AF360894

Forest Fire Risk Prediction

Forest Fire Risk Prediction

Editors

Rachael Nolan
Víctor Resco de Dios

MDPI • Basel • Beijing • Wuhan • Barcelona • Belgrade • Manchester • Tokyo • Cluj • Tianjin

Editors
Rachael Nolan
Hawkesbury Institute for the
Environment
Western Sydney University
Richmond
Australia

Víctor Resco de Dios
School of Life Sciences and
Engineering/Crop and Forest
Sciences
Southwest University of Science
and Technology/University of
Lleida-AGROTECNIO
Mianyang/Lleida
China

Editorial Office
MDPI
St. Alban-Anlage 66
4052 Basel, Switzerland

This is a reprint of articles from the Special Issue published online in the open access journal *Forests* (ISSN 1999-4907) (available at: www.mdpi.com/journal/forests/special_issues/Fire_Risk_Prediction).

For citation purposes, cite each article independently as indicated on the article page online and as indicated below:

LastName, A.A.; LastName, B.B.; LastName, C.C. Article Title. *Journal Name* **Year**, *Volume Number*, Page Range.

ISBN 978-3-0365-1474-1 (Hbk)
ISBN 978-3-0365-1473-4 (PDF)

Contents

About the Editors

Rachael Nolan

Dr Rachael Nolan obtained her Ph.D. from The University of Melbourne, studying ecohydrology and wildfire. She has two main areas of research. The first is focused on understanding how ecosystems respond to disturbance, and what this means for ecosystem services, such as carbon sequestration, habitat values, and water resources. The second research area is focused on developing models of forest flammability due to dynamics in fuel loads and fuel moisture content. She currently works at the Hawkesbury Institute for the Environment at Western Sydney University and is a member of the New South Wales Bushfire Risk Management Research Hub.

Víctor Resco de Dios

Víctor Resco de Dios obtained his Ph.D. at the University of Wyoming. His research is focused on understanding the reciprocal feedbacks between plant physiological ecology, fire activity and land management. He currently works at the Southwest University for Science and Technology and holds an affiliation with the University of Lleida and the Joint Research Unit CTFC-AGROTECNIO-CERCA Center.

Editorial

Some Challenges for Forest Fire Risk Predictions in the 21st Century

Víctor Resco de Dios [1,2,3,*] and Rachael H. Nolan [4,5]

1   School of Life Science and Engineering, Southwest University of Science and Technology, Mianyang 621010, China
2   Department of Crop and Forest Sciences, Universitat de Lleida, 25198 Lleida, Spain
3   Joint Research Unit CTFC-AGROTECNIO, Universitat de Lleida, 25198 Lleida, Spain
4   Hawkesbury Institute for the Environment, Western Sydney University, Locked Bag 1797, Penrith, NSW 2751, Australia; Rachael.Nolan@westernsydney.edu.au
5   Bushfire Risk Management Research Hub, Wollongong, NSW 2522, Australia
*   Correspondence: v.rescodedios@swust.edu.cn

Citation: Resco de Dios, V.; Nolan, R.H. Some Challenges for Forest Fire Risk Predictions in the 21st Century. *Forests* **2021**, *12*, 469. https://doi.org/10.3390/f12040469

Received: 7 April 2021
Accepted: 11 April 2021
Published: 12 April 2021

Global wildfire activity has experienced a dramatic surge since 2017. From Chile [1] to Indonesia [2], unprecedented fire behavior has occurred in many areas worldwide including, but not limited to, Portugal [3], Siberia [4], Australia [5,6], the Amazon and Orinoco basins [7], and the Western US [8]. This surge in global wildfire activity has led to a dramatic raise in human fatalities, and in socio-economic and ecological losses.

Wildfires have been ravaging through both fire-prone and non-fire-prone ecosystems. Although wildfires are regarded as a natural phenomenon in fire-prone ecosystems, and are necessary for the successful establishment and regeneration of many species, their positive effects are limited to instances in which the current fire regime resembles that under which the affected species evolved [9]. As the components of the fire regime change because of global change (including the frequency, intensity, and timing of forest fires), many species will experience a novel disturbance regime and, consequently, we can expect a reorganization in many ecosystems worldwide, potentially shifting towards an increase in ruderal species (ruderalization) [10].

Wildfires have also become increasingly common in many non-fire-prone ecosystems [11]. We have observed dramatic wildfires in the rainforests of the Amazon associated with deforestation, land use change, and other processes [12]. Additionally, increasing numbers of holdover fires beyond the Arctic circle, which resurface after burning underground during the winter, have been linked to a proliferation of spring and summer surface wildfires in the Arctic [4]. The effects of catastrophic wildfire on non-fire-prone ecosystems will be even more dramatic than in fire-prone ecosystems.

The currently ongoing, global change-induced, intensification of the fire regime has escalated from being primarily an ecological problem to also becoming a civil protection issue. For example, in Southern Europe, the number of wildfire fatalities over the last 13 years (473 fatalities) [13,14] has exceeded the number of fatalities in terrorist attacks in the entire European Union (448 victims) [15] (Figure 1). Wildfires also potentially pose a security threat by directly threatening defense infrastructure and diverting defense personnel to provide humanitarian assistance and disaster relief, as occurred during the 2019/2020 fire season in Australia [16].

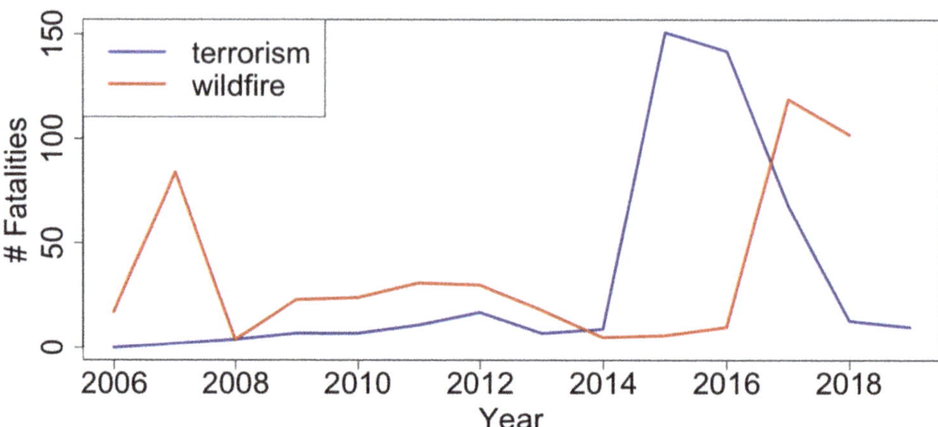

Figure 1. The number of fatalities resulting from wildfires is now larger than the number of terrorism victims in the European Union. We compiled data on people killed by terrorism in the EU since Europol TE-SAT reports [15] started in 2006 with data on wildfire fatalities for Southern Europe (Portugal, Spain, Greece, and Sardinia) [13,14].

Wildfires are also becoming an increasing public health issue. Smoke from wildfires is currently estimated to be the direct cause of death of 339,000 people annually [17], and there is a well-documented increase in hospital admissions due to smoke-enhanced cardiovascular and respiratory conditions, amongst others [18]. In interaction with the current pandemic, wildfire smoke could increase hospital admissions by 10% in areas affected by large wildfire [19]. Wildfires could also act as vectors for transporting airborne pathogens [20].

Extreme wildfires are also extending their impacts to an increasing range of sectors, including agriculture, infrastructure, transport, and tourism, to name a few, resulting in declines in consumer activity and large economic impacts [21].

Wildfires also alter the climatic system in a myriad of ways. Direct effects include enhanced carbon emissions, currently estimated at ~10% of fossil fuel emissions [22]. However, other contributions are more difficult to quantify. Wildfires alter albedo, with important consequences for the energy balance and fostering warming in tropical, but cooling in boreal ecosystems [23]. Wildfires also affect the water balance [24], and the capacity for future C sequestration in burned ecosystems may be reduced, relative to that prior to fire, when wildfires induce large-scale land degradation [9].

All in all, it appears that we are simply witnessing the "preview" of the havoc that climate change will bring for future fire regimes. In fact, a commonality across all the changes in the fire regime previously mentioned was a very marked increase in vapor pressure deficit, a very marked driver of fire activity [25,26], resulting from living in a 1 °C warmer world. As climate projections for the end of the century converge into a likely 3 °C warming scenario [27], we can expect, at least in the short term, a further intensification of the fire regime.

Increases in fire danger induced by climate change call for a re-evaluation of current approaches for assessing the likelihood of a catastrophic fire. Seasonal changes in wildfire likelihood result from changes in fuel moisture and in fire weather that, when an ignition source is provided, can lead to catastrophic wildfire if fuel build up is large enough [28]. The challenge lies in developing a quantitative and mechanistic understanding of fire danger that is deep enough to allow accurate forecasting, yet simple enough that can be used for operational purposes.

Traditional approaches for forecasting seasonal changes in fire weather relied on developing fire weather and danger indices that, often, seek to estimate fuel moisture and potential fire spread depending on past meteorological conditions. These indices have shown mixed success in predicting fire danger [29,30], and they have not been exempt from criticism [31]. Developing an early warning system for catastrophic fire, which predicts when and where critical wildfires will occur, as well as educating the populace on how to interpret these predictions, has become an increasing necessity [3].

Understanding changes in fire danger requires integration across multiple spatial and temporal scales: from leaves to regions, and from days to decades. This was the motivation to put forward this Special Issue on Forest Fire Risk Prediction: to provide the tools for improving fire danger forecasts in the 21st century. The Special Issue comprises 13 articles, touching upon different aspects of fire danger along a continuum of scales.

Broadly speaking, the articles contained in this Special Issue can be grouped into four topics. The first topic is the use of fire danger metrics and other approaches to understand variation in wildfire activity. The opening article by Fernandes [32] examines the thresholds in a widespread indicator of fire danger, the Fire Weather Index (FWI), associated with wildfires of different sizes across Portugal. Varela et al. [33] use the FWI to identify future vulnerability of archaeological and touristic sites in Greece. The study of Ma et al. [34] presents a broad assessment on the drivers of wildfire activity in China using random forest algorithms, and the study by Milanović et al. [35] additionally compares random forest with logistic regressions to estimate fire probability in Serbia. Zong et al. [36] uses a historical analysis of the drivers of forest fire in Central Asia to project changes under climate change in the 21st century.

The second topic covered by this Special Issue is devoted to understanding changes in the flammability of live fuel. Nolan et al. [37] review the mechanisms driving live fuel moisture content and propose a novel model that moves the field forward. Balaguer–Romano et al. [38] explore the hypothesis that needle senescence could affect wildfire behavior in Mediterranean pine forests. How to provide regional estimates of live fuel moisture content is explored by Luo et al. [39], who provide a case study for SW China. Della Rocca, et al. [40] address interactions between pathogen infections and flammability. Also along these lines, Collins et al. [41] address the effects of repeated fires on burn severity.

The third topic covered in this Special Issue is modeling dead fuel moisture content. Zhang and Sun [42] compare two methods for estimating diurnal changes in fine litter moisture. Log [43] further assesses moisture diffusion coefficients in *Calluna vulgaris* to better inform prescribed fire practices.

Finally, the study of Ma et al. [44] compares the emission factors of Chinese tree species ranging from Boreal to subtropical environments, and they present intriguing evidence indicating that wildfires could acidify forest ecosystems because of emissions. This varied collection of articles indicates that progress in fire danger predictions comes from a multidisciplinary and varied approach.

The cradle of Western civilization lies in Ancient Greece. They laid the foundations of science and democracy. As climate change intensifies worldwide, wildfires are even threatening the archaeological remains from Ancient Greece [33]. In his work *The Republic*, the Greek philosopher Plato wrote the Allegory of the Cave, where men living in the bottom of a cave perceived reality only through shadows and they were afraid of reaching out of the cave to see reality as it is. The only escapee from the cave who saw the actual world was the one who embraced knowledge. If we have any chance to prevent the worsening of the wildfire problem, we need now, more than ever, to resort back to logic and reason. This Special Issue is a testament to the fact that we have the knowledge and technical capability to anticipate the effects of global warming on wildfires. The challenge lies in convincing policy makers, managers, stakeholders, and all the other actors involved to follow through and take evidence-based decisions. The challenge lies in getting out of Plato's cave.

Author Contributions: Writing: V.R.d.D. and R.H.N. All authors have read and agreed to the published version of the manuscript.

Funding: V.R.d.D. acknowledges funding from the National Natural Science Foundation in China (U20A20179, 31850410483), the talent proposals in Sichuan Province (2020JDRC0065), from Southwest University of Science and Technology (18ZX7131), and the MICINN (RTI2018-094691-B-C31). R.H.N. was supported with funding from the New South Wales Department of Planning, Industry and Environment, via the NSW Bushfire Risk Management Research Hub.

Acknowledgments: We appreciate the comments that Dick Williams, Rob Whelan, and Matthias Boer provided on an earlier version of this article.

Conflicts of Interest: The authors declare no conflict of interest.

References

1. Gómez-González, S.; Ojeda, F.; Fernandes, P.M. Portugal and Chile: Longing for sustainable forestry while rising from the ashes. *Environ. Sci. Policy* **2018**, *81*, 104–107. [CrossRef]
2. Rochmyaningsih, D. Scientists in indonesia fear political interference. *Science* **2020**, *367*, 722. [CrossRef] [PubMed]
3. Boer, M.M.; Nolan, R.H.; Resco De Dios, V.; Clarke, H.; Price, O.F.; Bradstock, R.A. Changing weather extremes call for early warning of potential for catastrophic fire. *Earth's Future* **2017**, *5*, 1196–1202. [CrossRef]
4. McCarty, J.L.; Smith, T.E.L.; Turetsky, M.R. Arctic fires re-emerging. *Nat. Geosci.* **2020**, *13*, 658–660. [CrossRef]
5. Nolan, R.H.; Boer, M.M.; Collins, L.; Resco de Dios, V.; Clarke, H.; Jenkins, M.; Kenny, B.; Bradstock, R.A. Causes and consequences of eastern Australia's 2019–20 season of mega-fires. *Glob. Chang. Biol.* **2020**, *26*, 1039–1041. [CrossRef] [PubMed]
6. Boer, M.M.; Resco de Dios, V.; Bradstock, R.A. Unprecedented burn area of australian mega forest fires. *Nat. Clim Chang.* **2020**, *10*, 171–172. [CrossRef]
7. Barlow, J.; Berenguer, E.; Carmenta, R.; França, F. Clarifying amazonia's burning crisis. *Glob. Chang. Biol.* **2020**, *26*, 319–321. [CrossRef]
8. Pickrell, J.; Pennisi, E. Record U.S. and australian fires raise fears for many species. *Science* **2020**, *370*, 18–19. [CrossRef]
9. Resco de Dios, V. Plant-Fire Interactions. In *Applying Ecophysiology to Wildfire Management*; Springer: Cham, Switzerland, 2020; Volume 36.
10. Karavani, A.; Boer, M.M.; Baudena, M.; Colinas, C.; Díaz-Sierra, R.; Pemán, J.; de Luís, M.; Enríquez-de-Salamanca, Á.; Resco de Dios, V. Fire-induced deforestation in drought-prone mediterranean forests: Drivers and unknowns from leaves to communities. *Ecol. Monogr.* **2018**, *88*, 141–169. [CrossRef]
11. Boer, M.M.; Resco De Dios, V.; Stefaniak, E.Z.; Bradstock, R.A. A hydroclimatic model for the distribution of fire on earth. *Environ. Res. Commun.* **2021**, *3*, 035001. [CrossRef]
12. Armenteras, D.; González, T.M.; Vargas Ríos, O.; Meza Elizalde, M.C.; Oliveras, I. Incendios en ecosistemas del norte de Suramérica: Avances en la ecología del fuego tropical en Colombia, Ecuador y Perú. *Caldasia* **2020**, *42*, 1–16. [CrossRef]
13. Molina-Terrén, D.M.; Xanthopoulos, G.; Diakakis, M.; Ribeiro, L.; Caballero, D.; Delogu, G.M.; Viegas, D.X.; Silva, C.A.; Cardil, A. Analysis of forest fire fatalities in southern europe: Spain, Portugal, Greece and Sardinia (Italy). *Int. J. Wildland Fire* **2019**, *28*, 85. [CrossRef]
14. Haynes, K.; Short, K.; Xanthopoulos, G.; Viegas, D.; Ribeiro, L.M.; Blanchi, R. Wildfires and Wui Fire Fatalities. In *Encyclopedia of Wildfires and Wildland-Urban Interface (WUI) Fires*; Manzello, S.L., Ed.; Springer: Cham, Switzerland, 2020; pp. 1–16.
15. Europol. Eu Terrorism Situation & Trend Report (Te-Sat). Available online: http://www.europol.europa.eu/tesat-report?page=0,1 (accessed on 15 September 2020).
16. McDonald, M. After the fires? Climate change and security in australia. *Aust. J. Political Sci.* **2021**, *56*, 1–18. [CrossRef]
17. Johnston, F.H.; Henderson, S.B.; Chen, Y.; Randerson, J.T.; Marlier, M.; Defries, R.S.; Kinney, P.; Bowman, D.M.; Brauer, M. Estimated global mortality attributable to smoke from landscape fires. *Environ. Health Perspect.* **2012**, *120*, 695–701. [CrossRef]
18. Borchers Arriagada, N.; Palmer, A.J.; Bowman, D.M.; Morgan, G.G.; Jalaludin, B.B.; Johnston, F.H. Unprecedented smoke-related health burden associated with the 2019–20 bushfires in eastern australia. *Med. J. Aust.* **2020**, *213*. [CrossRef] [PubMed]
19. Henderson, S.B. The covid-19 pandemic and wildfire smoke: Potentially concomitant disasters. *Am. J. Public Health* **2020**, *110*, 1140–1142. [CrossRef]
20. Kobziar, L.N.; Thompson, G.R. Wildfire smoke, a potential infectious agent. *Science* **2020**, *370*, 1408–1410. [PubMed]
21. Filkov, A.I.; Ngo, T.; Matthews, S.; Telfer, S.; Penman, T.D. Impact of australia's catastrophic 2019/20 bushfire season on communities and environment. Retrospective analysis and current trends. *J. Saf. Sci. Resil.* **2020**, *1*, 44–56. [CrossRef]
22. Le Quéré, C.; Andrew, R.M.; Friedlingstein, P.; Sitch, S.; Hauck, J.; Pongratz, J.; Pickers, P.A.; Korsbakken, J.I.; Peters, G.P.; Canadell, J.G.; et al. Global carbon budget 2018. *Earth Syst. Sci. Data* **2018**, *10*, 2141–2194. [CrossRef]
23. Chapin, F.S.; Randerson, J.T.; McGuire, A.D.; Foley, J.A.; Field, C.B. Changing feedbacks in the climate–biosphere system. *Front Ecol. Environ.* **2008**, *6*, 313–320. [CrossRef]
24. Nolan, R.H.; Lane, P.N.J.; Benyon, R.G.; Bradstock, R.A.; Mitchell, P.J. Changes in evapotranspiration following wildfire in resprouting eucalypt forests. *Ecohydrology* **2014**, *7*, 1363–1377. [CrossRef]
25. Resco de Dios, V.; Fellows, A.W.; Nolan, R.H.; Boer, M.M.; Bradstock, R.A.; Domingo, F.; Goulden, M.L. A semi-mechanistic model for predicting the moisture content of fine litter. *Agric. For. Meteorol.* **2015**, *203*, 64–73.

26. Nolan, R.H.; Resco de Dios, V.; Boer, M.M.; Caccamo, G.; Goulden, M.L.; Bradstock, R.A. Predicting dead fine fuel moisture at regional scales using vapour pressure deficit from modis and gridded weather data. *Remote. Sens. Environ.* **2016**, *174*, 100–108. [CrossRef]
27. Hausfather, Z.; Peters, G.P. Emissions—The 'business as usual' story is misleading. *Nature* **2020**, *577*, 618–620. [CrossRef]
28. Bradstock, R.A. A biogeographic model of fire regimes in Australia: Current and future implications. *Glob. Ecol. Biogeogr.* **2010**, *19*, 145–158. [CrossRef]
29. Di Giuseppe, F.; Vitolo, C.; Krzeminski, B.; Barnard, C.; Maciel, P.; San-Miguel, J. Fire weather index: The skill provided by the european centre for medium-range weather forecasts ensemble prediction system. *Nat. Hazards Earth Syst. Sci.* **2020**, *20*, 2365–2378. [CrossRef]
30. Bedia, J.; Herrera, S.; Gutiérrez, J.M.; Benali, A.; Brands, S.; Mota, B.; Moreno, J.M. Global patterns in the sensitivity of burned area to fire-weather: Implications for climate change. *Agric. For. Meteorol.* **2015**, *214–215*, 369–379. [CrossRef]
31. Ruffault, J.; Martin-StPaul, N.; Pimont, F.; Dupuy, J.-L. How well do meteorological drought indices predict live fuel moisture content (lfmc)? An assessment for wildfire research and operations in mediterranean ecosystems. *Agric. For. Meteorol.* **2018**, *262*, 391–401. [CrossRef]
32. Fernandes, P.M. Variation in the canadian fire weather index thresholds for increasingly larger fires in Portugal. *Forests* **2019**, *10*, 838. [CrossRef]
33. Varela, V.; Vlachogiannis, D.; Sfetsos, A.; Politi, N.; Karozis, S. Methodology for the study of near-future changes of fire weather patterns with emphasis on archaeological and protected touristic areas in Greece. *Forests* **2020**, *11*, 1168. [CrossRef]
34. Ma, W.; Feng, Z.; Cheng, Z.; Chen, S.; Wang, F. Identifying forest fire driving factors and related impacts in china using random forest algorithm. *Forests* **2020**, *11*, 507. [CrossRef]
35. Milanović, S.; Marković, N.; Pamučar, D.; Gigović, L.; Kostić, P.; Milanović, S.D. Forest fire probability mapping in eastern Serbia: Logistic regression versus random forest method. *Forests* **2020**, *12*, 5. [CrossRef]
36. Zong, X.; Tian, X.; Yin, Y. Impacts of climate change on wildfires in central asia. *Forests* **2020**, *11*, 802. [CrossRef]
37. Nolan, R.H.; Blackman, C.J.; Resco de Dios, V.; Choat, B.; Medlyn, B.E.; Li, X.; Bradstock, R.A.; Boer, M.M. Linking forest flammability and plant vulnerability to drought. *Forests* **2020**, *11*, 779. [CrossRef]
38. Balaguer-Romano, R.; Díaz-Sierra, R.; Madrigal, J.; Voltas, J.; Resco de Dios, V. Needle senescence affects fire behavior in aleppo pine (*Pinus halepensis* Mill.) stands: A simulation study. *Forests* **2020**, *11*, 1054. [CrossRef]
39. Luo, K.; Quan, X.; He, B.; Yebra, M. Effects of live fuel moisture content on wildfire occurrence in fire-prone regions over southwest China. *Forests* **2019**, *10*, 887. [CrossRef]
40. Della Rocca, G.; Danti, R.; Hernando, C.; Guijarro, M.; Michelozzi, M.; Carrillo, C.; Madrigal, J. Terpenoid accumulation links plant health and flammability in the cypress-bark canker pathosystem. *Forests* **2020**, *11*, 651. [CrossRef]
41. Collins, L.; Hunter, A.; McColl-Gausden, S.; Penman, T.D.; Zylstra, P. The effect of antecedent fire severity on reburn severity and fuel structure in a resprouting eucalypt forest in Victoria, Australia. *Forests* **2021**, *12*, 450. [CrossRef]
42. Zhang, Y.; Sun, P. Study on the diurnal dynamic changes and prediction models of the moisture contents of two litters. *Forests* **2020**, *11*, 95. [CrossRef]
43. Log, T. Modeling drying of degenerated *Calluna vulgaris* for wildfire and prescribed burning risk assessment. *Forests* **2020**, *11*, 759. [CrossRef]
44. Ma, Y.; Tigabu, M.; Guo, X.; Zheng, W.; Guo, L.; Guo, F. Water-soluble inorganic ions in fine particulate emission during forest fires in chinese boreal and subtropical forests: An indoor experiment. *Forests* **2019**, *10*, 994. [CrossRef]

 *forests*

Article

Variation in the Canadian Fire Weather Index Thresholds for Increasingly Larger Fires in Portugal

Paulo M. Fernandes

Centro de Investigação e Tecnologias Agroambientais e Biológicas, CITAB, Universidade de Trás-os-Montes e Alto Douro (UTAD), Quinta dos Prados, Apartado 1013, 5000-801 Vila Real, Portugal; pfern@utad.pt; Tel.: +351-259-350-484

Received: 21 August 2019; Accepted: 23 September 2019; Published: 24 September 2019

Abstract: Forest fire management relies on the fire danger rating to optimize its suite of activities. Limiting fire size is the fire management target whenever minimizing burned area is the primary goal, such as in the Mediterranean Basin. Within the region, wildfire incidence is especially acute in Portugal, a country where fire-influencing anthropogenic and landscape features vary markedly within a relatively small area. This study establishes daily fire weather thresholds associated to transitions to increasingly larger fires for individual Portuguese regions (2001–2011 period), using the national wildfire and Canadian fire weather index (FWI) databases and logistic regression. FWI thresholds variation in relation to population density, topography, land cover, and net primary production (NPP) metrics is examined through regression and cluster analysis. Larger fires occur under increasingly higher fire danger. Resistance to fire spread (the fire-size FWI thresholds) varies regionally following biophysical gradients, and decreases under more complex topography and when NPP and occupation by flammable forest or by shrubland increase. Three main clusters synthesize these relationships and roughly coincide with the western north-central, eastern north-central and southern parts of the country. Quantification of fire-weather relationships can be improved through additional variables and analysis at other spatial scales.

Keywords: fire danger rating; fire management; fire regime; fire size; fire weather; Portugal

---

1. Introduction

Exposure to increasingly severe and anomalous fire weather makes fire disasters a growing global concern, especially where flammable vegetation and developed areas intermingle [1]. Such is the case of the Mediterranean Basin, where forest fires are prominent and an average of 360 kha burn every year in the Iberian Peninsula, southern France, Italy, and Greece, with northern and central Portugal being the hotspot of fire incidence [2]. Combination of an oceanic-influenced Mediterranean climate favoring high plant productivity and fast fuel build-up, prevalence of flammable vegetation types, rough terrain, and particularly high ignition density are credited to account for the high annual burn rates (up to 6%) observed in Portugal [3].

Forest fire management comprises a suite of activities that seek to minimize the socioeconomic and environmental impacts of fire, namely ignition prevention, fire detection, initial attack resource deployment and dispatch, large fire management, strategic planning, and fuel management (e.g., through prescribed burning) [4]. Fire danger rating supports all these activities [5] through indices calculated from current and past weather that express the individual and combined effects of atmospheric conditions and drought, as in the systems developed in Canada [6], the USA [7], and Australia [8] that describe how easily a fire will ignite and spread, and how difficult its control will be.

The Canadian forest fire weather index system (CFFWIS) [9] is commonly used in research and management applications across the world and consists of six fuel moisture and fire behavior indices calculated from ambient temperature, relative humidity, wind speed, and 24-h rainfall. Operational fire danger rating in Portugal relies on the fire weather index (FWI) of the CFFWIS, calibrated to indicate fire suppression difficulty as determined by fireline intensity in maritime pine (*Pinus pinaster* Ait.) stands, and considers five classes, respectively low (FWI < 8.4), moderate (8.5 < FWI < 17.1), high (17.2 < FWI < 24.5), very high (24.6 < FWI < 38.2), and extreme (FWI > 38.2) [10]. This approach offers an objective and quantitative indication of fire potential, albeit generic as it does not consider variation in fuel and terrain conditions. Consequently, outputs of fire weather systems benefit from regional interpretation and statistical evaluation against historical fire activity that lead to probabilistic assessments of fire likelihood and threshold values to define fire danger classes [5,11]. In Europe, the CFFWIS has been calibrated to depict distinct fire activity levels [12–15] and has been used to model fire activity (number of fires, burned area) across various spatiotemporal scales [16–23] and to assess the likelihood and characteristics of large fires [24–28].

Fire management often seeks to minimize the area burned by unplanned fire, as it equates to abating its environmental and societal negative impacts. Since fire size distribution is highly uneven, most of the burned area is accounted for by a small fraction of the total number of fires, e.g., [29], those that escape initial control efforts, are driven by extreme weather conditions, and can become catastrophic if wildland-urban interfaces are impacted [1,30]. Fire size distribution is thus useful to evaluate fire management policies and for strategic planning [31], and operational planning of pre-suppression and suppression activities can be guided by fire danger rating such that individual fires do not exceed predefined size targets [32]. Fire size depends of fire behavior characteristics, namely rate of spread (determined by weather, topography, and fuel), landscape barriers, and fire suppression effort and effectiveness, which are affected respectively by the amount of simultaneous wildfires and fire behavior [31]. The relative importance of the numerous factors involved is variable and can change depending on fire-size class or across the fire size continuum [24,26,27,33–35]. The need for achieving progress on the understanding of fire weather (and other influences) roles on fire growth to increasingly larger sizes is obvious, given the impacts associated to larger fires [36].

Fire activity and fire regimes, including fire-size distribution [31], vary along climatic and land use gradients [37] and result from the interaction of biophysical and social factors, hence are subject to substantial anthropogenic-induced change, through modified vegetation and ignition patterns and fire suppression [37–39] capable of overriding background fire-climate [40] and fire-weather [41] relationships. The regional scale context is thus relevant to better understand how fire activity (including fire size) is connected to fire weather [21,42] and, more generally, to climate [40,43].

Marked spatial contrasts in fire activity in Portugal [44] and the concomitant influence of biophysical and socioeconomic factors [45–47] motivates this study examination of regional fire activity in relation to fire weather. I hypothesize that larger fire development in response to increasingly severe fire weather is dependent on land use and vegetation, topography, and variables affecting human fire initiation and control, which should be reflected in regionally variable fire size-weather relationships.

2. Materials and Methods

2.1. Data

This study relates to mainland Portugal (89,089 km^2) and the 2001–2011 period, which is relatively short but was selected to ensure consistency in record keeping standards and minimize variability in fire management policies and land-use changes that could affect fire activity, e.g., [48]. The study period includes both extremely high (2003, 2005) and extremely low (2007, 2008) fire years in terms of burned area, ranging from 20.0 kha to 471.8 kha. Individual fire data (date, duration, location, and size) was sourced from the Portuguese rural fire database [49] and comprised 333,378 records for 2001–2011, corresponding to a burned area of 1671.4 kha (Figure 1a), according to the national fire atlas [50].

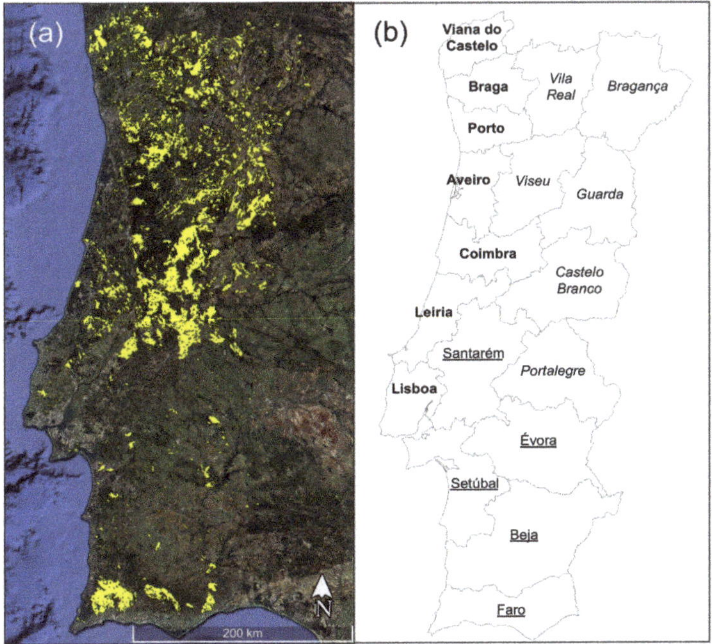

Figure 1. Cumulative burned area in Portugal (2001–2011) overlaid on Google Earth imagery (map data courtesy of ©2018 Google) (**a**) and administrative regions (*distritos*) (**b**). Bold, italic, and underlined lettering indicates regions belonging respectively to the North and Central West (NCW), North and Central East (NCE), and South (S) supra regions.

Fire weather was described through the FWI of the CFFWIS, calculated daily from noon weather observations acquired by the Portuguese Sea and Atmosphere Institute network of automatic weather stations. *Distrito* ($n = 18$, Figure 1b) is the administrative regional scale that underpins fire preparedness and suppression organization and decision-making in Portugal. The FWI at the region (i.e., *distrito*) level was obtained by averaging the FWI values of the stations (between 2 and 7) located within each region. The fraction of fire starts that attained or exceeded increasingly larger size thresholds were calculated for each combination of region and day of the time series, respectively 0.01, 0.1, 1, 10, 100, 500, and 1000 ha. Those fractions were then recoded as 0 or 1, respectively for days with none and one or more fires belonging to a given size class; no-fire days were also attributed 0. Thus, each fire size-based partition divided the whole dataset in two subsets.

A number of variables expected to influence fire-weather relationships were calculated for each region. Population density, as a strong determinant of ignition density [51], correlation with unburnable land and road network density, and a relevant factor in fire control response time and available resources, was obtained by averaging the 2001 and 2011 national census.

Topography and terrain characteristics influence ignition patterns, fire behavior, and fire suppression operations and were derived from a 25-m resolution digital elevation model: Elevation, slope, and the topographic ruggedness index (TRI) [52] that expresses the average elevation change between any point on a digital grid and the surrounding area.

Net primary productivity (NPP, t C year^{-1}) was used as a generic indicator of potential fuel load and was calculated for each region from the gridded (0.25° resolution) spatial data product global patterns in net primary productivity [53,54]. Data from the 5th National Forest Inventory [55] was used to calculate the regional fractions of land covered by agriculture, shrubland, and forest. Additionally, a "flammable forest" category was calculated as the fraction of forest cover comprised of maritime pine or eucalypt (mostly *Eucalyptus globulus* Labill.). These species tend to occupy large and continuous tracts of the landscape, often in the form of naturally-regenerated and highly-flammable low and dense stands [56], and are more prone to fire than any other forest type [57,58].

2.2. Data Analyses

Summary statistics were calculated for the regional variation of daily FWI and percentages of days with fires ≥0.01, 0.1, 1, 100, 500, and 1000 ha. Most fires in Portugal are short-lived, as expected in an anthropogenic landscape under a full fire suppression policy, e.g., only 4.1% of the fires in the database lasted more than 8 h, and only 0.9% extended beyond 24 h; Table S1 gives percentiles of fire duration for each of the binned fire-size classes. The duration of active fire spread is even smaller, as fire duration in the database is the time elapsed between fire detection and fire extinction. Consequently, I reasonably assumed that fire weather on the day a fire starts determines whether it will grow to a size larger than the thresholds considered [59], which is further supported by the fact that in Portugal fires larger than 2500 ha typically (interquartile range) attain 1000 ha in 2 h to 10 h [25].

Logistic regression [60] with the FWI of the fire-start day as the independent variable was employed to model the daily likelihood of fires ≥0.01, 0.1, 1, 10, 100, 500, and 1000 ha at the regional level. The χ^2 likelihood ratio test was used to assess statistical significance ($p < 0.05$). Model classification performance was assessed with the area under the receiver operating characteristic (ROC) curve (AUC) and the misclassification rate (fraction of the observations incorrectly classified by the model) was calculated. A probability of 0.5 was the cut-off assumed for a fire-day event and the corresponding FWI was calculated from the fitted models for each region and fire size class. Correlations between the FWI thresholds for the various fire size classes were inspected, as well as between those and the regional median and 90th percentile of the FWI.

A correlation analysis was carried out between the variables expected to explain patterns in fire-FWI relationships, i.e., population density and the biophysical variables previously described. The regional variation in FWI thresholds allowing increasingly larger fires was examined as a function of those variables, but only the classes more relevant for fire management were retained, i.e., ≥1, 100, and 1000 ha, respectively FWI_1, FWI_100 and FWI_1000. Univariate least-squares regression analysis was carried out on log-transformed variables, to achieve more Gaussian distributions and because visual inspection indicated adequate fit by power functions. Residuals were checked for normality and homoscedasticity. To aid in interpretation, and anticipating common fire-FWI relationships, the 18 regions were allocated to three higher-level regions, respectively North and Central West (NCW), North and Central East (NCE), and South (S).

Combining the independent context variables with the FWI thresholds and applying hierarchical cluster analysis enabled the identification of similar regions in respect to fire-FWI relationships; as fires ≥1000 ha were very scarce in some regions, the corresponding FWI threshold was excluded from the analysis.

3. Results

Regional diversity in the putative factors related with fire activity in Portugal was manifest (Table 1). In particular, variation in population density approached two orders of magnitude. The remaining variables varied at least approximately two-fold (NPP and agriculture land cover) and up to 10- and 12-fold, respectively shrubland and flammable forest cover.

Table 1 Regional means (standard deviations) of variables expected to influence the relationships between fire activity and fire weather in Portugal. TRI is the topographic ruggedness index and NPP is net primary productivity.

Region	Popul. Density	Elevation	Slope	TRI	NPP	Land Cover Fractions			
	(no. km^{-2})	(m)	(°)		(t C year^{-1})	Agriculture	Shrubland	Forest	Flammable Forest *
Aveiro	261.2	200 (215)	4.8 (5.0)	36.0 (33.0)	4.2 (0.4)	0.25	0.08	0.53	0.95
Beja	15.0	177 (77)	2.5 (2.1)	23.2 (18.8)	2.9 (0.8)	0.50	0.13	0.35	0.17
Braga	323.2	356 (276)	7.7 (5.9)	53.8 (35.5)	4.3 (0.3)	0.30	0.27	0.32	0.82
Bragança	21.9	610 (186)	6.4 (4.5)	47.6 (29.8)	2.6 (0.5)	0.35	0.34	0.29	0.39
Castelo Branco	30.2	408 (192)	5.1 (4.6)	37.1 (29.7)	3.1 (0.5)	0.26	0.32	0.40	0.78
Coimbra	109.5	270 (257)	5.3 (5.4)	41.5 (38.3)	3.7 (0.4)	0.21	0.17	0.52	0.96
Évora	22.8	218 (74)	1.7 (1.2)	12.1 (7.7)	3.2 (0.4)	0.42	0.05	0.50	0.08
Faro	87.9	176 (137)	4.7 (3.7)	46.1 (30.1)	3.3 (1.1)	0.26	0.38	0.26	0.26
Guarda	30.6	698 (232)	6.0 (5.4)	44.5 (35.7)	2.6 (0.4)	0.28	0.48	0.19	0.58
Leiria	135.0	199 (157)	3.7 (3.2)	27.3 (20.5)	3.7 (0.5)	0.27	0.16	0.45	0.95
Lisboa	801.1	116 (82)	3.2 (2.8)	23.5 (17.4)	3.4 (0.4)	0.48	0.13	0.18	0.77
Portalegre	19.8	272 (107)	2.3 (2.0)	16.2 (13.6)	3.2 (0.4)	0.41	0.11	0.44	0.22
Porto	769.7	253 (215)	6.2 (5.2)	45.4 (32.7)	4.0 (0.3)	0.29	0.17	0.31	0.87
Santarém	66.7	133 (94)	2.4 (2.2)	17.8 (14.7)	3.4 (0.3)	0.33	0.12	0.49	0.53
Setúbal	164.4	78 (56)	1.8 (1.9)	14.1 (12.9)	3.1 (0.7)	0.29	0.07	0.56	0.29
Viana do Castelo	110.3	386 (3249)	10.1 (6.3)	68.0 (36.8)	4.5 (0.2)	0.21	0.38	0.30	0.74
Vila Real	49.5	698 (245)	8.3 (5.4)	58.2 (32.4)	3.2 (0.4)	0.25	0.39	0.32	0.74
Viseu	77.3	562 (239)	7.5 (5.8)	53.5 (33.5)	3.3 (0.7)	0.24	0.26	0.45	0.84

* As a fraction of total forest cover.

Correlation analysis indicates that regions characterized by rougher terrain tended to have less agriculture, more flammable forest, and, especially, more shrubland; forest and shrubland fractions were negatively correlated; and the flammable forest fraction increased with NPP (Table S2). Northwestern and southern Portugal were dissimilar in population density, terrain, NPP, and land cover patterns, indicating potential enhancement of ignition density and fire spread and intensity in the former.

Inter-regional fire weather variability was relevant (Figure 2), given altitudinal differences and the east to west and south to north gradients of increasing oceanic influence that moderates the fire-prone Mediterranean climate. Northwestern Portugal regions (Viana do Castelo, Braga, Porto, Aveiro), with more temperate climate, were characterized by lower FWI values and narrower FWI ranges in contrast with the drier and warmer south of the country (Évora, Beja). Regardless of these differences, and noting that annual fire activity concentrates on a short number of days [61], the 90th percentile of the FWI was higher than the threshold for extreme fire behavior (FWI = 38) in all regions but one (Viana do Castelo).

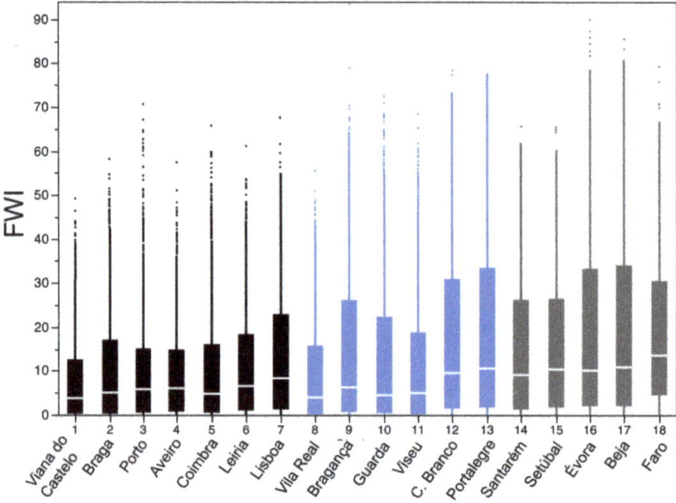

Figure 2. Boxplots (with outliers) of the fire weather index (FWI) daily variation for the regions (*distritos*) of Portugal (2001–2011). The box spans the interquartile range and the whiskers extend to the 90th and 10th percentiles. Black, blue, and grey boxplots correspond respectively to the NCW, NCE and S supra regions.

Increasingly larger fires consistently occurred on an increasingly smaller number of days (Table 2). Such a decreasing trend was particularly noticeable for fires surviving to ≥10 ha and ≥100 ha. Comparatively with other regions, large fires (≥100 ha) were a greater portion (≥3%) of the total number of fires in a few regions of the northern half of the country, namely Guarda, Vila Real, Bragança, and Viseu. Increasingly larger fires corresponded to increasingly higher FWI values (Table 3). The relative increase in FWI thresholds was especially relevant for fires ≥1 ha and ≥10 ha. Variation in the FWI thresholds among regions was quite high up to 10–99 ha fires, ranging from 26 to 76, but declined for the subsequent fire-size partition classes. While the FWI threshold for fires ≥1 ha corresponded to quite mild fire weather conditions in some regions, particularly Viana do Castelo, Vila Real, and Guarda, it increased to very high and extreme fire danger levels in others, namely Portalegre, Évora, and Faro.

Table 2. Regional percentages of days with fires ≥0.01, 0.1, 1, 100, 500, and 1000 ha during the study period (2001–2011).

Region	Fire Size (ha)						
	0.01	0.1	1	10	100	500	1000
Aveiro	59.4	45.1	23.7	4.9	0.9	0.3	0.2
Beja	28.6	24.8	18.3	6.2	1.3	0.3	0.1
Braga	57.3	51.1	38.2	11.9	2.3	0.4	0.1
Bragança	51.3	47.7	36.6	11.7	3.3	0.8	0.4
Castelo Branco	44.5	36.0	24.1	5.9	2.3	1.1	0.7
Coimbra	45.2	32.1	14.9	3.3	1.3	0.6	0.4
Évora	23.9	20.8	13.5	4.1	1.0	0.3	0.2
Faro	34.3	23.3	13.8	2.6	0.9	0.4	0.3
Guarda	54.4	49.2	41.1	15.6	5.2	1.5	0.8
Leiria	50.2	36.3	18.0	3.4	1.2	0.5	0.2
Lisboa	62.8	47.6	35.6	3.6	0.5	0.1	0.1
Portalegre	25.9	20.9	13.1	3.2	0.7	0.3	0.2
Porto	62.0	51.3	35.5	9.4	1.9	0.5	0.1
Santarém	55.1	42.4	25.9	5.0	1.9	0.8	0.3
Setúbal	50.9	33.8	21.1	2.9	0.6	0.1	0.1
Viana do Castelo	51.0	44.8	34.4	11.5	2.6	0.6	0.2
Vila Real	58.3	48.4	40.9	13.1	3.9	1.1	0.4
Viseu	59.5	52.4	39.4	11.5	3.6	1.1	0.5

The probability of distinguishing between fire and no-fire days as per the AUC varied between 0.67 and > 0.99 among the fitted logistic models, and misclassification rates ranged from < 1% to 36% (Table 3). AUC and misclassification rate tended to respectively increase and decrease with fire size, primarily reflecting the increasingly higher prevalence of no-fire days.

All fire-size FWI thresholds were significantly ($p < 0.05$) correlated, but the strength of the correlations diminished with fire-size dissimilarity (Table 4), e.g., $r = 0.93$ between the thresholds for fires ≥1 ha and fires ≥10 ha, but only $r = 0.50$ between the former and the threshold for fires ≥1000 ha. Nearly all FWI thresholds were strongly and positively coupled with the median FWI for all sample days (both fire and no-fire days), more closely associated with the thresholds for fires <100 ha, and with the 90th percentile of the FWI, better correlated with large-fire thresholds (Table 4).

The daily FWI threshold for fires ≥1 ha in size decreased with TRI, elevation, and the fractions of shrubland, flammable forest, and forest in general, by decreasing order of importance (Table 5). Flammable forest fraction had the strongest association with the FWI threshold for fires ≥100 ha, followed by TRI, NPP, and population density, but the p-value for the latter was just 0.052; increases in these variables implied lower thresholds. The FWI threshold for fires ≥1000 ha increased with agricultural land cover and decreased with the fraction of flammable forest and with NPP, but the amount of variation accounted for by these variables was inferior to what was attained for fires ≥1 ha and ≥100 ha using the best univariate relationships. Figure 3 shows the best fitting relationships.

Three major groups of regions emerged from the cluster analysis that mostly conformed to the a priori supra regions (Figure 4). Low FWI thresholds, the highest population density, NPP and flammable forest fraction, and intermediate shrubland cover characterize the NCW cluster ($n = 7$). The NCE cluster ($n = 6$) did not differ from the former regarding FWI_1 but fires ≥100 ha occurred under more severe fire weather conditions, in landscapes with lower NPP, intermediate cover of flammable forest types and the highest incidence of shrubland. The uppermost FWI thresholds corresponded to the S cluster ($n = 5$), in association with the lowest elevation, slope, TRI, NPP and fractions of flammable forest types and shrubland, and the highest proportion of farmland. The bottom end of the clustering hierarchy was composed of contiguous regions, with two exceptions (Lisboa, Porto; and Castelo Branco, Faro).

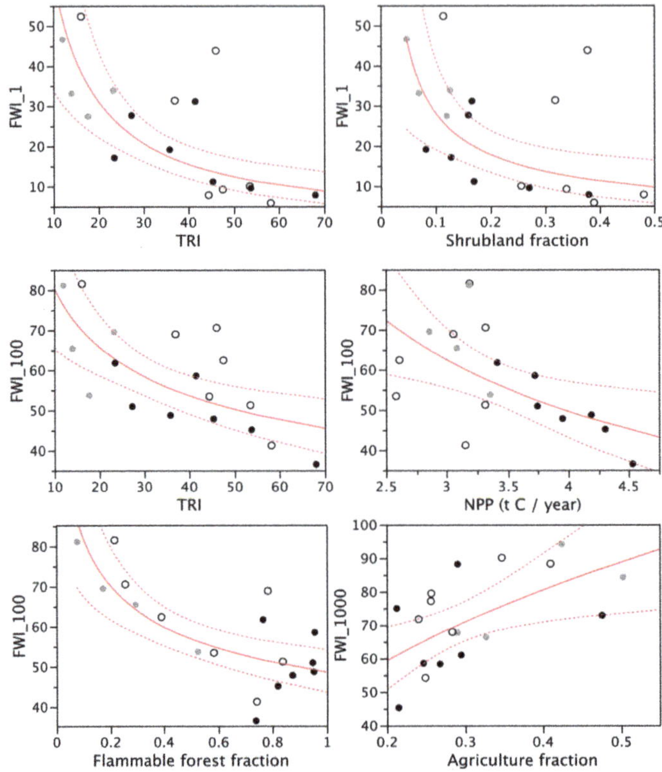

Figure 3. Selected relationships of the form $y = a \times x^b$ between regional FWI thresholds for the occurrence of fires ≥ 1 (FWI_1), ≥ 100 (FWI_100), and ≥ 1000 (FWI_1000) ha and biophysical variables, with 95% confidence intervals. Models were fitted on log-transformed variables. Black, white, and grey circles correspond to the NCW, NCE, and S supra regions of Portugal.

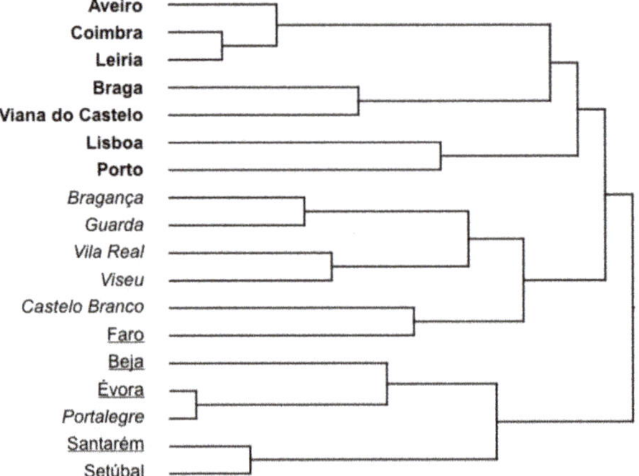

Figure 4. Regional clustering dendrogram as a function of FWI thresholds (≥ 1 ha, and ≥ 100 ha), population density, and biophysical variables (TRI, elevation, NPP, land cover fractions of agriculture, shrubland, forest, and flammable forest). Bold, italic, and underlined lettering indicates regions belonging respectively to the NCW, NCE, and S supra regions of Portugal.

Table 3. FWI thresholds resulting from logistic regression for the daily occurrence of increasingly larger fires in Portugal regions. Numbers within brackets respectively indicate the area under the curve (AUC) and misclassification rate. Italics denote thresholds exceeding the maximum observed FWI. Missing FWI values correspond to thresholds lower than the previous fire-size threshold.

Region	Fire Size (ha)						
	0.01	0.1	1	10	100	500	1000
Aveiro	0.6 (0.88; 0.11)	6.6 (0.83; 0.24)	19.1 (0.76; 0.29)	38.5 (0.80; 0.07)	48.7 (0.89; 0.01)	53.7 (0.93; <0.01)	*58.8 (0.91; <0.01)*
Beja	10.0 (0.73; 0.24)	18.5 (0.71; 0.29)	33.8 (0.71; 0.35)	63.7 (0.73; 0.16)	69.5 (0.89; 0.03)	84.3 (0.89; 0.01)	-
Braga	0.0 (0.89; 0.11)	2.3 (0.87; 0.14)	9.5 (0.80; 0.28)	30.8 (0.84; 0.14)	45.1 (0.92; 0.04)	54.6 (0.94; 0.01)	*61.0 (0.36; <0.01)*
Bragança	0.0 (0.76; 0.10)	0.0 (0.75; 0.16)	9.2 (0.72; 0.32)	45.8 (0.79; 0.20)	62.4 (0.86; 0.06)	79.9 (0.83; 0.01)	90.1 (0.86; 0.01)
Castelo Branco	0.0 (0.74; 0.12)	9.3 (0.72; 0.28)	31.3 (0.70; 0.35)	61.2 (0.81; 0.18)	68.9 (0.87; 0.05)	76.9 (0.89; 0.02)	79.5 (0.90; 0.01)
Coimbra	0.0 (0.77; 0.22)	12.2 (0.75; 0.33)	31.1 (0.76; 0.22)	50.3 (0.84; 0.06)	58.5 (0.83; 0.02)	*66.0 (0.88; 0.01)*	*75.0 (0.85; 0.01)*
Évora	23.7 (0.75; 0.29)	29.6 (0.74; 0.32)	46.6 (0.73; 0.30)	68.5 (0.80; 0.10)	81.1 (0.88; 0.02)	*93.1 (0.93; 0.01)*	*94.2 (0.96; 0.01)*
Faro	14.1 (0.76; 0.24)	29.1 (0.72; 0.34)	43.8 (0.73; 0.26)	65.2 (0.80; 0.05)	78.5 (0.86; 0.02)	-	-
Guarda	0.0 (0.76; 0.17)	0.0 (0.78; 0.24)	7.8 (0.75; 0.31)	38.9 (0.79; 0.21)	53.4 (0.87; 0.08)	63.6 (0.92; 0.02)	77.9 (0.93; 0.01)
Leiria	1.7 (0.82; 0.17)	11.4 (0.77; 0.25)	27.6 (0.76; 0.24)	43.5 (0.89; 0.06)	50.9 (0.93; 0.02)	56.3 (0.94; 0.01)	58.3 (0.97; <0.01)
Lisboa	0.5 (0.88; 0.12)	9.5 (0.86; 0.23)	17.1 (0.83; 0.24)	47.4 (0.88; 0.04)	61.7 (0.94; 0.01)	*72.9 (0.93; <0.01)*	-
Portalegre	16.8 (0.68; 0.33)	30.5 (0.68; 0.36)	52.3 (0.67; 0.32)	75.6 (0.77; 0.08)	*81.5 (0.85; 0.02)*	-	*88.3 (0.92; 0.01)*
Porto	0.2 (0.91; 0.09)	4.4 (0.90; 0.15)	11.1 (0.87; 0.21)	31.4 (0.87; 0.12)	47.8 (0.94; 0.03)	62.7 (0.96; 0.01)	*88.2 (0.97; <0.01)*
Santarén	0.4 (0.84; 0.13)	11.3 (0.78; 0.26)	27.4 (0.77; 0.29)	47.8 (0.88; 0.08)	53.7 (0.32; 0.03)	58.9 (0.93; 0.01)	*66.3 (0.93; 0.01)*
Setúbal	8.2 (0.86; 0.19)	20.9 (0.80; 0.27)	33.1 (0.77; 0.27)	59.3 (0.82; 0.04)	65.4 (0.91; 0.01)	65.5 (0.98; <0.01)	*67.7 (>0.99; <0.01)*
Viana do Castelo	0.0 (0.82; 0.12)	1.7 (0.79; 0.20)	7.8 (0.77; 0.32)	26.0 (0.78; 0.17)	36.5 (0.88; 0.04)	43.0 (0.94; 0.01)	45.3 (0.99; 0.01)
Vila Real	0.0 (0.85; 0.07)	0.6 (0.79; 0.19)	5.8 (0.76; 0.29)	30.5 (0.80; 0.18)	41.2 (0.88; 0.06)	50.6 (0.91; 0.02)	54.2 (0.95; 0.01)
Viseu	0.0 (0.83; 0.10)	0.6 (0.81; 0.19)	10.0 (0.77; 0.30)	38.5 (0.82; 0.16)	51.2 (0.90; 0.05)	62.6 (0.92; 0.02)	71.8 (0.91; 0.01)

Table 4. Correlation matrix (*r* values) between FWI thresholds for each fire size class, and between the latter and the median and 90th percentile of the FWI ($n = 18$). All correlations are significant at $p < 0.001$, except * ($p < 0.05$) and ** ($p < 0.01$).

	Fire-Size Thresholds						FWI	
	0.1	**1**	**10**	**100**	**500**	**1000**	**Median**	**90th Perc.**
0.01	0.902	0.819	0.789	0.794	0.673 **	0.541 *	0.719	0.666 **
0.1		0.963	0.889	0.827	0.655 **	0.483 *	0.853	0.620 **
1			0.928	0.851	0.677 **	0.501 *	0.828	0.668 **
10				0.969	0.834	0.659 **	0.875	0.859
100					0.922	0.786	0.819	0.888
500						0.879	0.697 **	0.839
1000							0.525 *	0.717

Table 5. Univariate explanation (r^2) of regional variability in daily FWI thresholds for fires attaining 1, 100, and 1000 ha through regional descriptors and relationships of the form log(y) = a + b log(x). Missing values indicate non-significant ($p > 0.05$) effects. Negative and positive influences are represented respectively by (−) and (+).

Variable	1 ha	100 ha	1000 ha
Popul. density		*0.22 (−)	
Elevation	0.44 (−)		
TRI	0.56 (−)	0.48 (−)	
NPP		0.33 (−)	0.29 (−)
Land cover fractions			
Agriculture			0.33 (+)
Shrubland	0.39 (−)		
Forest	0.24 (−)		
Flammable forest	0.32 (−)	0.54 (−)	0.32 (−)

*$p = 0.052$.

4. Discussion

4.1. FWI Thresholds for Increasingly Larger Fires and Fire Danger Rating

As expected, increasingly larger fires are gradually restricted to a lesser number of days (Table 2), those characterized by increasingly higher FWI (Table 3). Ability of the fire danger classification [10] in operational use in Portugal to match fire activity depends on whether its representation of fire behavior can be reasonably generalized and, overall, to what extent it captures the compounded influences involved. While regional FWI thresholds for the occurrence of fires ≥10 ha were either within the very high or the extreme fire danger classes [10], all thresholds for fires ≥100 ha were within the extreme class. This operational fire danger classification indicates fire control difficulty in case a fire occurs, rather than whether it will occur and overall fire activity. Still, the current nationwide fire danger thresholds were able to signal daily conditions for significant fires (≥10 ha) and, especially, for large fires (≥100 ha), despite the existing regional variability in FWI thresholds. Lower relative differences among regions were also found in the FWI thresholds required for the occurrence of large fires versus fires <100 ha. Both findings might arise from an overriding effect of fire weather, through which fire behavior is gradually equalized among the existing forest and shrubland vegetation types [62]; differential fire preference for vegetation types disappear [57]; and fire development to significant dimensions is fast if the initial attack fails when fire behavior is above the extinction capacity, partially cancelling interregional differences in ignition density, vegetation, and terrain.

Within-region discrimination between FWI thresholds was lower for fires ≥100, ≥500, and ≥1000 ha in comparison with smaller fires. This indicates that, within a given landscape context, the response in large fire size is less than proportional to increasingly severe weather conditions reflected in the

FWI. A limitation, as with any fire danger rating system, is that the FWI does not encapsulate all weather influences on fire growth, namely atmosphere structure and stability [25,63] and rapidly changing conditions or fine-scale variability resulting from thunderstorm outflows, frontal passages, or wind–terrain interactions [59]. However, results align with previous work revealing control of fire size by landscape structure [27,64]. This does not preclude weather-fuel synergies: Towards the upper end of the fire size range the response to increased meteorological fire danger is greater when fuel hazard is uniformly high across the landscape [25] and the largest fires coincide with extreme fire weather coupled with high fuel build-up [65,66].

Contemporary fire-weather relationships have been extensively examined in Portugal and the Mediterranean Basin, as described in the Introduction section. Nonetheless, there is a paucity of operationally oriented work with straightforward application to fire management decision-making. The FWI thresholds for the daily occurrence of fires of different sizes determined in this study can be viewed as an initial step in the development of a predictive system tailored to regional specificities. Additional variables are probably warranted for improved predictive performance in an operational context, namely seasonal patterns of fire occurrence, a descriptor of fire load, i.e., the suppression effort needed to tackle the fires occurring in a given region during a given time interval [4], and the ability to predict short-duration strong winds [67]. I used *distrito* as the regional stratification unit, to conform to the current and merely political fire-management zoning of the country, but usage of more coherent land units, e.g., phytoclimatic zones [42] or ecoregions [65], would probably result in more robust fire–weather relationships [20].

4.2. Regional Variation in Fire-Size Response to the FWI

Mainland Portugal offers an interesting opportunity to examine the interaction of natural and anthropogenic influences in determining the fire regime, given their variability within a relatively small area, roughly 600 km by 200 km. The substantial regional variation in fire weather thresholds required for wildfire development to different sizes was consistent with gradients in biophysical factors (Table 5, Figure 3). The FWI threshold for fire growth to a given size can be seen as the resistance of the human-landscape system to fire spread. Thus, resistance to fire spread decreased under rougher terrain, higher plant productivity, or higher coverage by shrubland, forest, and flammable forest, and was enhanced by the extent of farming, with the relative relevance of these individual influences depending on the fire-size partition considered. Generally, larger fires were fostered by the concomitance of more and better-connected fuel in topographically complex landscapes; these features are correlated, and are expected to match with less inhabited regions, but correlation analysis overlooks it (Table S2), possibly because of substantial intra-regional variation in population density and human influence [47].

Decreased correlation between FWI thresholds for increasingly distant fire-size partitions (Table 4) suggests that, as one example, changes in the relative importance of fire drivers across the fire size range [68] that may be a consequence of interactions between population density and landscape structure. The regions of Porto and Bragança offer clear-cut examples in this respect. Both have a low FWI_1 threshold, in Porto due to extremely high population density within landscapes dominated by wildland-urban interfaces (hence very high ignition pressure), in Bragança because of landscape dominance by farmland and shrubland, which presumes the use of fire in land management. These regions also share very high FWI_1000 thresholds, as the burnable landscape is either limited in extent or is substantially fragmented and particularly large fires will only develop under particularly severe and uncommon fire weather.

Data scatter in Figure 3, along with Table 1, allow some discussion on the relative role of human versus natural influences on FWI thresholds. For example, the Faro *distrito* deviates from the relationships between TRI and shrubland fraction with FWI_1, and from the FWI_100-TRI relationship, exhibiting high resistance to fire in relation to what would be expected. Vila Real shares similar NPP with Portalegre and Évora, but their respective FWI_100 are in the opposite extremes of the distribution, respectively corresponding to low and high resistance to fire spread. Population density is the distinguishing factor in both cases, presumably conferring higher resistance to fire when it is lower. Likewise, Viana do Castelo and Castelo Branco are similar in their fractions of flammable forest but contrast in FWI_100, respectively denoting low and high resistance to fire, possibly because of higher population density and marginally lower agricultural land use in the former. This suggests that population density, as a surrogate for ignition density [45,51,69], contributes to distinguish FWI-fire thresholds among similarly fire-prone landscapes.

The interaction of biophysical and human factors with the likelihood of weather conditions conducive to significant fire activity should introduce further complexity and nuance in fire–weather relationships. However, strong correlations between regional FWI thresholds and the corresponding median and 90th percentile of the FWI (Table 4) suggest that regional distinctions in the prevalence of more extreme fire weather (Figure 2) did not affect the fire-size FWI thresholds, likely because in Mediterranean climates the NPP gradient controls how fire activity responds to fire-inducing weather [70]. A study in southern France found that only 25% of the variation in wildfire spatial patterns was attributed to spatial variation in fire weather [35].

Cluster analysis revealed three broad pyroregions (Figure 4), roughly corresponding to the NCW, NCE, and S supra regions of the country, and representing differential resistance to fire spread, from low to high, as determined by land use, topography, NPP, and population density. This supplements previous analyses examining regional or spatial patterns of fire activity metrics that highlighted the role of climate [44], recognized the existence of synergistic effects [69], and carried out explicit modeling [25]. However, within-cluster variation was substantial down to adjacent regions. Further advances in understanding and systematizing fire-weather relationships from a fire management perspective and, more generally, in typifying fire regimes, can be achieved by working at finer spatial scales and using landscape-level metrics of fuel structure [27,71,72].

5. Conclusions

Fires grow in the landscape to increasingly large sizes as allowed by the composite influences of atmospheric and fuel dryness conditions, which in this study were expressed by the FWI of the Canadian forest fire weather index system. Larger fires require progressively more extreme fire weather, which occurs on a limited set of days. However, fire–weather relationships, as expressed by FWI-thresholds associated to the fire-size classes considered, show important regional variation in Portugal, even if the operationally-used fire danger rating classification is satisfactory at discriminating days with large fire (≥100 ha) occurrence. Relative differences between FWI thresholds for large fire-size classes variants decreased both within and among regions as compared with smaller fires, possibly signaling a greater control of fire spread exerted by landscape structure and a decrease in the anthropogenic-related ability to restrain fire development, but also fire–atmosphere–terrain influences and interactions unaccounted for by the analysis.

FWI thresholds as a function of fire size are an expression of resistance to fire spread. Such levels of resistance followed biophysical gradients. Coincidence of higher NPP and flammable forest fraction enabled fire development to larger sizes under comparatively mild fire weather. Decrease in NPP and flammable forest cover and high shrubland cover signaled intermediate resistance. Finally, decreased terrain roughness with less forest and shrubland and more agriculture corresponded with the highest resistance to fire spread. Population density is both a source of ignitions and a facilitator of fire suppression and its potential role warrants further research. Future improvements in fire-weather relationships analysis will likely result from the inclusion of additional variables, examined at the scales most coherent with the objective, i.e., fire management versus understanding of processes.

Supplementary Materials: The following are available online at http://www.mdpi.com/1999-4907/10/10/838/s1, Table S1: Percentiles of fire duration (days) for fires ≥0.01, 0.1, 1, 100, 500, and 1000 ha during the study period (2001–2011); Table S2: Correlation matrix between regional population density and biophysical variables among Portugal regions.

Funding: This work is financed by the ERDF—European Regional Development Fund through the COMPETE Program and by national funds through the FCT (Fundação para a Ciência e a Tecnologia) within project FIRE ENGINE—Flexible Design of Forest Fire Management Systems (MIT/FSE/0064/2009) and project UID/AGR/04033/2019.

Acknowledgments: I am indebted to Marco Magalhães and Tiago Monteiro-Henriques, who respectively derived the terrain metrics and NPP at regional level.

Conflicts of Interest: The author declares no conflict of interest.

References

1. Bowman, D.M.J.S.; Williamson, G.J.; Abatzoglou, J.T.; Kolden, C.A.; Cochrane, M.A.; Smith, A.M.S. Human exposure and sensitivity to globally extreme wildfire events. *Nat. Ecol. Evol.* **2017**, *1*, 0058. [CrossRef] [PubMed]
2. Turco, M.; Herrera, S.; Tourigny, E.; Chuvieco, E.; Provenzale, A. A comparison of remotely-sensed and inventory datasets for burned area in Mediterranean Europe. *Int. J. Appl. Earth Obs.* **2019**, *82*, 101887. [CrossRef]
3. Mateus, P.; Fernandes, P.M. Forest Fires in Portugal: Dynamics, Causes and Policies. In *Forest Context and Policies in Portugal*; Reboredo, F., Ed.; Springer International Publishing: Cham, Switzerland, 2014; pp. 97–115.
4. Martell, D.L. Forest Fire Management. In *Handbook of Operations Research In Natural Resources*; Weintraub, A., Romero, C., Bjørndal, T., Epstein, R., Miranda, J., Eds.; Springer: Boston, MA, USA, 2007; pp. 489–509.
5. Wotton, B.M. Interpreting and using outputs from the Canadian Forest Fire Danger Rating System in research applications. *Environ. Ecol. Stat.* **2009**, *16*, 107–131. [CrossRef]
6. Stocks, B.J.; Lynham, T.J.; Lawson, B.D.; Alexander, M.E.; Wagner, C.V.; McAlpine, R.S.; Dube, D.E. Canadian forest fire danger rating system: An overview. *For. Chron.* **1989**, *65*, 258–265. [CrossRef]
7. Burgan, R.E. *1988 Revisions to the 1978 National Fire-Danger Rating System*; US Department of Agriculture, Forest Service, Southeastern Forest Experiment Station: Asheville, NC, USA, 1988; p. 144.
8. Matthews, S. A comparison of fire danger rating systems for use in forests. *Aust. Meteorol. Ocean.* **2009**, *58*, 41. [CrossRef]
9. Van Wagner, C.E. *Development and Structure of the Canadian Forest Fire Weather Index System*; Forestry Technical Report 35; Canadian Forest Service: Ottawa, ON, Canada, 1987; p. 35.
10. Palheiro, P.M.; Fernandes, P.; Cruz, M.G. A fire behaviour-based fire danger classification for maritime pine stands: Comparison of two approaches. *For. Ecol. Manag.* **2006**, *234*, S54. [CrossRef]
11. Andrews, P.L.; Loftsgaarden, D.O.; Bradshaw, L.S. Evaluation of fire danger rating indexes using logistic regression and percentile analysis. *Int. J. Wildland Fire* **2003**, *12*, 213–226. [CrossRef]
12. DaCamara, C.C.; Calado, T.J.; Ermida, S.L.; Trigo, I.F.; Amraoui, M.; Turkman, K.F. Calibration of the Fire Weather Index over Mediterranean Europe based on fire activity retrieved from MSG satellite imagery. *Int. J. Wildland Fire* **2014**, *23*, 945–958. [CrossRef]
13. Davies, G.M.; Legg, C.J. Regional variation in fire weather controls the reported occurrence of Scottish wildfires. *PeerJ* **2016**, *4*, e2649. [CrossRef]
14. De Jong, M.C.; Wooster, M.J.; Kitchen, K.; Manley, C.; Gazzard, R.; McCall, F.F. Calibration and evaluation of the Canadian Forest Fire Weather Index (FWI) System for improved wildland fire danger rating in the United Kingdom. *Nat. Hazard Earth Sys.* **2016**, *16*, 1217–1237. [CrossRef]
15. Šturm, T.; Fernandes, P.M.; Šumrada, R. The Canadian fire weather index system and wildfire activity in the Karst forest management area, Slovenia. *Eur. J. For. Res.* **2012**, *131*, 829–834. [CrossRef]
16. Amatulli, G.; Camia, A.; San-Miguel-Ayanz, J. Estimating future burned areas under changing climate in the EU-Mediterranean countries. *Sci. Total Environ.* **2013**, *450–451*, 209–222. [CrossRef] [PubMed]
17. Carvalho, A.; Flannigan, M.D.; Logan, K.; Miranda, A.I.; Borrego, C. Fire activity in Portugal and its relationship to weather and the Canadian Fire Weather Index System. *Int. J. Wildland Fire* **2008**, *17*, 328–338. [CrossRef]
18. Giannakopoulos, C.; LeSager, P.; Moriondo, M.; Bindi, M.; Karali, A.; Hatzaki, M.; Kostopoulou, E. Comparison of fire danger indices in the Mediterranean for present day conditions. *iForest* **2012**, *5*, 197. [CrossRef]

19. Jiménez-Ruano, A.; Mimbrero, M.R.; Jolly, W.M.; de la Riva Fernández, J. The role of short-term weather conditions in temporal dynamics of fire regime features in mainland Spain. *J. Environ. Manag.* **2019**, *241*, 575–586. [CrossRef] [PubMed]

20. Padilla, M.; Vega-García, C. On the comparative importance of fire danger rating indices and their integration with spatial and temporal variables for predicting daily human-caused fire occurrences in Spain. *Int. J. Wildland Fire* **2011**, *20*, 46–58. [CrossRef]

21. Papakosta, P.; Straub, D. Probabilistic prediction of daily fire occurrence in the Mediterranean with readily available spatio-temporal data. *iForest* **2016**, *10*, 32. [CrossRef]

22. Urbieta, I.R.; Zavala, G.; Bedia, J.; Gutiérrez, J.M.; Miguel-Ayanz, J.S.; Camia, A.; Keeley, J.E.; Moreno, J.M. Fire activity as a function of fire—weather seasonal severity and antecedent climate across spatial scales in southern Europe and Pacific western USA. *Environ. Res. Lett.* **2015**, *10*, 114013. [CrossRef]

23. Venäläinen, A.; Korhonen, N.; Hyvärinen, O.; Koutsias, N.; Xystrakis, F.; Urbieta, I.R.; Moreno, J.M. Temporal variations and change in forest fire danger in Europe for 1960–2012. *Nat. Hazards Earth Sys.* **2014**, *14*, 1477–1490. [CrossRef]

24. Ager, A.A.; Preisler, H.K.; Arca, B.; Spano, D.; Salis, M. Wildfire risk estimation in the Mediterranean area. *Environmetrics* **2014**, *25*, 384–396. [CrossRef]

25. Fernandes, P.M.; Barros, A.M.G.; Pinto, A.; Santos, J.A. Characteristics and controls of extremely large wildfires in the western Mediterranean Basin. *J. Geophys. Res. Biogeo* **2016**, *121*, 2141–2157. [CrossRef]

26. Fernandes, P.M.; Pacheco, A.P.; Almeida, R.; Claro, J. The role of fire-suppression force in limiting the spread of extremely large forest fires in Portugal. *Eur. J. For. Res.* **2016**, *135*, 253–262. [CrossRef]

27. Fernandes, P.M.; Monteiro-Henriques, T.; Guiomar, N.; Loureiro, C.; Barros, A.M.G. Bottom-up variables govern large-fire size in Portugal. *Ecosystems* **2016**, *19*, 1362–1375. [CrossRef]

28. Lahaye, S.; Curt, T.; Fréjaville, T.; Sharples, J.; Paradis, L.; Hély, C. What are the drivers of dangerous fires in Mediterranean France? *Int. J. Wildland Fire* **2018**, *27*, 155–163. [CrossRef]

29. Strauss, D.; Bednar, L.; Mees, R. Do one percent of the forest fires cause ninety-nine percent of the damage? *For. Sci.* **1989**, *35*, 319–328.

30. Lannom, K.O.; Tinkham, W.T.; Smith, A.M.S.; Abatzoglou, J.; Newingham, B.A.; Hall, T.E.; Morgan, P.; Strand, E.K.; Paveglio, T.B.; Anderson, J.W.; et al. Defining extreme wildland fires using geospatial and ancillary metrics. *Int. J. Wildland Fire* **2014**, *23*, 322–337. [CrossRef]

31. Cui, W.; Perera, A.H. What do we know about forest fire size distribution, and why is this knowledge useful for forest management? *Int. J. Wildland Fire* **2008**, *17*, 234–244. [CrossRef]

32. Arienti, M.C.; Cumming, S.G.; Boutin, S. Empirical models of forest fire initial attack success probabilities: The effects of fuels, anthropogenic linear features, fire weather, and management. *Can. J. For. Res.* **2006**, *36*, 3155–3166. [CrossRef]

33. Fang, L.; Yang, J.; Zu, J.; Li, G.; Zhang, J. Quantifying influences and relative importance of fire weather, topography, and vegetation on fire size and fire severity in a Chinese boreal forest landscape. *For. Ecol. Manag.* **2015**, *356*, 2–12. [CrossRef]

34. Moreira, F.; Catry, F.X.; Rego, F.; Bacao, F. Size-dependent pattern of wildfire ignitions in Portugal: When do ignitions turn into big fires? *Landsc. Ecol.* **2010**, *25*, 1405–1417. [CrossRef]

35. Ruffault, J.; Mouillot, F. Contribution of human and biophysical factors to the spatial distribution of forest fire ignitions and large wildfires in a French Mediterranean region. *Int. J. Wildland Fire* **2017**, *26*, 498–508. [CrossRef]

36. Price, O.F.; Penman, T.; Bradstock, R.; Borah, R. The drivers of wildfire enlargement do not exhibit scale thresholds in southeastern Australian forests. *J. Environ. Manag.* **2016**, *181*, 208–217.

37. Whitman, E.; Batllori, E.; Parisien, M.-A.; Miller, C.; Coop, J.D.; Krawchuk, M.A.; Chong, G.W.; Haire, S.L. The climate space of fire regimes in north-western North America. *J. Biogeogr.* **2015**, *42*, 1736–1749. [CrossRef]

38. Bowman, D.M.J.S.; Balch, J.; Artaxo, P.; Bond, W.J.; Cochrane, M.A.; D'Antonio, C.M.; DeFries, R.; Johnston, F.H.; Keeley, J.E.; Krawchuk, M.A.; et al. The human dimension of fire regimes on Earth: The human dimension of fire regimes on Earth. *J. Biogeogr.* **2011**, *38*, 2223–2236. [CrossRef]

39. Parisien, M.-A.; Miller, C.; Parks, S.A.; DeLancey, E.R.; Robinne, F.-N.; Flannigan, M.D. The spatially varying influence of humans on fire probability in North America. *Environ. Res. Lett.* **2016**, *11*, 075005. [CrossRef]

40. Syphard, A.D.; Keeley, J.E.; Pfaff, A.H.; Ferschweiler, K. Human presence diminishes the importance of climate in driving fire activity across the United States. *Proc. Natl. Acad. Sci. USA* **2017**, *114*, 13750–13755. [CrossRef]

41. Ruffault, J.; Mouillot, F. How a new fire-suppression policy can abruptly reshape the fire-weather relationship. *Ecosphere* **2015**, *6*, art199. [CrossRef]

42. Bedia, J.; Herrera, S.; Gutiérrez, J.M. Assessing the predictability of fire occurrence and area burned across phytoclimatic regions in Spain. *Nat. Hazards Earth Sys.* **2014**, *14*, 53–66. [CrossRef]

43. Jiménez-Ruano, A.; Mimbrero, M.R.; de la Riva Fernández, J. Understanding wildfires in mainland Spain. A comprehensive analysis of fire regime features in a climate-human context. *Appl. Geogr.* **2017**, *89*, 100–111. [CrossRef]

44. Trigo, R.M.; Sousa, P.M.; Pereira, M.G.; Rasilla, D.; Gouveia, C.M. Modelling wildfire activity in Iberia with different atmospheric circulation weather types. *Int. J. Climatol.* **2016**, *36*, 2761–2778.

45. Costa, L.; Thonicke, K.; Poulter, B.; Badeck, F.-W. Sensitivity of Portuguese forest fires to climatic, human, and landscape variables: Subnational differences between fire drivers in extreme fire years and decadal averages. *Reg. Environ. Change* **2011**, *11*, 543–551. [CrossRef]

46. Marques, S.; Borges, J.G.; Garcia-Gonzalo, J.; Moreira, F.; Carreiras, J.M.B.; Oliveira, M.M.; Cantarinha, A.; Botequim, B.; Pereira, J.M.C. Characterization of wildfires in Portugal. *Eur. J. For. Res.* **2011**, *130*, 775–784. [CrossRef]

47. Sá, A.C.L.; Turkman, M.A.A.; Pereira, J.M.C. Exploring fire incidence in Portugal using generalized additive models for location, scale and shape (GAMLSS). *Model. Earth Syst. Environ.* **2018**, *4*, 199–220. [CrossRef]

48. Moreno, M.V.; Conedera, M.; Chuvieco, E.; Pezzatti, G.B. Fire regime changes and major driving forces in Spain from 1968 to 2010. *Environ. Sci. Policy* **2014**, *37*, 11–22. [CrossRef]

49. ICNF. Available online: http://www2.icnf.pt/portal/florestas/dfci/inc/estat-sgif (accessed on 20 August 2019).

50. ICNF—Cartografia da área Ardida. Available online: http://www2.icnf.pt/portal/florestas/dfci/inc/mapas (accessed on 20 August 2019).

51. Catry, F.X.; Rego, F.C.; Bação, F.L.; Moreira, F. Modeling and mapping wildfire ignition risk in Portugal. *Int. J. Wildland Fire* **2010**, *18*, 921–931. [CrossRef]

52. Riley, S.J.; DeGloria, S.D.; Elliot, R. A terrain ruggedness index that quantifies topographic heterogeneity. *Intermt. J. Sci.* **1999**, *5*, 23–27.

53. Imhoff, M.L.; Bounoua, L.; Ricketts, T.; Loucks, C.; Harriss, R.; Lawrence, W.T. *HANPP Collection: Global Patterns in Net Primary Productivity (NPP)*; SEDAC: Palisades, NY, USA, 2004.

54. Imhoff, M.L.; Bounoua, L. Exploring global patterns of net primary production carbon supply and demand using satellite observations and statistical data. *J. Geophys. Res.-Atmos.* **2006**, *111*, D22. [CrossRef]

55. ICNF, Relatório Final IFN5—FloreStat. Available online: http://www2.icnf.pt/portal/florestas/ifn/ifn5/rel-fin (accessed on 20 August 2019).

56. Fernandes, P.M. Combining forest structure data and fuel modelling to classify fire hazard in Portugal. *Ann. For. Sci.* **2009**, *66*, 415. [CrossRef]

57. Barros, A.M.G.; Pereira, J.M.C. Wildfire selectivity for land cover type: Does size matter? *PLoS ONE* **2014**, *9*, e84760. [CrossRef]

58. Moreira, F.; Vaz, P.; Catry, F.; Silva, J.S. Regional variations in wildfire susceptibility of land-cover types in Portugal: Implications for landscape management to minimize fire hazard. *Int. J. Wildland Fire* **2009**, *18*, 563–574. [CrossRef]

59. Jolly, W.M.; Freeborn, P.H.; Page, W.G.; Butler, B.W. Severe fire danger index: A forecastable metric to inform firefighter and community wildfire risk management. *Fire* **2019**, *2*, 47. [CrossRef]

60. Hosmer, D.W.; Lemeshow, S. *Applied Logistic Regression*, 2nd ed.; Wiley: New York, NY, USA, 2000; p. 375.

61. Amraoui, M.; Pereira, M.G.; DaCamara, C.C.; Calado, T.J. Atmospheric conditions associated with extreme fire activity in the Western Mediterranean region. *Sci. Total Environ.* **2015**, *524–525*, 32–39.

62. Cruz, M.G.; Alexander, M.E. The 10% wind speed rule of thumb for estimating a wildfire's forward rate of spread in forests and shrublands. *Ann. For. Sci.* **2019**, *76*, 44.

63. Potter, B.E. Atmospheric interactions with wildland fire behaviour—II. Plume and vortex dynamics. *Int. J. Wildland Fire* **2012**, *21*, 802–817.

64. Viedma, O.; Angeler, D.G.; Moreno, J.M. Landscape structural features control fire size in a Mediterranean forested area of central Spain. *Int. J. Wildland Fire* **2009**, *18*, 575–583. [CrossRef]

65. Fernandes, P.M.; Loureiro, C.; Magalhães, M.; Ferreira, P.; Fernandes, M. Fuel age, weather and burn probability in Portugal. *Int. J. Wildland Fire* **2012**, *21*, 380–384. [CrossRef]
66. Koutsias, N.; Arianoutsou, M.; Kallimanis, A.S.; Mallinis, G.; Halley, J.M.; Dimopoulos, P. Where did the fires burn in Peloponnisos, Greece the summer of 2007? Evidence for a synergy of fuel and weather. *Agr. For. Meteorol.* **2012**, *156*, 41–53.
67. Abatzoglou, J.T.; Balch, J.K.; Bradley, B.A.; Kolden, C.A. Human-related ignitions concurrent with high winds promote large wildfires across the USA. *Int. J. Wildland Fire* **2018**, *27*, 377–386. [CrossRef]
68. Slocum, M.G.; Beckage, B.; Platt, W.J.; Orzell, S.L.; Taylor, W. Effect of climate on wildfire size: A cross-scale analysis. *Ecosystems* **2010**, *13*, 828–840. [CrossRef]
69. Nunes, A.N. Regional variability and driving forces behind forest fires in Portugal an overview of the last three decades (1980–2009). *Appl. Geogr.* **2012**, *34*, 576–586. [CrossRef]
70. Pausas, J.G.; Paula, S. Fuel shapes the fire—climate relationship: Evidence from Mediterranean ecosystems. *Global Ecol. Biogeogr.* **2012**, *21*, 1074–1082. [CrossRef]
71. Curt, T.; Borgniet, L.; Bouillon, C. Wildfire frequency varies with the size and shape of fuel types in southeastern France: Implications for environmental management. *J. Environ. Manag.* **2013**, *117*, 150–161. [CrossRef]
72. Loepfe, L.; Martinez-Vilalta, J.; Oliveres, J.; Piñol, J.; Lloret, F. Feedbacks between fuel reduction and landscape homogenisation determine fire regimes in three Mediterranean areas. *Forest Ecol. Manag.* **2010**, *259*, 2366–2374. [CrossRef]

 *forests*

Article

Methodology for the Study of Near-Future Changes of Fire Weather Patterns with Emphasis on Archaeological and Protected Touristic Areas in Greece

Vassiliki Varela *[ID], Diamando Vlachogiannis[ID], Athanasios Sfetsos[ID], Nadia Politi and Stelios Karozis[ID]

Environmental Research Laboratory (EREL), INRASTES, NCSR "Demokritos", 15341 Agia-Paraskevi, Greece; mandy@ipta.demokritos.gr (D.V.); ts@ipta.demokritos.gr (A.S.); nadiapol@ipta.demokritos.gr (N.P.); skarozis@ipta.demokritos.gr (S.K.)
* Correspondence: vvarela@ipta.demokritos.gr; Tel.: +30-210-6503403

Received: 14 September 2020; Accepted: 30 October 2020; Published: 31 October 2020

Abstract: This work introduces a methodology for assessing near-future fire weather pattern changes based on the Canadian Fire Weather Index system components (Fire Weather Index (FWI), Initial Spread Index (ISI), Fire Severity Rating (FSR)), applied in touristic areas in Greece. Four series of daily raster-based datasets for the fire seasons (May–October), concerning a historic (2006 to 2015) and a future climatology period (2036–2045), were created for the areas under consideration, based on high-resolution climate modelling with the Representative Concentration Pathway (RCP), PCR 4.5 and RCP 8.5 scenarios. The climate model data were obtained from the European Coordinated Downscaling Experiment (EURO-CORDEX) climate database and consisted of atmospheric variables as required by the FWI system, at 12.5 km spatial resolution. The final datasets of the abovementioned variables used for the study were processed at 5 km spatial resolution for the domain of interest after applying regridding based on the nearest neighbour interpolating process. Geographic Information Systems (GIS) spatial operations, including spatial statistics and zonal analyses, were applied on the series of the derived daily raster maps in order to provide a number of output thematic layers. Moreover, historic FWI percentile values, which were estimated for Greece in the frame of a past research study of the Environmental Research Laboratory (EREL), were used as reference data for further evaluation of future fire weather changes. The straightforward methodology for the assessment of the evolution of spatial and temporal distribution of Fire weather Danger due to climate change presented herewith is an essential tool for enhancing the knowledge for the decision support process for forest fire prevention, planning and management policies in areas where the fire risk both in terms of fire hazard likelihood and expected impact is quite important due to human presence and cultural prestige, such as archaeological and tourist protected areas.

Keywords: fire danger; fire weather patterns; climate change; RCP; FWI system; SSR

---

1. Introduction

Fire plays an important role in ecosystems structure and function in forested and non-forested lands worldwide and in the Mediterranean region as well, where the climate favours noticeable ecological diversity and wildland fire occurrence. All over the Mediterranean areas, fuel moisture level is expected to decrease due to climate change. Thus, the meteorological danger of wildland fires is likely to increase as the region becomes drier [1] with extended low moisture areas northwards [2]. In addition, the presence of human population is intense in many of these areas, making fire hazard and risk management a major concern and priority. This need emanates from the fact that wildfire activity is projected to increase under future climate conditions and in conjunction with ongoing land use change, forest fires are becoming a reoccurring hazard of forested landscapes globally, posing significant risks to local and regional communities [3,4].

Consequently, new territories start facing increased fire risk but the long term interactions between environmental factors, the social context and the fire regime, as well as the changing fire behaviour spatial patterns, are still largely unknown [5], despite that the relationship of wildfire occurrence to physical and social factors is a widely-researched topic [6,7]. Usually, for the prediction of wildfire occurrence, the factors considered include meteorological data [8,9] and physical indices from fire-danger rating systems [9,10]. Such prediction forms the basis for costly wildfire pre-suppression activities, such as aircraft fire detection flights and pre-fire distribution of firefighting means [11].

The potential for climate change to cause "novel" or "no analogue" environmental conditions in some ecosystems presents new challenges for management, policy and planning [4]. In addition to the day-by-day fire-danger mapping, spatial information related to the fire history and physiognomic characteristics, such as maps of fire frequency, severity, size, and pattern, are useful for planning fire and natural resource management at a strategic level for an area of interest. This type of information can also be used for the assessment of risk and ecological conditions and for the study of fire regimes and their changes as a function of the specific characteristics/features of a territory, such as climate, topography, vegetation and land use [12].

Furthermore, Geographic Information Systems (GIS) can be used for the integration of spatial layers of information for the identification and analysis of spatial patterns of wildfire occurrence and for deriving fire risk at different scales. Different spatial analysis techniques can be applied to answer the questions of "where" and "why" these wildfires are occurring [6,13]. It is also well known that the comparison among regions (e.g., of different geographic orientation and context; northwest, southwest, intermountain region), within regions (across biophysical settings), and across time is a powerful way to understand the factors that determine and constrain the fire patterns [14–16].

It is clearly justified that the main causes of fire have to be minimized, a process which needs to include the investigation of the social and economic factors that lead people to start fires, increasing awareness of the danger, encouraging good behaviour and sanctioning offenders. In particular, the importance of the wildland-urban interface in potentially catalysing fire impacts should be focused on a context where wildfires are genuinely understood as a natural hazard and defensible space is considered even from a social and policy perspective [7].

Montiel et al., 2016 [6], stated that zoning and characterizing a fire-prone territory require special spatial units to analyse the variability of attributes previously defined as relevant to wildfires, both structurally and dynamically. They proposed the use of landscape units obtained from landscape character assessments as the most appropriate ones for a smaller scale approach and drainage basins, which can then be easy to define using GIS, for a larger scale approach.

In addition to the above zoning approach, which is based on physical characteristics, the authors of the current study consider that social and cultural aspects should not be underestimated for the delineation of fire-prone territories and for the definition of respective fire management units. According to Ryan et al., (2012) [17], a landscape approach provides the tools for organizing and understanding intellectual and practical issues engaged by the topic of fire effects on cultural resources.

Any management decisions that affect cultural resources, also affect people and local communities—sometimes in direct and damaging ways. Understanding fuels, fire behaviour, and heat transfer mechanisms is a key to predicting, managing, and monitoring the effects of fire on cultural resources. Cultural resources are important resources that bind those of us living today with our ancestors, traditions, and histories. They are generally viewed as non-renewable resources. They are often fragile tangible objects, susceptible to thermal damage during wildland fires (wildfires and prescribed fires) and physical damage from management-related disturbances [17].

Recognizing these particularities, as well as the necessity of both rational and effective fire management in changing fire regime conditions in Greece due to climate change, the methodology for the study of near-future fire weather changes, proposed in the current paper, focuses on specific areas which cover a number of criteria that combine social and cultural aspects with fire regime and physical characteristics. Greece has suffered from significant forest fires during the last few decades, which, among others, has affected forested archaeological sites [18,19]. As these areas accept thousands of tourists every year, their effective protection and management is essential and imperative.

A number of previous studies elaborated, most of them by using the FWI system indices, on the impact of climate change on the fire regime in the Mediterranean region and particularly in Greece, and indicated that fire danger and risk would increase in the near future in these areas [3,20–22]. Likewise, in a 2011 research work [23], where the FWI system indices and other climatic indices were used with the the Intergovernmental Panel on Climate Change (IPCC) Special Report Emission Scenarios (SRES), A1B, meteorological data and the Regional Atmospheric Climate Model (RACMO2) version 2 of KNMI (Royal Netherlands Meteorological Institute: De Bilt, Netherlands) a substantial increase was found in the number of fire risk days in the near future for a number of areas of agricultural and touristic importance in Greece. Many of the abovementioned studies have used older versions of climatic models, nowadays considered rather less accurate and reliable; however, all the research results have clearly shown the tendency of fire danger and risk to increase in the region, highlighting its importance on economical and societal domains. Moreover, they have put on the table the necessity for studying the fire regime future projection in view of climate change, as an essential prerequisite for fire management, in every area under consideration. Thus, the previous research works mainly focused on the quantitative results obtained for the specific considered geographical region, as well as on the grade of changes in fire danger and risk indices. The current study aims at the provision of indicators and methodological tools for the quantitative assessment of fire weather in "Areas of Interest (AoI)", applicable in any geographical region, that can be considered as distinctive units in terms of all levels of fire management. Accordingly, the AoI selected for Greece, consist of extended areas of the Natura 2000 network, which include archaeological sites or sites of natural beauty with intense touristic load. It is worth mentioning that, according to de Rigo et al., 2017 [5] for specific typologies of forests, increasing the size of protected areas, such as Natura 2000 sites, might even be considered as a potential option for adaptation if other strategies are considered in parallel.

2. Materials and Methods

In this section, the developed methodology to obtain the stated objectives of the study are described. The research approach, based on ArcGIS 10.8 (Environmental Systems Research Institute –ESRI: Charlotte, NC, USA), comprised a number of flow processes, which are depicted in the diagram of Figure 1. In a first step, the fire weather calculation for the whole country was carried out, based on the Canadian FWI system components and the input climate datasets of the studied historic and future periods, for the two different emission scenarios. Then, a number of auxiliary thematic layers were selected concerning the country, such as the Natura 2000 map, the bioclimatic map that was part of the ESRI living Atlas and the FWI extreme class thresholds map as derived according to recent past research of the authors [23], Varela et al., 2018. These thematic layers served to select the AoI, taking into consideration a number of criteria, discussed later. In a next step, the selected AoI and the thematic layers were used with GIS functions and tools (e.g., cell by cell analysis, zonal analysis and statistics, spectral profile analysis) to derive the final products which were daily maps and spatial datasets for a number of components of FWI that express the fire spread rate and fire danger. In the sub-sections below, the applied methodology is presented in detail together with the datasets used and the procedure for the analysis of the results.

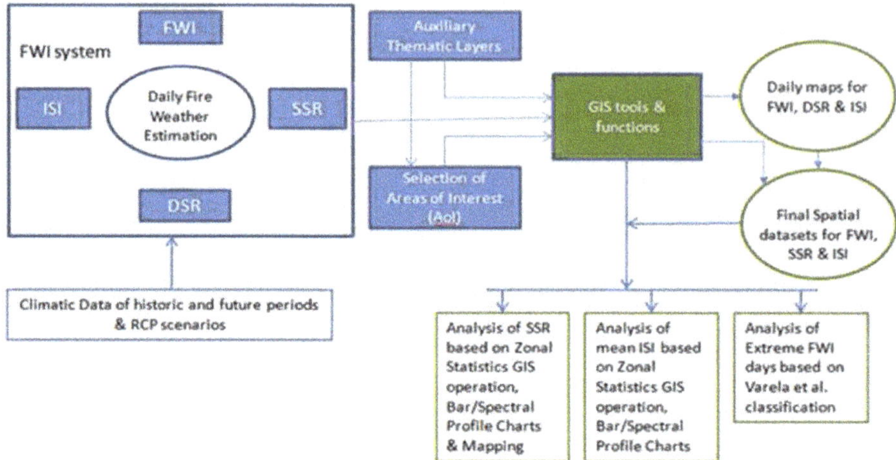

Figure 1. Flow process of the developed methodology for the derivation of fire weather patterns for selected Areas of Interest (FWI—Fire Weather Index, ISI—Initial Spread Index, DSR—Daily Severity Rating, SSR—Seasonal Severity Rating, RCP—Representative Concentration Pathway, GIS—Geographic Information System).

2.1. Fire Weather Estimation and Climatic Data

The Canadian FWI System, apart from its classical use as a daily fire weather rating system, is widely used for the projection and study of fire weather due to climate change [24,25]. The FWI System is comprised of six components: five of them are intermediate outputs of the system, namely the Fine Fuel Moisture Code, the Duff Moisture Code, the Drought Code, the Initial Spread Index and the BuildUp Index. The final output is the Fire Weather Index (FWI). An extension of the FWI output is the Daily Severity Rating (DSR) [24,25], which is averaged through a fire season for the calculation of the Seasonal Severity Rating (SSR), allowing for the objective comparison of fire danger from year to year and from region to region. It has been used for expressing the projection of Fire danger changes by the European Environment Agency [26].

The parameters of the FWI system that are employed for the purposes of this study are:

- The Initial Spread Index (ISI) which expresses the expected rate of fire spread.
- The Fire Weather Index (FWI) as the main indicator of fire danger representing the potential fire line intensity [14].
- The Daily Severity Rating (DSR) and fire season SSR, which represent the difficulty of controlling fires and reflect the expected fire suppression expected efforts [27].

Daily meteorological values at noon (12:00) of near surface temperature, relative humidity and 10-m wind speed, as well as 24-h cumulative precipitation, are used for the calculation of the components of the system for the whole country. The values of FWI vary from 0 to above 100. The ISI parameter is also used for operational purposes as an indicator of fire spread rate [28]. According to Viegas et al., 1999 [10], ISI was found to be an interesting index for the prediction of extreme fire conditions, as a result of an extended drought and strong wind conditions.

The above parameters of FWI system are classified into 46 classes for operation purposes, using varying classification thresholds, depending on the area of application [23,24,29], and they are an important and effective decision support tool used for forecasting the levels of preparedness needed for an area, the determination of the appropriate mitigation measures, the definition of organizational requirements and the support of fire control measures that are appropriate for the area [24].

In the current study, the climatic data, used as input to the FWI system, were derived from the state-of-the-art Global–Regional Climate modelling system ICHEC-EC-EARTH v2.3 /SMHI-RCA4 v4 (Irish Centre for High-End Computing (ICHEC)-European Consortium-Earth (EC-Earth) version 2.3/Swedish Meteorological and Hydrological Institute (SMHI)-Rossby Centre Atmospheric model version 4) simulations [30] of the openly accessible SMHI archives of the European Coordinated Downscaling Experiment (EURO-CORDEX) at spatial resolution of 12 km. The Regional Climate Model (RCM) simulations were retrieved for the domain of Greece, for the historic (2006 to 2015) and future (2036 to 2045) periods and for the latest IPCC RCP scenarios of climate forcing of 4.5 and 8.5 W/m^2 [31,32]. The retrieved variables were values of temperature at 2 m, wind speed at 10 m and relative humidity at 12:00 UTC (Universal Time Coordinated) as well as total precipitation on a 24 h basis.

The RCA4 model performance was evaluated for the recent past climate by Strandberg et al., 2014 [33]. An uncertainty assessment due to systematic bias of the model future climate simulation RCA4 has already been carried out by Sørland et al., 2018 [34]. The excellent performance of the SMHI model has been assessed for Greece by comparing the SMHI-RCA4 EUROCORDEX historic simulations with available meteorological data from the Hellenic National Meteorological Service (HNMS), by Katopodis et al., 2019 [35].

2.2. Auxiliary Thematic Layers and Selection of AoI

As previously mentioned, specific thematic layers were used as auxiliary data for the derivation of AoI and the analysis of the results. In particular, the thematic layers employed were the Natura 2000 map for Greece, the Bioclimatic World map and the FWI extreme class thresholds.

The Natura 2000 dataset consists of Special Areas of Conservation (SACs) and Special Protection Areas (SPAs) designated, respectively, under the Habitats Directive and Birds Directive. The Habitats Directive requires Sites of Community Importance (SCIs), which, upon the agreement of the European Commission, become Special Areas of Conservation (SACs) to be designated for species other than birds, and for habitat types (e.g., particular types of forest, grasslands, wetlands, etc.). In Greece, 202 areas have been registered as SPAs and 241 as SCIs. The area covered by the above 443 Greek Natura 2000 areas, cover about 19% of the country [36].

The Bioclimatic World map is part of the ESRI living Atlas [37] and provides access to a 250 m cell-sized raster with a bioclimatic stratification into classes based on factors that influence the distribution of plants and animals [38]. This layer was used to select distribution across the country AoI of various bioclimatic zones reflecting different fire prone areas.

The FWI extreme class thresholds map, according to the methodology of Varela et al., 2018 [23] (Figure 2), was necessary for the analysis and interpretation of the results. According to this methodology, the FWI classification takes into consideration the environmental variety of the country, which highly influences the significance of FWI values and consequently their interpretation as reasonable and functional fire danger classes. The classification approach, applied in the Greek Local Forest Service Office (LFSO) areas, is based on Percentile Indices and provide suitably varying FWI boundaries of classes based on the specific physical characteristics of the study area.

In compliance with the aim of the current study, the AoI were selected in order to satisfy a combination of the following criteria:

- Forest Fire prone areas
- Areas with high cultural and/or touristic interest
- Areas with high ecological/environmental interest
- Distribution in a variety of Bioclimatic zones within Greece

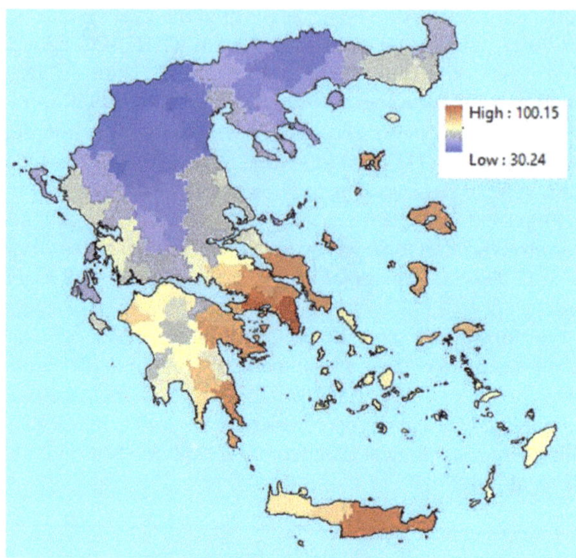

Figure 2. Map of Fire Weather Index (FWI) values thresholds for Extreme FWI class for each Forest Office Area.

Thus, the Natura 2000 network of areas was considered as a very important and appropriate basis, since it covers a significant part of the country and includes areas with the above defined criteria. A brief preliminary study of the Natura 2000 areas in conjunction with additional data regarding archaeological sites and touristic load, taking into account the Bioclimatic zones, resulted in the final list of the selected 19 areas (Table 1).

Table 1. Catalogue of the selected Natura 2000 areas.

CODE	HECTARES	NAME
GR1150012	17,592.2	Thasos (Ypsario Mountain and Seaside Zone) and Koinyra, Xironisi Islands
GR2130011	53,407.8	Central Zagori and Eastern Part Of Mitsikeli Mountain
GR4210005	27,696.2	Rodos Island: Akramytis, Armenistis, Attavyros, Streams and Seaside Zone (Karavola-Ormos Glyfada)
GR1250001	19,139.5	Olympos Mountain
GR3000001	14,902.4	Parnitha Mountain
GR1270014	23,451.1	Sithonias Peninsula
GR4110011	14,787.9	Olympos Lesvou Mountain
GR2550009	48,785.9	Taygetos-Lagkada Trypis Mountain
GR4210029	13,441.9	Eastern Rodos Island: Profitis Ilias-Epta Piges-Ekvoli Loutani-Katergo, Stream Gadoura-Lindou Stream-Pentanisa and Tetrapolis Islands, Psalidi Hill
GR3000013	5392.5	Kythira And Related Islands: Prasonisi, Dragonera, Antidragonera, Avgo, Kapello, Koufo Kai Fidonisi
GR4340014	13,979.8	Samaria National Park-Trypitis Canyon-Psilafi-Koustogerako
GR2220006	20,715.2	Kefalonia Island: Ainos, Agia Dynati Kai Kalon Oros
GR2410002	34,384.0	Parnassos Mountain
GR1270003	33,567.8	Athos Peninsula
GR1430001	31,112.2	Pilio Mountain and Seaside Zone
GR2330004	314.8	Olympia
GR2550006	53,367.5	Taygetos Mountain
GR3000005	5374.3	Sounio-Patroklou Island and Seaside Zone
GR3000006	8819.2	Ymittos-Kaisariani Aesthetical Forest-Vouliagmenis Lake

For the analytical and methodological purposes of the study, taking into account the extended regime of the Fire Weather influence, as well as the spatial resolution of the climatic data, a buffer of 10 km was applied on the polygons of the selected NATURA areas to create the Areas of Interest (AoI). The buffered areas which comprise the AoI are presented in Figure 3.

Figure 3. Map of the Areas of Interest (selected Natura 2000 areas buffered by 10 km).

2.3. Application of GIS Tools and Functions

The study of Fire Weather patterns and changes was elaborated using tools and functions of ARC-GIS v.10.8 software [39].

The software applications, tools and functions adopted for the implementation of the methodology are described below:

i.   FWI System Raster Calculator

For the daily calculation of the FWI system map series, for the historic and future time period the FWI_G.FMIS module of the proprietary software GeographicalFire Management Information System—G.FMIS v.1, (Varela Vassiliki & Eftychidis Georgios, Attika, Greece), has been used [40,41]. The software is developed in C++ programming language, based on the structure and equations of the Canadian FWI system [42] and supports the massive calculation of maps for all the FWI system intermediate and final parameters in ARCGIS Grid ASCII format.

ii.   Buffering

Euclidean buffers measure distance in a two-dimensional Cartesian plane, where straight-line or Euclidean distances are calculated between two points on a flat surface (the Cartesian plane).

iii.   Cell by cell analysis

This function is used for the calculation of per-cell statistics from multiple rasters. The statistics used for the current study are maximum, mean, minimum, range and standard deviation.

iv.   Zonal analyses and statistics

The Zonal tools allow for performing analysis where the output is a result of computations carried out on all cells that belong to each input zone. A zone can be defined as being one single area, but it can also be composed of multiple disconnected elements, or regions.

With the Zonal Statistics tool, a statistic is calculated for each zone defined by a zone dataset, based on values from another dataset (a value raster). A single output value is computed for every zone in the input zone dataset. The Zonal Statistics as a Table tool calculates all—a subset or a single statistic that is valid for the specific input but returns the result as a table instead of an output raster.

v.   Spectral profile analysis

Spectral profile charts allow us to select areas of interest or ground features on the image and review the spectral information of all bands in a chart format [43]. The boxes' chart type of the spectral profile analysis allows us to visualize and compare the distribution and central tendency of the values of the set of pixels for each of the AoI, collected through their quartiles, which are a way of categorizing the examined parameter values into four equal groups based on five key values: minimum, first quartile, median, third quartile, and maximum. A quartile is a type of quantile that divides the number of data points into four more or less equal parts, or quarters. This type of analysis has been used as a supervisory method for the comparison of the distribution and frequency of the values of SSR and ISI parameters within each of the AoI.

In order to study the Fire Weather patterns and their changes in the near future due to climate change, a number of analyses were performed, using the above described GIS functions and operations, for the selected FWI system parameters. The first phase of the work concerned a generic part of application of the FWI system equations for the calculation of daily maps and the creation of the basic spatial datasets. Then, different analyses were considered as appropriate for each parameter, for the provision of easy to interpret and meaningful outputs to be used as quantitative indicators of the fire weather characteristics and their future changes for the areas of interest.

Firstly, the calculation of daily maps for the selected FWI parameters (i.e., FWI, DSR, ISI) was carried out for the historic and future periods for RCP 4.5 and RCP 8.5 scenarios for the AoI, using raster calculation techniques and the algorithm and equations of the FWI system. Then, the FWI system's final spatial datasets were created for the studied periods of time for all the AoI, based on the above daily calculated maps and using the GIS "cell by cell" analyses. More particularly, the groups of spatial datasets that were created concerned:

(a)   The Seasonal Severity Rating (SSR) for the historic and future fire periods for the climatic scenarios RCP 4.5 and RCP 8.5;
(b)   The mean values of Initial Spread Index for the historic and future fire periods for RCP 4.5 and RCP 8.5;
(c)   The number of days with extreme FWI, per fire period for the historic and future fire periods for RCP 4.5 and RCP 8.5. The FWI thresholds for the extreme FWI class for the whole Greece were defined by Varela et.al., 2018 [23].

For each group of the above spatial datasets, i.e., (a–c), further analyses were performed as detailed below.

2.4. Analyses of SSR Spatial Datasets for the Historic and Future Scenarios

The Zonal Statistics GIS operation has been used for the calculation of Mean, Maximum, Minimum, STD and Range of values of SSR within each of the AoI for the historic and future period and the respective tables have been calculated. The tables with the calculated values for the historic and the two future period scenarios are presented in Tables S1–S3.

In addition, combined Bar Charts were created for the presentation and comparison of the SSR mean values within each of the AoI for the historic and future period scenarios (Figure 4). Moreover, Spectral Profile Charts of SSR values distribution within each of the AoI, for the historic and future period climatic scenarios were created. Indicative profile charts of two AoI are presented in Figures 5 and 6.

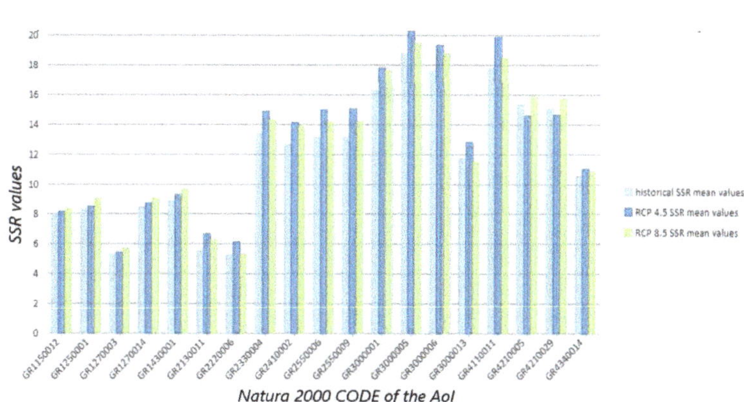

Figure 4. Bar chart of historic Seasonal Severity Rating (SSR) value, RCP 4.5 SSR value and RCP 8.5 SSR value by Area of Interest (AoI).

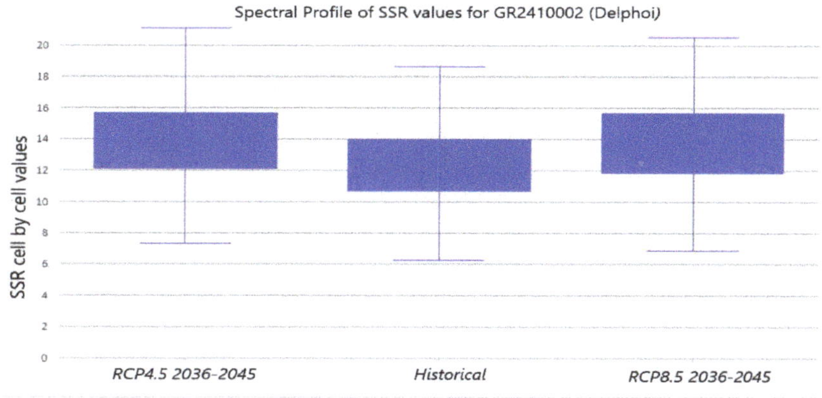

Figure 5. Spectral Profile of Seasonal Severity Rating (SSR) values for GR2410002 (Delphoi), for: RCP 4.5 2036–2045 (**left**), historic period (**centre**), and RCP 8.5 2036–2045 (**right**).

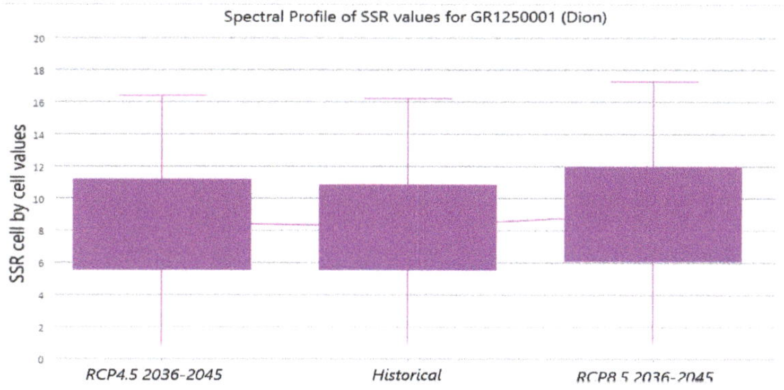

Figure 6. Spectral Profile of Seasonal Severity Rating (SSR) values for: GR1250001 (Dion), for RCP 4.5 2036–2045 (**left**), historic period (**centre**), and RCP 8.5 2036–2045 (**right**).

Finally, the mapping of SSR Range and Mean parameters of each of the AoI for the historic and future time-periods, derived from the Zonal Statistics, were classified in categories (Figures 7 and 8). The above selected parameters were considered as essential indicators of Fire weather diversity and intensity for each of the AoI and their mapping provided an informative portrayal and comparison of Fire Weather conditions.

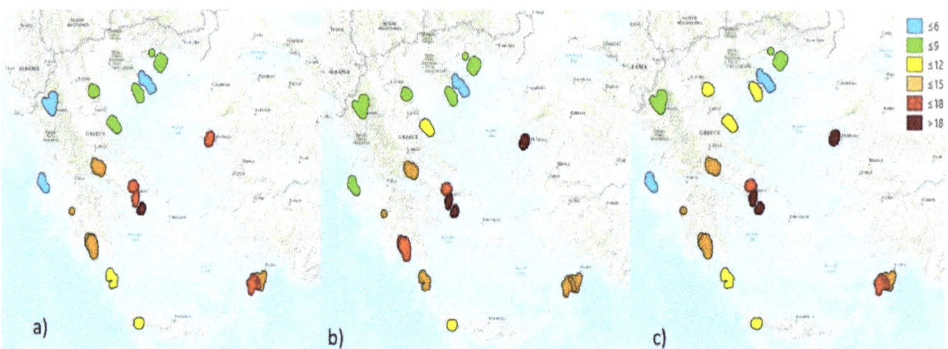

Figure 7. Maps of coloured Areas of Interest (AoI) depicting classified Seasonal Severity Rating (SSR) Mean values, for: (**a**) the historic period, (**b**) RCP 4.5 2036–2045, (**c**) RCP 8.5 2036–2045.

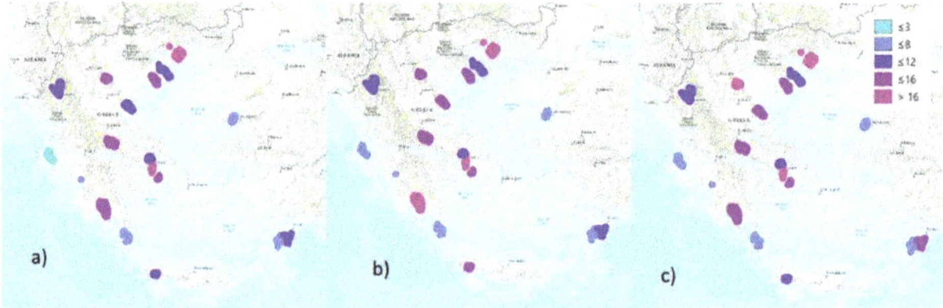

Figure 8. Maps of Seasonal Severity Rating (SSR) Range coloured classes within the Areas of Interest (AoI) for: (**a**) the historic period, (**b**) RCP 4.5 2036–2045, (**c**) RCP 8.5 2036–2045.

2.5. Analyses of Mean ISI Spatial Datasets for the Historic and Future Scenarios

Similar to the above approach, the Zonal Statistics GIS operation was applied for the calculation of the Mean, Maximum, Minimum STD and Range of values of ISI within each of the AoI for both periods and respective tables have been calculated. The tables with the calculated values for the historic and the two future scenarios are presented in Tables S4–S6.

The combined Bar Charts were then calculated for the presentation and comparison of the mean ISI within each of the AoI, for the historic and future period scenarios (Figure 9). Finally, the Spectral Profile Charts of ISI value distribution within each of the AoI, for the historic and future period scenarios were created. Indicative profile charts of two AoI are shown in Figures 10 and 11.

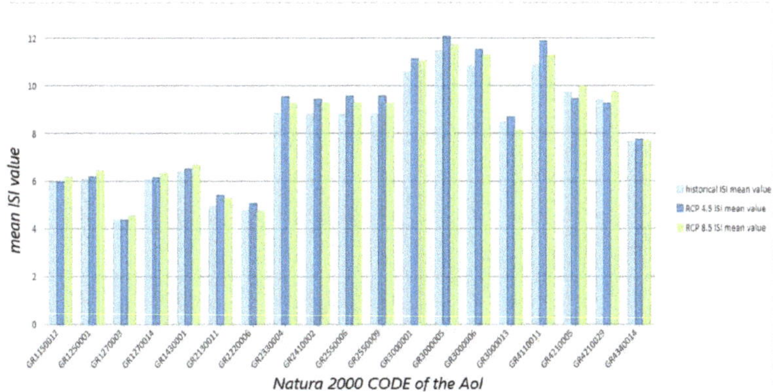

Figure 9. Bar chart of historic Initial Spread Index (ISI) Mean value, RCP 4.5 ISI Mean value and RCP 8.5 ISI Mean value by Area of Interest.

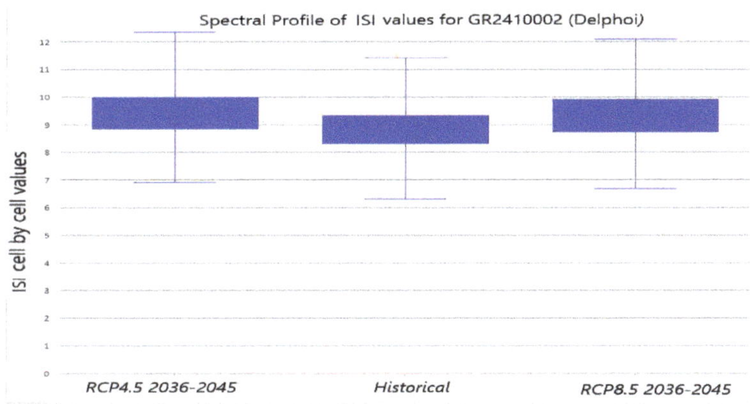

Figure 10. Spectral Profile of Initial Spread Index (ISI) values for GR2410002 (Delphoi), for RCP 4.5 2036–2045 (**left**), historic period (**centre**), and RCP 8.5 2036–2045 (**right**).

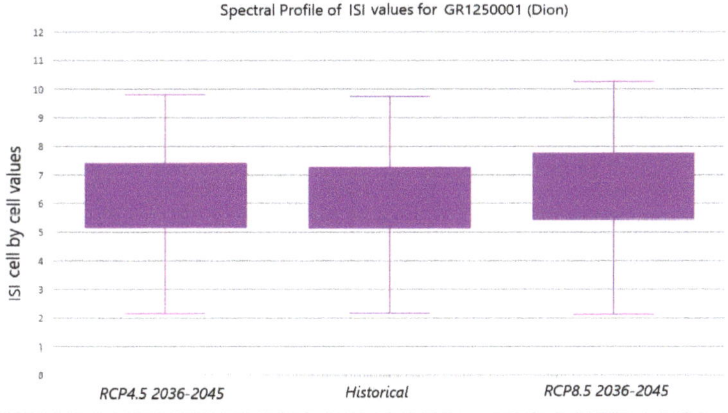

Figure 11. Spectral Profile of Initial Spread Index (ISI) values for: GR1250001 (Dion), for RCP 4.5 2036–2045 (**left**), historic period (**centre**), and RCP 8.5 2036–2045 (**right**).

2.6. Analyses of "Extreme FWI Days" Spatial Datasets for the Historic and Future Scenarios

The mapping of the difference among the RCP 4.5 and RCP 8.5 future periods and the historic period, of the maximum days with extreme FWI calculated in the cells within each area of interest was performed. Two maps were created as a result of this analysis, each one corresponding to the difference among RCP 4.5 and RCP 8.5 for 2036 to 2045 and the historic period, respectively (Figure 12).

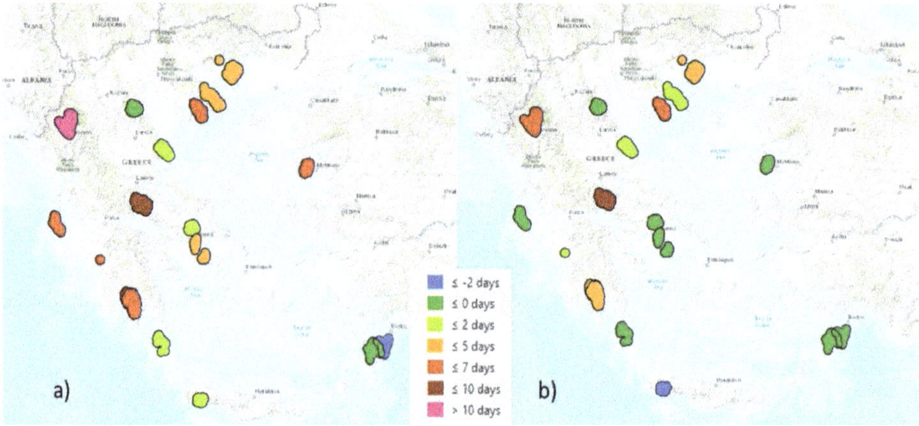

Figure 12. Differences in the results of maximum number of days with extreme Fire Weather Index (range depicted with colours) within the areas of interest between: (**a**) RCP 4.5 and historic data sets, (**b**) RCP 8.5 and historic datasets.

3. Results

The results of this study, which are based on the methodology described above, are presented in the form of tables, bar charts, box charts and maps, aiming to provide different views of fire weather parameters, in order to allow their study both at a local and national level.

3.1. Seasonal Severity Rating (SSR) Mapping and Analysis

The tables with the calculated values derived from Zonal statistics analyses, which are described in Section 2.4 (Tables S1–S3) in Supplementary Material), show that there is a variety of Seasonal Severity Rating (SSR) values for all the statistical parameters, among the AoI for the historic and future periods.

The results indicate that the values of SSR increase everywhere under future scenarios, while this increase is higher for RCP4.5 than for RCP8.5 for a number of areas. However, in three areas, SSR is found to be lower for RCP4.5 than for the historical period. More particularly, the mean values vary from 5.28 for GR1270003 up 18.78 for GR3000005 for the historic period, from 5.44 for GR1270003 up 20.33 for GR3000005 for the future RCP 4.5; and 5.28 for GR2220006 up 20.33 for GR3000005 for RCP 8.5. The range of SSR values within each of the AoI, which shows the diversity of fire weather within the area, is found between 2.79 and 16.39 for the historic period, while for RCP 4.5 and RCP 8.5, SSR values vary from 7.91 to 24.98 and from 6.83 to 24.16, respectively.

Figure 4, which represents the SSR mean values as a bar chart showcases in a more concise manner the variety of the SSR values among the AoI and for the studied periods and scenarios. On the other hand the box plot diagrams in Figures 5 and 6, which are based on Spectral Profile analyses for two (2) selected AoI that were selected as an example for presenting this type of information, provide a more detailed picture of the profile of the SSR values within each of the AoI for both fire periods.

Mapping of the mean and the range of SSR values in Figures 7 and 8, respectively, at the national level, using color-coded classes, allow for visual comparison among the AoI and among the studied time periods. The results show the variety in SSR among the AoI for the historical time, which are expected due to the diversity of the physical characteristics of the selected AoI, as well as the different levels of changes due to the two future climate scenarios. As an example, in Figure 7, it is obvious that the changes in fire weather due to climate change are more significant for three AoI in northern Greece, which are classified in a higher class and their colour code changes from green to yellow, compared to other AoI, which are classified in the same class as in the historic period map.

3.2. Initial Spread Index (ISI) Analysis

ISI parameter is an important intermediate parameter of FWI and similarly to the SSR parameter, Tables S4–S6 (Supplementary Material) and Figures 9–11 provide different analysis views of this parameter for the historical and future periods. The resulting values indicate that the values of ISI also increase in all areas for both future scenarios, while this increase is higher for RCP 4.5 than for RCP 8.5 for most areas.

3.3. Difference of the Maximum Number of Days with Extreme Fire Weather Mapping and Analysis

The maps that were derived from the calculation of the difference of the maximum number of days with extreme FWI between the future and historic periods are depicted in Figure 12. The results obtained show quite clearly that, in some areas (i.e., blue, green colour coded AoI), the number of extreme days slightly decreases or does not change significantly, while it is found to increase by more than seven days in other areas (i.e., red, purple colour coded AoI). The greatest difference of more than 10 days is found in the case of RCP 4.5 (Figure 12a).

4. Discussion

The analysis described above, is proposed as a methodology for the provision of information for all the areas of interest (AoI), in order to study the various perspectives of fire weather due to climate change.

The resulting values of the fire weather parameters for the selected areas and for the historical and future periods for two climate scenarios of the applied climate model, indicate that significant changes are expected in fire weather due to climate change. Moreover, these changes also have an important spatial variation, which needs to be highlighted, in order to be taken into account for rational fire management adaptation.

The accuracy of the results in terms of the estimated values of fire weather parameters, depends mainly on the inherent uncertainties of the input model climatic data. However, a further analysis on these uncertainties as well on the specific resulting values obtained for the areas under consideration, for the two future scenarios, are considered beyond the scope of the current study, which focuses on the methodology for studying fire weather patterns for distinctive fire management entities. Thus, the results obtained are discussed below mainly from the methodological point of view and the fire management perspective.

Previous research results on climate change impact on forest fires presented the grade of changes by applying a variety of available climate models on different geographical environments. Those studies indicated clearly that climate change should be considered as an important factor affecting future fire regime worldwide and thus as an essential subject for the adaptation of fire management policy and actions [44]. On the other hand, fire management is a complex task, demanding a concise picture of the situation at all operational levels—i.e., at national, sub-national and community (local) levels [45]. The analyses and indicators, which are proposed here, can be applied in local management units and, at the same time, can provide meaningful and easily interpreted results at the sub-national (regional) and national level in a consistent manner. For the presentation and evaluation of the proposed methodology, two climatic models were chosen to be applied in the selected areas in Greece, which were considered as discrete management units. The following discussion of the results aspires to stand out in terms of the operational usefulness of the methodology.

The areas of interest are faced not only as natural protected units but also as landscape entities specifically characterized in terms of Fire Weather, which in turn is the main component of fire danger and risk evaluation for each of the area, at the entity level. Moreover, at a higher level of geographical/administrative analysis (i.e., regional, national), mapping of the classified entities according to specific fire weather related variables, provides a concise portrayal of the spatial distribution and interrelation of fire weather levels at a specific time "instance".

The comparison among the historic and future time "instances", which, in our study, were represented by RCP 4.5 and RCP 8.5 climate change scenarios for the years 2036–2045, permitted the in depth study of the fire weather patterns and their anticipated changes through time. The approach followed included two levels of information, the first covering the needs of an extensive scientific analysis (i.e., tables of variables, graphs) and the other made suitable for direct operational purposes/application (i.e., single and simple Indicators, maps of classified areas of interest).

The basic climatic datasets for the application of the proposed methodology were based on two IPCC RCP scenarios of 4.5 and 8.5 W/m^2, from the openly accessible SMHI-RCA4 archives. The data were spatially downscaled from 12 to 5 km resolution by applying a regridding method based on the nearest neighbour interpolating process.

The zonal statistics for Seasonal Severity Rating (SSR) and Initial Spread Index (ISI) parameters are presented in Tables S1–S6 (Supplementary material). The Range of Seasonal Severity Range (SSR) and Initial Spread Index (ISI) values within each area of interest, as well as their Mean value and Standard Deviation (STD) for the historic and the future scenarios were the descriptive indicators of these two important parameters of fire weather. The calculated indicators allowed for both the comparison among the areas for a specific time lapse and the study of fire weather evolution in time. From the Range and STD values in these Tables S1–S6, it was deduced that some areas were characterized by intrinsic homogeneity in both parameters (e.g., GR2220006, GR23300004) while others showed high diversity in fire weather levels (e.g., GR1150012, GR 1250011).

The bar charts analyses of the Mean values of both parameters (SSR and ISI) for the areas of interest, which were shown in Figures 4 and 9, respectively, constituted a comprehensive means of presentation and comparison of the level of historic and future fire weather for the areas. According to this analysis, both future scenarios lead to an increase in the two parameters in all the areas under examination.

Furthermore, spectral profiles of each of the area of interest (Figures 5, 6, 10 and 11) could provide additional information about the distribution of mean and extreme values of parameters within each of the area for all the time periods, and could be used in conjunction with the bar charts.

This type of information, which is available at the "area of interest" level of analysis, can be useful as a decision support background for the fire management actions of the agency that is responsible for the area (i.e., local forest office, management body), as it provides a brief overview of the current and future fire weather conditions within the area. In addition to the above outputs, classification and mapping of the areas of interest according to historic and future SSR Mean and SSR Range (Figures 7 and 8), can provide a comprehensive and informative view for operational purposes at a regional or national level. It is worth mentioning at this point that, for the classification thresholds and the number of classes for SSR and ISI parameters, further research is essential for the definition of appropriate values, customized at a regional or national level.

The classification thresholds of these two indicators, for the purposes of the current paper, were defined based on the range and distribution of values that were obtained for the examined areas in the historic period in a way to accommodate a satisfactory for the analysis number of classes. The classification of the mean SSR values lead to a map of six classes of areas of interest. For the future scenarios, the classes were found to change for some of the areas of interest. More particularly, for RCP4.5 in total six areas of interest were found at a higher class and two areas of interest at a lower class, while for RCP 8.5, seven areas of interest were shown at a higher class.

Similarly, the classification for the SSR range of the historic period lead to five classes of areas of interest. For the future RCP 4.5 scenario, three areas changed to a higher class and one areas to a lower class, while for RCP 8.8, four areas were found at a higher class.

The estimation and mapping of the difference of the maximum number of days with extreme fire weather between historic and future periods (Figure 12) is another informative output of the proposed methodology, to be used for operational purposes, as it is a valid indicator of the potential changes of the preparedness planning actions.

Further information can be figured out about the patterns of changes. For example, some areas that are characterized by humid and cold climate (e.g., GR21300011, GR2410002) are expected to suffer significant changes in the number of extreme fire weather days, while others which belong to the classical Mediterranean zone show very slight changes (e.g., GR3000013, GR43400014). This is considered an interesting finding, since it indicates that climate change tends to affect, more severely, areas less adapted to forest fires. In those areas, ecosystems are less resilient on wildland fires and the recovery after a fire occurrence will take much longer or even may not occur at all. Future research on specific characteristics of the areas in terms of the climatic zone and other physical descriptors may provide interesting conclusions about the fire regime due to climate change aspects.

The above output indicators and physical descriptors of the methodology presented as maps and tables for the selected areas in Greece, are easily applicable to other geographical areas, either at a local or at a national level, since they are based on classical GIS functions. Besides, the mapping of FWI system parameters for the study of current and future fire weather, which constitutes the underlying dataset for the development and application of the methodology in any geographical area, is a common practice for all the categories of relevance to forest fire stakeholders. Moreover, as implied above, the introduced methodology can be the basis for further enhancement of indicators and descriptors related to forest fire regimes, aiming to facilitate fire management to a greater extent.

5. Conclusions

A straight forward methodology was presented for the estimation of fire weather indicators of the current fire weather and for the near future for areas which are considered as management units in terms of fire management. The proposed methodology, easy to apply using simple GIS functionality, can be used as an informative decision support tool for operational purposes in any geographical area, at a local and national level. The application of the methodology provided interesting, easy to interpret results for the area of Greece for the anticipated changes in fire weather danger, due to near future climate changes. The outcomes of the methodology, which are provided both as indicators for individual areas and as maps at a regional or national level, could be used in conjunction with other thematic layers and information for the areas of interest. The combination and mapping of the various indicators provide a concise view of the fire weather characterizing each area and also allow for comparison among the management units and the extraction and study of spatiotemporal patterns. This in turn constitutes a valid approach for providing in depth knowledge of the current and future fire weather regime which would not be obvious otherwise.

Supplementary Materials: The following are available online at http://www.mdpi.com/1999-4907/11/11/1168/s1, Table S1: Statistics for the areas of interest for SSR calculated for the historic dataset, Table S2: Statistics for the Areas of interest for SSR calculated for the scenario RCP 4.5 for the period 2036–2045, Table S3: Statistics for the Areas of interest for SSR calculated for the scenario RCP 8.5 for the period 2036–2045, Table S4: Statistics for the AoI for mean ISI cell values calculated for the historic dataset, Table S5: Statistics for the AoI for mean ISI cell values calculated for the scenario RCP 4.5 for the period 2036–2045, Table S6: Statistics for the AoI for mean ISI cell values calculated for the scenario RCP 8.5 for the period 2036–2045.

Author Contributions: Author Contributions: Individual contributions for this paper were as follows: conceptualization, V.V.; methodology, V.V. and D.V.; software, V.V. and S.K.; validation, V.V., A.S. and D.V., formal analysis, V.V., D.V. and A.S.; investigation, V.V. and N.P.; resources, V.V., A.S., N.P.; data curation, V.V. and S.K.; writing—original draft preparation, V.V.; writing—review and editing, D.V.; visualization, V.V.; supervision, V.V. and D.V.; project administration, A.S.; funding acquisition, A.S. All authors have read and agreed to the published version of the manuscript.

Funding: This research was funded by project XENIOS T1EΔK-02219 of the Greek Operational Programme Competitiveness, Entrepreneurship and Innovation (EPAnEK) of NSRF 2014-2020

Conflicts of Interest: The authors declare no conflict of interest.

References

1. Spinoni, J.; Naumann, G.; Vogt, J.V. Pan-European seasonal trends and recent changes of drought frequency and severity. *Glob. Planet. Chang.* **2017**, *148*, 113–130. [CrossRef]
2. Hertig, E.; Tramblay, Y. Regional downscaling of Mediterranean droughts under past and future climatic conditions. *Glob. Planet. Chang.* **2017**, *151*, 36–48. [CrossRef]
3. Boer, M.M.; Nolan, R.H.; Resco De Dios, V.; Clarke, H.; Price, O.F.; Bradstock, R.A. Changing Weather Extremes Call for Early Warning of Potential for Catastrophic Fire. *Earth's Future* **2017**, *5*, 1196–1202. [CrossRef]
4. Moritz, M.A.; Batllori, E.; Bradstock, R.A.; Gill, A.M.; Handmer, J.; Hessburg, P.F.; Leonard, J.; McCaffrey, S.; Odion, D.C.; Schoennagel, T.; et al. Learning to coexist with wildfire. *Nature* **2014**, *515*, 58–66. [CrossRef] [PubMed]
5. De Rigo, D.; Libertà, G.; Houston Durrant, T.; Artés Vivancos, T.; San-Miguel-Ayanz, J. *Forest Fire Danger Extremes in Europe under Climate Change: Variability and Uncertainty*; JRC Technical Reports. PESETA III Project-Climate Impacts and Adaptation in Europe, Focusing on Extremes, Adaptation and the 2030s. Task 11-Forest Fires. Final Report; Luxembourg Publications Office of the European Union: Luxembourg, 2017; ISBN 978-92-79-77046-3. [CrossRef]
6. Montiel Molina, C.; Galiana-Martín, L. Fire Scenarios in Spain: A Territorial Approach to Proactive Fire Management in the Context of Global Change. *Forests* **2016**, *7*, 273. [CrossRef]
7. Feltman, J.A.; Straka, T.J.; Post, C.J.; Sperry, S.L. Geospatial Analysis Application to Forecast Wildfire Occurrences in South Carolina. *Forests* **2012**, *3*, 265–282. [CrossRef]
8. Keane, R.E.; Burgan, R.; van Wagtendonk, J. Mapping wildland fuels for fire management across multiple scales: Integrating remote sensing, GIS, and biophysical modeling. *Int. J. Wildland Fire* **2001**, *10*, 301–319. [CrossRef]
9. Crimmins, M.A. Synoptic climatology of extreme fire-weather conditions across the southwest United States. *Int. J. Climatol.* **2006**, *26*, 1001–1016. [CrossRef]
10. Viegas, D.X.; Bovio, G.; Ferreira, A.; Sol, B. Comparative study of various methods of fire danger evaluation in southern Europe. *Int. J. Wildland Fire* **1999**, *9*, 235–246. [CrossRef]
11. Haines, D.A. *Where to Find Weather and Climate Data for Forest Research Studies and Management Planning (General Technical Report NC-27)*; USDA Forest Service, North Central Forest Experiment Station: St. Paul, MN, USA, 1977.
12. Krusel, N.; Packham, D.; Tapper, N. Wildfire activity in the Mallee Shrubland of Victoria, Australia. *Int. J. Wildland Fire* **1993**, *3*, 217–227. [CrossRef]
13. Morgan, P.; Hardy, C.C.; Swetnam, T.W.; Rollins, M.G.; Long, D.G. Mapping fire regimes across time and space: Understanding coarse and fine-scale fire patterns. *Int. J. Wildland Fire* **2001**, *10*, 329–342. [CrossRef]
14. Miguel, J.S.; Chuvieco, E.; Handmer, J.; Moffat, A.; Montiel-Molina, C.; Sandahl, L.; Viegas, D. Climatological Risk: Wildfires, Chapter 3 UNDERSTANDING DISASTER RISK: HAZARD RELATED RISK ISSUES-SECTION III. Available online: https://drmkc.jrc.ec.europa.eu/portals/0/Knowledge/ScienceforDRM/ch03_s03/ch03_s03_subch0310.pdf (accessed on 1 April 2020).
15. Rollins, M.G.; Swetnam, T.W.; Morgan, P. Evaluating a century of fire patterns in two Rocky Mountain wilderness areas using digital fire atlases. *Can. J. For. Res.* **2001**, *31*, 2107–2123. [CrossRef]
16. Swetnam, T.W.; Allen, C.D.; Betancourt, J.L. Applied historic ecology: Using the past to manage for the future. *Ecol. Appl.* **1999**, *9*, 1189–1206. [CrossRef]
17. Ryan, K.C.; Jones, A.T.; Koerner, C.L.; Lee, K.M. (Eds.) *Wildland Fire in Ecosystems: Effects of Fire on Cultural Resources and Archaeology*; U.S. Department of Agriculture, Forest Service, Rocky Mountain Research Station: Fort Collins, CO, USA, 2012.
18. Olympia Saved as Greek Fires Continue to Rage. 2007. Available online: https://www.euronews.com/2007/08/27/olympia-saved-as-greek-fires-continue-to-rage (accessed on 12 June 2020).
19. Wildfire Breaks out Near Tomb of Agamemnon in Greece. 2020. Available online: https://www.theguardian.com/world/2020/aug/30/wildfire-breaks-out-near-tomb-of-agamemnon-greece (accessed on 1 June 2020).

20. Giannakopoulos, C.; Le Sager, P.; Bindi, P.; Moriondo, M.; Kostopoulou, E.; Goodess, C.M. Climatic changes and associated impacts in the Mediterranean resulting from a 2 °C global warming. *Glob. Planet. Change* **2009**. [CrossRef]

21. National Bank of Greece. *The Environmental, Economic and Social Impacts of Climate Change in Greece*; National Bank of Greece: Athens, Greece, 2011; p. 494. ISBN 978-960-7032-58-4.

22. Giannakopoulos, C.; Kostopoulou, E.; Varotsos, K.V.; Tziotziou, K. An integrated assessment of climate change impacts for Greece in the near future. *Reg. Environ. Chang.* **2011**, *11*, 829–843. [CrossRef]

23. Varela, V.; Sfetsos, A.; Vlachogiannis, D.; Gounaris, N. Fire Weather Index (FWI) classification for fire danger assessment applied in Greece. *Tethys* **2018**, *15*, 31–40. [CrossRef]

24. Flannigan, M.; Cantin, A.S.; de Groot, W.J.; Wotton, M.; Newbery, A.; Gowman, L.M. Global wildland fire season severity in the 21st century. *Forest Ecol. Manag.* **2013**, *294*, 54–61. [CrossRef]

25. Van Wagner, C.E. Development and structure of the canadian forest fire weather index system. In *Can. For. Serv., Forestry Tech. Rep*; Canadian Forestry Service, Headquarters: Ottawa, ON, Canada, 1987.

26. European Environment Agency. Available online: https://www.eea.europa.eu/data-and-maps/figures/projected-meteorological-forest-fire-danger (accessed on 10 April 2020).

27. Schoenberg, F.P.; Chang, C.; Keeley, J.E.; Pompa, J.; Woods, J.; Xu, H. A critical assessment of the burning index in Los Angeles County, California. *Int. J. Wildland Fire* **2007**, *16*, 473–483. [CrossRef]

28. New Zealand Fire Service. *Apply Fire Weather (FWI) Index System, Study Guide*; New Zealand Fire Service: Wellington, New Zealand, 2016; p. 87.

29. Dimitrakopoulos, A.P.; Bemmerzouk, A.M.; Mitsopoulos, I.D. Evaluation of the Canadian fire weather index system in an eastern Mediterranean environment. *Meteorol. Appl.* **2011**, *18*, 83–93. [CrossRef]

30. Samuelsson, P.; Gollvik, S.; Kupiainen, M.; Kourzeneva, E.; van de Berg, W.J. *The Surface Processes of the Rossby Centre Regional Atmospheric Climate Model (RCA4)*; Meteorologi Nr 157 (SMHI): Norrköping, Sweden, 2015.

31. Moss, R.; Edmonds, J.; Hibbard, K.; Manning, M.R.; Rose, S.K.; Van Vuuren, D.P.; Carter, T.R.; Emori, S.; Kainuma, M.; Kram, T.; et al. The next generation of scenarios for climate change research and assessment. *Nature* **2010**, *463*, 747–756. [CrossRef]

32. EURO-CORDEX. Coordinated Downscaling Experiment-European Domain. 2019. Available online: https://www.euro-cordex.net/ (accessed on 2 May 2019).

33. Strandberg, G.; Bärring, L.; Hansson, U.; Jansson, C.; Jones, C.; Kjellström, E.; Kupiainen, M.; Nikulin, G.; Samuelsson, P.; Ullerstig, A. *CORDEX Scenarios for Europe from the Rossby Centre Regional Climate Model RCA4*; Swedish Meteorological and Hydrological Institute: Norrköping, Sweden, 2014.

34. Sørland, S.L.; Schär, C.; Lüthi, D.; Kjellström, E. Bias patterns and climate change signals in GCM-RCM model chains. *Environ. Res. Lett.* **2018**, *13*, 074017. [CrossRef]

35. Katopodis, T.; Vlachogiannis, D.; Politi, N.; Gounaris, N.; Karozis, S.; Sfetsos, A. Assessment of climate change impacts on wind resource characteristics and wind energy potential in Greece. *J. Renew. Sustain. Energy* **2019**, *11*. [CrossRef]

36. The Natura 2000 Areas of Greece. Available online: https://www.geogreece.gr/natura_en.php (accessed on 1 April 2020).

37. Esri Living Atlas of the World. Available online: https://livingatlas.arcgis.com/en/home/ (accessed on 12 June 2020).

38. Metzger, M.J.; Bunce, R.G.H.; Jongman, R.H.G.; Sayre, R.; Trabucco, A.; Zomer, R. High-resolution bioclimate map of the world. *Glob. Ecol. Biogeogr.* **2013**, *22*, 630–638. [CrossRef]

39. ArcGIS. The Mapping and Analytics Platform. Available online: https://www.esri.com/en-us/arcgis/about-arcgis/overview (accessed on 12 June 2020).

40. Varela, V.; Eftichidis, G.; Margaritis, M. Design and Implementation of a User Interface for a Forest Fire Management Information System. In Proceedings of the 2nd International Conference on Forest Fire Research, Coimbra, Portugal, 21–24 November 1994; Volume 1, pp. 113–121.

41. Eftichidis, G.; Margaritis, E.; Sfiris, A.; Varela, V. Fire management information systems: FMIS. In Proceedings of the III Int. Conference on Forest Fire Research/14th Conference on Fire and Forest Meteorology, Luso, Coimbra, Portugal, 16–20 November 1998; Volume 2, pp. 2641–2642.

42. Van Wagner, C.E.; Pickett, T. Equations and FORTRAN Program for the Canadian Forest fire Weather Index System. *Ont. For. Tech. Rep.* **1985**, *2*, 18.

43. Spectral Profile Analysis in ArcGIS Pro. Available online: https://pro.arcgis.com/en/pro-app/help/data/imagery/spectral-profile-chart.htm (accessed on 12 June 2020).

44. Association for Fire Ecology. The San Diego Declaration on climate change and Fire Management. In Proceedings of the Third International Fire Ecology and Management Congress, San Diego, CA, USA, 13–17 November 2006.
45. FAO Strategy on Forest Fire Management. Available online: http://www.fao.org/forestry/firemanagement/strategy/en/ (accessed on 15 October 2020).

Publisher's Note: MDPI stays neutral with regard to jurisdictional claims in published maps and institutional affiliations.

Article

Identifying Forest Fire Driving Factors and Related Impacts in China Using Random Forest Algorithm

Wenyuan Ma [1], **Zhongke Feng [1,*]**, **Zhuxin Cheng [1]**, **Shilin Chen [1] and Fengge Wang [2]**

1   Precision Forestry Key Laboratory of Beijing, Beijing Forestry University, Beijing 10083, China; imawenyuan@126.com (W.M.); chengzhuxin1993@foxmail.com (Z.C.); chenshilin@bjfu.edu.cn (S.C.)
2   Forest Fire Prevention and Monitoring Center in Ministry of Emergency Management of China, Beijing 100054, China; slgaj9204@sina.com
*   Correspondence: zhongkefeng@bjfu.edu.cn; Tel.: +86-138-1030-5579

Received: 22 March 2020; Accepted: 28 April 2020; Published: 1 May 2020

Abstract: Reasonable forest fire management measures can effectively reduce the losses caused by forest fires and forest fire driving factors and their impacts are important aspects that should be considered in forest fire management. We used the random forest model and MODIS Global Fire Atlas dataset (2010~2016) to analyse the impacts of climate, topographic, vegetation and socioeconomic variables on forest fire occurrence in six geographical regions in China. The results show clear regional differences in the forest fire driving factors and their impacts in China. Climate variables are the forest fire driving factors in all regions of China, vegetation variable is the forest fire driving factor in all other regions except the Northwest region and topographic variables and socioeconomic variables are only the driving factors of forest fires in a few regions (Northwest and Southwest regions). The model predictive capability is good: the AUC values are between 0.830 and 0.975, and the prediction accuracy is between 70.0% and 91.4%. High fire hazard areas are concentrated in the Northeast region, Southwest region and East China region. This research will aid in providing a national-scale understanding of forest fire driving factors and fire hazard distribution in China and help policymakers to design fire management strategies to reduce potential fire hazards.

Keywords: forest fire driving factors; forest fire occurrence; random forest; forest fire management; China

1. Introduction

Forests are ecosystems with rich biodiversity [1–3], and they play an important role in soil and water conservation, climate regulation, the carbon cycle and other aspects [4,5]. Fire, which affects the biodiversity, species composition and ecosystem structure of forest ecosystems, is the dominant disturbance factor in many forest ecosystems [6–9]. Moreover, fire also affects human lives, regional economies and environmental health [10–12]. In short, forest fires threaten the sustainable development of modern forestry and human security [13]. Therefore, as an important component of global environmental change, forest fires have become the focus of forestry and ecological research [14,15]. An important aspect of forest fire management and prevention is studying forest fire driving factors and their impacts, which can help fire prevention departments to accurately assess forest fire hazards and effectively implement forest fire prevention strategies [11,16]. Forest fires are affected complexly by many driving factors, so it is very important to select appropriate forest fire driving factors and prediction models.

Forest fire driving factors have generally been divided into four types, namely, climate, vegetation, topography and socioeconomic [17,18], which vary at different temporal and spatial scales [19]. Regarding impact modes, climate factors control the accumulation and water content of forest fuels [20,21], which are usually considered as the major determinants of forest fire occurrence [22]. Vegetation is a source of forest fuel and directly affects the ignition capacity [23,24]. Topography can affect the structure and distribution of vegetation, thus affecting the possibility of forest fires as well as the spread speed and direction of forest fires [25]. Socioeconomic factors affect forest fire occurrence via building expansion, traffic network construction and human-related activities, which increase pressure on wildlands, bringing ignition sources close to forests [23,26]. In terms of impact scope, climate affects forest fires on a larger scale while the vegetation, topography and socioeconomic factors affect forest fires on a smaller scale [27]. In terms of impact relationship, there are nonlinear relationships and thresholds between forest fire driving factors and forest fire occurrence [28–30]. Random forest is a machine learning algorithm, which can automatically select important variables and flexibly evaluate the complex interaction between variables. In recent years, random forest has been used in the study of forest fire driving factors and has shown better prediction ability than multiple linear regression [31] and logistic regression [18].

Previous studies have analysed forest fire distribution, forest fire frequency and burnt area at the national scale in China. Tian [32] analysed the spatial and temporal distribution characteristics of wildfires for 2008–2012 in mainland China. Chang [33] explored the environmental factors influencing the spatial variation in the mean number of fires and mean burned forest area from 1987 to 2007 at a provincial level using cluster analysis and redundancy analysis. Zhong [34] analysed the changes in fire frequency and burnt area during 1992–1999 in China. Lu [35] analysed the impacts of annual temperature and precipitation on the burnt area dynamics in China. Ying used ground-based data to analyse the environmental and social factor contributions to the spatial variation of forest fire frequency and burnt area summarized at a county level during 1989–1991 in China [36]. However, previous studies have used models to analyse the driving factors and their impacts of forest fire occurrence in China, mainly at the provincial scale, such as in Fujian province [18], Heilongjiang province [29,37] and Shanxi province [38]. There is still a lack of nationwide research on forest fire driving factors and their influence on recent forest fires. The value of this study lies in conducting the nationwide research which can provide a detailed analysis and practical information of the forest fire hazard and would help governments to formulate more accurate forest fire prevention strategies and allocate resources rationally. In this study, we used the random forest model and forest fire ignitions for 2010~2016 (obtained from MODIS Global Fire Atlas dataset) to evaluate the impact of four types of forest fire driving factors and the regional differences of these factors in China. This study has three objectives: (1) to determine the forest fire driving factors in various geographical regions of China and analyse how they affect forest fire occurrence; (2) to map the likelihood of forest fire occurrence in China and (3) to discuss forest fire prevention strategies in different geographical areas of China.

2. Materials and Methods

2.1. Study Area

The study area covered mainland China (Hong Kong, Macao and Taiwan were not analysed due to a lack of data). The driving factors of forest fires and their effect were analysed in 6 geographical regions: Northeast region (NE), North China region (N), East China region (E), Northwest region (NW), Southwest region (SW) and Mid-south region (MS). Each region is an aggregation of provinces with adjacent locations and similar topography, economy and climate. The details of each region are shown in Figure 1 and Table 1.

Table 1. Basic description of the six geographical regions in China.

Study Area	Province	Main Climate Types	Topography	Dominant Vegetation Types	Socioeconomic Conditions
Northeast region	Heilongjiang, Jilin, and Liaoning	Middle temperate monsoon climate	Dominant terrain is plains and mountains. The elevation in most areas is below 500 m.	Temperate coniferous broadleaved mixed forests and cold temperate coniferous forest. Forest coverage is 41.59% [39].	The total population is 108.75 million. The per capita GDP is ¥49,891 yuan [40].
North China region	Inner Mongolia, Shanxi, Beijing, Tianjin, and Hebei	Middle temperate continental climate and warm temperate monsoon climate	The dominant terrain types are plateaus and hills. The elevation in most areas is below 2000 m.	Warm temperate deciduous broadleaved forest and temperate grassland. Forest coverage is 21.09% [39].	The total population is 174.79 million. The per capita GDP is ¥64,194 yuan [40].
East China region	Shandong, Jiangsu, Anhui, Zhejiang, Shanghai, Jiangxi, and Fujian	Warm temperate monsoon climate and subtropical monsoon climate	The dominant terrain types are plains and mountains. The elevation in most areas is below 1000 m.	Warm temperate deciduous broadleaved forest and subtropical evergreen broadleaved forest. Forest coverage is 40.64% [39].	The total population is 408.98 million. The per capita GDP is ¥78,271 yuan [40] (China Statistical Yearbook, 2018).
Northwest region	Xinjiang, Gansu, Ningxia, Qinghai, and Shaanxi	Warm temperate continental climate	The dominant terrain types, which fluctuate greatly, are deserts and high mountains. The elevation in most areas is between 500 and 5000 m.	Temperate desert and alpine vegetation on the Qinghai-Tibetan plateau. Forest coverage is 8.21% [39].	The total population is 101.86 million. The per capita GDP is ¥45,463 yuan [40].
Southwest region	Tibet, Sichuan, Chongqing, Yunnan, and Guizhou	Subtropical monsoon climate and alpine climate	The terrain is complex and consists of basins, plateaus and mountains. The elevation in most areas is between 500 and 6000 m.	Alpine vegetation on the Qinghai-Tibetan plateau, subtropical evergreen broadleaved forest and tropical rainforest. Forest coverage is 25.75% [39].	The total population is 200.95 million. The per capita GDP is ¥43,609 yuan [40].
Mid-south region	Henan, Hubei, Hunan, Guangxi, Guangdong, and Hainan	Subtropical monsoon climate and tropical monsoon climate	The dominant terrain types are plains and mountains. The elevation in most areas is below 1000 m.	Subtropical evergreen broadleaved forest and tropical rainforest. Forest coverage is 44.63% [39].	The total population is 393.01 million. The per capita GDP is ¥57,664 yuan [40].

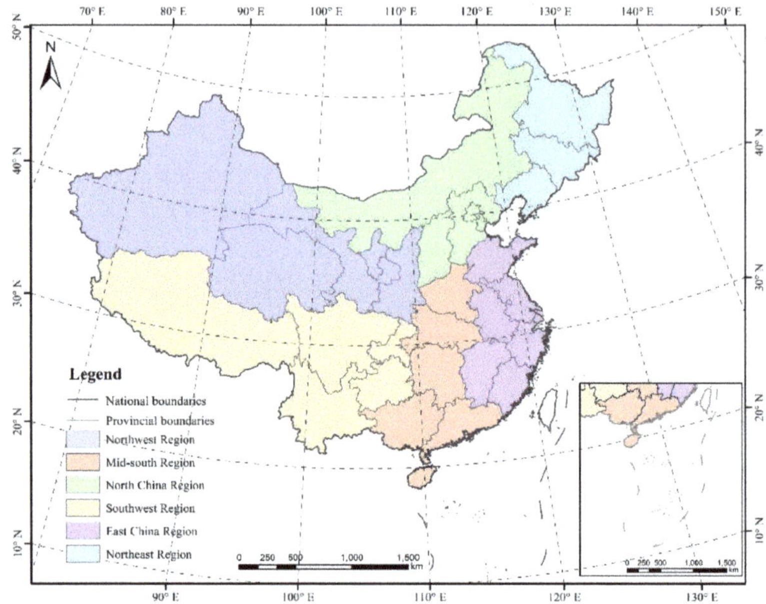

Figure 1. Six geographical regions in China.

2.2. Data Preparation

2.2.1. Dependent Variables

We identified 17,466 forest fires (ignitions) between 2010 and 2016 across mainland China with the Global Fire Atlas dataset (downloaded from the Oak Ridge National Laboratory (ORNL) Distributed Active Archive Center (DAAC), https://daac.ornl.gov) and Chinese land-use type dataset (downloaded from the Resource and Environment Data Cloud Platform, http://www.dsac.cn). The timing and location of the fire ignitions were provided by the Global Fire Atlas, which is a global dataset that records the daily dynamics of individual fires based on the Global Fire Atlas algorithm and estimated burn dates from the Moderate Resolution Imaging Spectroradiometer (MODIS) [41]. A Chinese land-use type dataset provided the forest land range in mainland China for 2015 at a 1000-m spatial resolution. According to this range, we identified Chinese forest fire ignitions for 2010–2016 from the Global Fire Atlas dataset in ArcGIS10.2 software (Environmental Systems Research Institute, RedLands, CA, USA). Figure 2 shows the distribution of the forest fire ignitions for six geographical regions in China.

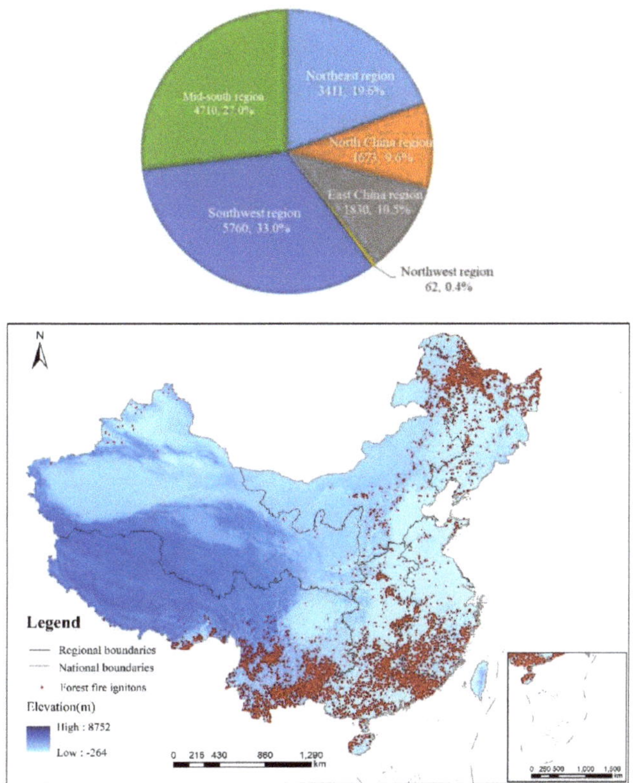

Figure 2. Distribution of the number, proportion and location of forest fire ignitions in six geographical regions from 2010 to 2016.

Modelling forest fire occurrence requires a binary target variable, so a certain percentage of control points (nonfire points) were generated randomly according to three principles: (1) the ratio of forest fire ignition points to control points was 1:1.5 [29], (2) the control points were located within the forest land range in mainland China and (3) the points were random in both time and space. ArcGIS10.2 software was used to randomly generate the control points, and the dates of the control points were randomly selected during 2010–2016 to meet the randomness of time.

2.2.2. Explanatory Variables

A total of 21 variables, grouped into climate, topography, vegetation and socioeconomic categories, were selected as the initial forest fire driving factors (Table 2). All variables were integrated in ArcGIS10.2 software and extracted to the forest fire ignition points and control points.

Table 2. Initial explanatory variables that were collected for inclusion as forest fire drivers.

Variable Type	Variable Name	Code	Source	Resolution, Units	References
Climatic	Annual precipitation in the year before the fire	Pre_year0	National Earth System Science Data Sharing Infrastructure, National Science & Technology Infrastructure of China (http://www.geodata.cn)	0.25°, 0.1 mm	[18,31,42]
	Annual precipitation in the year of the fire	Pre_year1		0.25°, 0.1 mm	
	Annual soil moisture in the year before the fire	Soil_mois0		0.25°, $m^3\ m^{-3}$	
	Annual soil moisture in the year of the fire	Soil_mois1		0.25°, $m^3\ m^{-3}$	
	Daily average ground surface temperature	GST_avg		0.1 °C	
	Daily precipitation	Pre_daily		0.1 mm	
	Daily average air pressure	Pres_avg		0.1 hPa	
	Daily average relative humidity	RH_avg	Daily Data Set of China's Surface Climate Data (V3.0), National Meteorological Information Centre (http://data.cma.cn)	%	[18,31,37,43,44]
	Daily minimum relative humidity	RH_min		%	
	Sunshine hours	SSD		0.1 h	
	Daily average temperature	Tem_avg		0.1 °C	
	Daily average wind speed	Win_avg		0.1 m/s	
	Daily maximum wind speed	Win_max		0.1 m/s	
Topographic	Elevation	Elevation	Geospatial Data Cloud site, Computer Network Information Center, Chinese Academy of Sciences (http://www.gscloud.cn)	90 m, m	[28,29,45,46]
	Slope	Slope		90 m, °	
	Aspect	Aspect		90 m	
Vegetation	Fractional vegetation cover	FVC	Resource and Environment Data Cloud Platform (http://www.resdc.cn)(Xu,2018)	1 km	[18]
Socioeconomic	The distance to road and railway	Dis_road	National Catalogue Service for Geographic Information website (http://www.webmap.cn)	1:1,000,000, m	[18,28,29,46,47]
	The distance to settlement	Dis_sett		1:1,000,000, m	
	Population density	POP	National Earth System Science Data Center (http://www.geodata.cn)	1 km, number/km^2	
	Per capita gross national product	GDP		1 km, RMB/km^2	

Climate Variables

Climate variables affect fuel accumulation and moisture which largely determine the time, location and occurrence probability of forest fires [31]. In this study, the initial climate variables include annual variables and daily variables. The annual variables are precipitation and soil moisture. As climate factors in the period before the fire can also affect vegetation accumulation and the fuel moisture content, precipitation and soil moisture in the year before individual forest fire ignition during 2010–2016 were also taken into consideration [29,31]. We downloaded precipitation data with a 1-km spatial and monthly temporal resolution [42] and soil moisture data with a 0.25° spatial and monthly temporal resolution from the National Earth System Science Data Sharing Infrastructure, National Science & Technology Infrastructure of China (http://www.geodata.cn). Based on these data, we calculated the annual cumulative precipitation and the annual average soil moisture for 2009–2016.

The daily initial climate variables include daily average temperature, daily average ground surface temperature, daily average relative humidity, daily minimum relative humidity, daily precipitation, daily average atmospheric pressure, sunshine hours, daily average wind speed and daily maximum wind speed. The daily humidity, precipitation, wind speed and sunshine hours affect the possibility of forest fire occurrence by reflecting fuel moisture. Daily temperature is the key condition triggering fire ignition. Atmospheric pressure can affect the oxygen content in the air, and the pressure obviously differs due to significant altitude differences and the complex terrain in China; therefore, atmospheric pressure was also considered as an initial climate variable. Daily climate data were obtained from the Daily Data Set of China's Surface Climate Data (V3.0) of the National Meteorological Information Centre (http://data.cma.cn), and we included daily data from 824 national weather stations in China. The daily climate variable values for each fire ignition and control point were provided by the weather station nearest that point.

Topographic Variables

Topography influences the possibility of forest fire occurrence by affecting the vegetation composition and distribution and local microclimate [25]. In this study, the initial topographic variables include elevation, slope and aspect. Data for these variables were extracted from digital elevation model (DEM) data in China with 90-m spatial resolution (obtained from Geospatial Data Cloud site, Computer Network Information Center, Chinese Academy of Sciences, http://www.gscloud.cn). Aspect was divided into 8 categories according to the criteria in Table 3.

Table 3. Aspect classification criteria.

Aspect	Azimuth (°)
North	337.5~22.5
Northeast	22.5~67.5
East	67.5~112.5
Southeast	112.5~157.5
South	157.5~202.5
Southwest	202.5~247.5

Vegetation Variable

The initial vegetation variable is the fractional vegetation cover (FVC), which is the percentage of the vertical projection of vegetation area to the ground surface within a unit area [48] and can well represent the amount of forest fuel [18,20]. The normalized difference vegetation index (NDVI) is significantly better than other vegetation indices in estimating FVC [49,50], so we calculated FVC based on the annual NDVI dataset for 2010–2016. The calculation formula is as follows:

$$FVC = (NDVI - NDVI_{soil}) / (NDVI_{veg} - NDVI_{soil}) \tag{1}$$

where $NDVI_{soil}$ is the NDVI of bare soil and $NDVI_{veg}$ is the NDVI of dense vegetation canopy. The annual NDVI dataset for 2010–2016 was from the Resource and Environment Data Cloud Platform (http://www.resdc.cn), and the resolution was 1 km [51].

Socioeconomic Variables

Socioeconomic variables affect the probability of forest fire occurrence by affecting human activities. Human travel and engaging in production activities in or around forests will increase the occurrence probability of forest fires. In this study, the initial socioeconomic variables include the distance to the road and railway, the distance to the settlement, gross national product (GDP) and population density. Collectively, these variables can reflect the accessibility of a forest and the possibility of people engaging in fire-prone behaviours in forests [23,26]. The road, railway and settlement datasets were from the National Basic Geographic Database of 1:1 million, which was published on the National Catalogue Service for Geographic Information website (http://www.webmap.cn). The distance between the forest fire ignitions and control points to the nearest road and railway and settlement areas was calculated using the ArcGIS 10.2 "near analysis tool." The population density dataset and GDP dataset were downloaded by the National Earth System Science Data Center (http://www.geodata.cn), and the resolution was 1 km.

2.3. Model

The random forest model was used to identify the forest fire driving factors and their corresponding impacts on forest fire occurrence in 6 geographical regions of China and the whole study area. Random forest is an ensemble learning technique that is derived from classification or regression trees (CARTs). Random forest has a high prediction accuracy and high tolerance to outliers and "noise," and it has shown good prediction ability in forest fire forecasting [30,52]. The random forest model is composed of a combination of various classification trees, which are individually generated by bootstrap samples. Two-thirds of the data are used to train the random forest model and one-third of the data (the out-of-bag samples, OOB) for model validation [53]. Variable importance can also be measured by OOB, which compares increases in OOB error with that variable randomly permuted and all others unchanged [54,55]. The importance score of a variable is as follows [56]:

$$VI\left(X^{j}\right) = \frac{1}{ntree}\sum_{t}\left(err'OOB_{t}^{j} - errOOB_{t}^{j}\right) \tag{2}$$

where X^{j} is the jth variable, $ntree$ is the number of trees, $errOOB_{t}^{j}$ is the OOB error of each tree t and $err'OOB_{t}^{j}$ is the OOB error when X^{j} is permuted, while all other variables remain unchanged among OOB data. For regression, the OOB error is the mean square error; meanwhile, for classification, the OOB is misclassification probability.

In this study, RF was used for classification, which divided dependent variables into two categories: forest fire occurrence and forest fire nonoccurrence. When using an RF model, the number of trees to run (ntree) and the number of variables to try at each split (mtry) need to be defined. According to previous experience [56,57], the value of mtry was set as $\sqrt{number\ of\ variables}$ and the value of ntree was set to 2000. The varSelRF package in R statistical software was applied to select significant variables from the initial variables. Then, we measured and ranked the variable importance of these variables. The partialPlot function was used to draw partial dependence plots which can describe the relationship between the dependent variables and explanatory variables.

To eliminate bias, in each study region and the whole study area, we selected 80% of the original dataset (training dataset) to build the model, and the remaining 20% of the original dataset (independent validation dataset) was used to assess the performance of the final model. Each training dataset was divided into an inner training dataset (60%) and an inner validation dataset (40%) randomly [52]; this procedure was repeated 5 times, and 5 random subsamples of data in each study region and the whole study area were obtained. Each subsample contained an inner training dataset and an inner validation dataset, and each subsample generates an intermediate model. The variables that were selected as significant variables in at least 3 of the 5 intermediate model were considered as the forest fire driving factors in a region. In each region, the driving factors and training dataset were used to build a final model, and the independent validation dataset was used to validate the model [30].

2.4. Prediction Accuracy of the Models

The receiver operating characteristic (ROC) curve, a coordinate schema analysis method, was applied to measure the predictive capability of the RF models using the area under the curve (AUC) [28,58,59]. The AUC values ranged from 0.5 to 1, with values closer to 1 indicating a relatively higher accuracy, while an AUC value of >0.8 usually indicates good predictive capability [18,60]. We used the Youden criterion, calculated according to the sensitivity and specificity of ROC (Youden criterion = sensitivity + specificity − 1) [28,61], to determine the cut-off point, which was the threshold for judging whether a fire occurred in the models [62]. If the predicted probability was higher than the cut-off point, it was assumed that a forest fire had occurred and vice versa. The prediction accuracy of the model was calculation based on the cut-off point.

2.5. Mapping Forest Fire Occurrence Likelihood

Based on the fire occurrence probability calculated by the random forest model for fire ignitions and nonfire points, we used ordinary kriging interpolation to map the forest fire occurrence likelihood in mainland China in ArcGIS 10.2 [30].

3. Results

3.1. Identification of Forest Fire Driving Factors and Their Importance Ranks

Table 4 and Figure 3 show the forest fire driving factors and their importance rank in six regions and the whole study area. Table A1 and Figure A1 show the significant variables and their importance rank of each intermediate model.

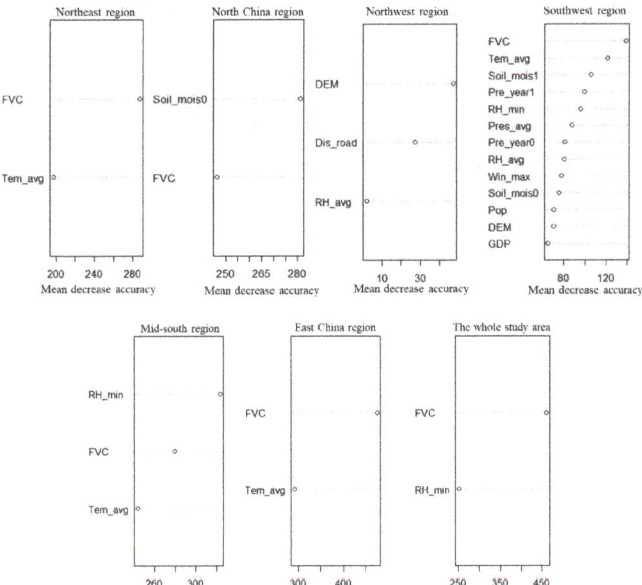

Figure 3. Importance rankings of the forest fire driving factors according to the mean decrease accuracy in six geographical regions and the whole study area. The abbreviated variable names are the same as in Table 4.

Table 4. The results when identifying the forest fire driving factors in six geographical regions and the whole study area.

Variable Type	Variable	NE	N	NW	SW	MS	E	The Whole Study Area
	Pre_year0				+			
	Pre_year1				+			
	Soil_mois0		+		+			
	Soil_mois1	/	/	/	+	/	/	/
	Tem_avg	+			+	+	+	
	GST_avg	/	/	/	/	/	/	/
Climatic	RH_avg			+	+			
	RH_min				+	+		+
	Pre_daily							
	Pres_avg				+			
	SSD							
	Win_avg							
	Win_max				+			
	DEM			+	+			
Topographic	Aspect							
	Slope							
Vegetation	FVC	+	+		+	+	+	+
	Dis_road			+				
Socioeconomic	Dis_sett							
	Pop				+			
	GDP				+			

VIF (variance inflation factor) was used to measure the amount of multicollinearity in the explanatory variables. When VIF > 10, then collinearity in the explanatory variable exists and is excluded in the random forest model. "+" indicates that the variable was identified as being a forest fire driving factor in a given region, and "/" indicates that the variable is excluded due to multicollinearity. NE: Northeast region; N: North China region; NW: Northwest region; SW: Southwest region; MS: Mid-south region; E: East China region; Pre_year0: annual precipitation in the year before the fire; Pre_year1: annual precipitation in the year of the fire; Soil_mois0: annual soil moisture in the year before the fire; Soil_mois1: annual soil moisture in the year of the fire; Tem_avg: daily average temperature; GST_avg: daily average ground surface temperature; RH_avg: daily average relative humidity; RH_min: daily minimum relative humidity; Pre_daily: daily precipitation; Pres_avg: daily average air pressure; SSD: sunshine hours; Win_avg: daily average wind speed; Win_max: daily maximum wind speed; DEM: elevation; FVC: fractional vegetation cover; Dis_road: the distance to road and railway; Dis_sett: the distance to settlement; Pop: population density; GDP: gross national product.

3.2. Influence of the Forest Fire Driving Factors on Forest Fire Occurrence in Different Regions

Partial dependence plots of each forest fire driving factor in each region were drawn to analyse the variables' influence intervals and trends on the probability of forest fire occurrence, where x is the variable value and y is logit of the probability of forest fire occurrence/2 [30]. The markers on the x-axis show the data distribution, where fewer marks indicate less training data and inaccurate model predictions; therefore, only the impact trends within the dense data range are discussed in this study.

Figure 4 shows a nonlinear relationship between each forest fire driving factor and the probability of forest fire occurrence. The vegetation variable shows the same influence trend on the forest fire occurrence probability in each region, and the overall trend is fluctuating. When the fractional vegetation cover is approximately 0.9, the probability of forest fire occurrence shows a peak value and then shows a sharp decline trend, while the probability is lowest when the fractional vegetation cover is approximately 0.98. The impact of climate variables is complex. The daily average temperature shows the same influence trend in the Northeast region and Southwest region: it was positively correlated with the probability of forest fire occurrence initially and negatively correlated after the values exceeded thresholds (12 °C in the Northeast region and 21 °C in the Southwest region). However, it shows another influence trend in the Mid-south region and East China region: the probability of forest fire occurrence is stable at higher values within 20 °C and decreases sharply when the daily average temperature exceeds 20 °C. The average daily relative humidity and the minimum daily relative humidity are generally negative correlated with the probability of forest fire occurrence in the respective regions. The annual soil moisture shows different influence trends in the North China and Southwest regions: in the North China region, it shows a fluctuating trend, while in the Southwest region, the probability of forest fire occurrence increases initially and then decreases as the annual soil moisture increases. For other climate variables, annual precipitation in the year before the fire and the year of the fire, daily average air pressure and daily maximum wind speed were generally positively correlated with the probability of forest fire occurrence. The elevation shows similar influence trends in the Northwest region and Southwest region and is negatively correlated with the probability of forest fire occurrence. Among socioeconomic variables, the probability of forest fire occurrence decreases with increasing distance from roads and increases initially and then declines with increasing population density and GDP.

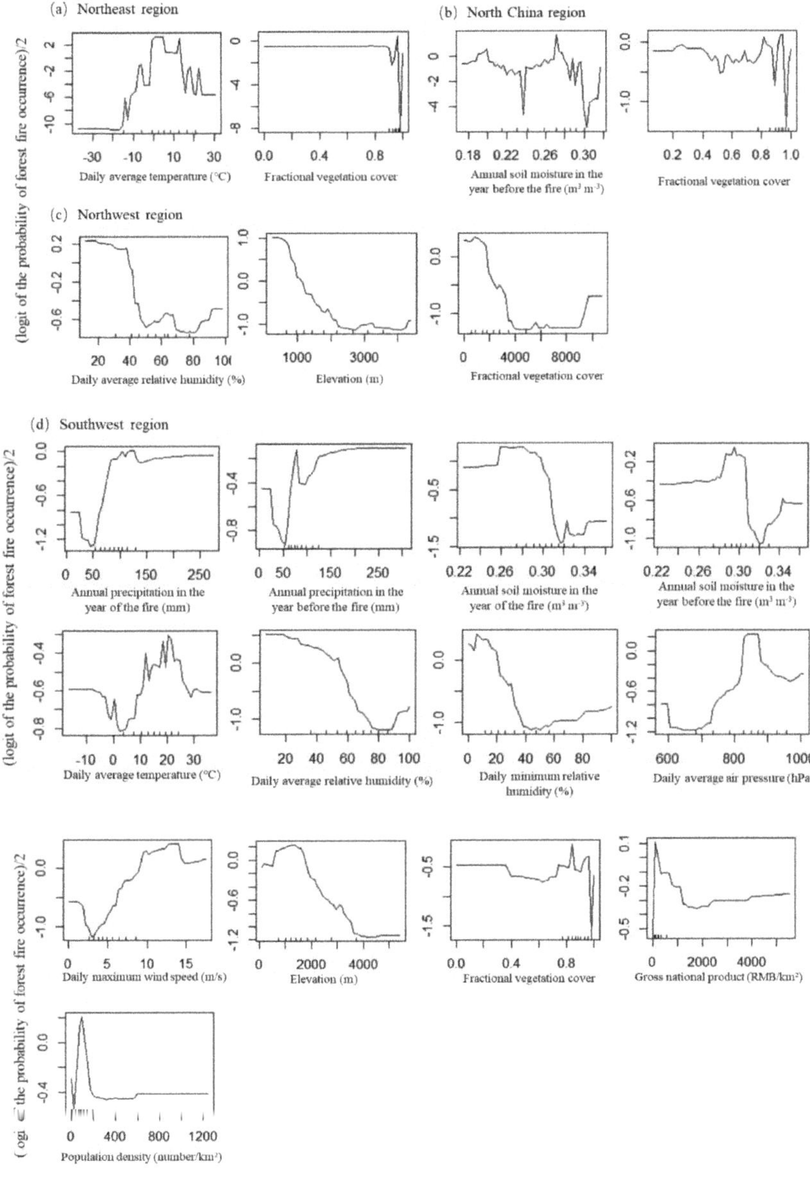

Figure 4. *Cont.*

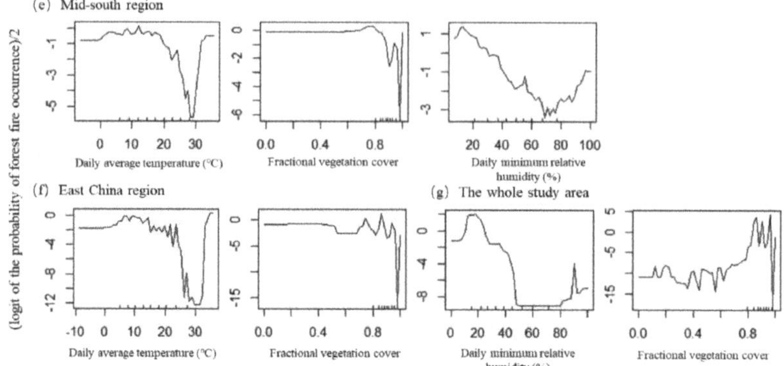

Figure 4. Partial dependence plots of each forest fire driving in six geographical regions and the whole study area.

3.3. Model Prediction Accuracy in Different Regions

The AUC values of each final model and intermediate model are greater than 0.85, and the prediction accuracy is between 70.0% and 91.4% (Table 5), which indicates that the model predictive capability is good. In the final models, the AUC (0.974) and prediction accuracy (91.4% for training and 89.3% for testing) in the East China region were the highest. The AUC (0.871) and prediction accuracy (81.75% for training and 70.52% for testing) in the Northwest region were the lowest, which may be due to the too-few fire ignition in the Northwest region.

Table 5. Comparisons of prediction accuracy of random forest models.

Regions	Model	AUC Value	Cut-Off	Prediction Accuracy (%)	
		(Intermediate Model 1/2/3/4/5)	(Intermediate Model 1/2/3/4/5)	Training (Subtraining 1/2/3/4/5)	Validation (Subvalidation 1/2/3/4/5)
Northeast region	Intermediate model	0.978/0.981/0.971/0.969/0.981	0.466/0.494/0.488/0.466/0.420	92.8/93.0/90.3/89.2/92.5	93.2/92.4/89.4/87.5/93.1
	Final model	0.976	0.493	92	91.5
North China region	Intermediate model	0.974/0.969/0.969/0.970/0.971	0.396/0.426/0.435/0.433/0.390	92.7/92.5/92.7/91.8/92.7	91.6/92.2/91.5/93.5/91.6
	Final model	0.971	0.402	92.8	92.2
East China region	Intermediate model	0.963/0.956/0.963/0.957/0.955	0.470/0.397/0.415/0.457/0.462	90.9/86.8/89.4/89.5/88.8	88.4/86.3/86.7/88.0/89.6
	Final model	0.955	0.467	89	86.2
Northwest region	Intermediate model	0.974/0.981/0.959/0.979/0.960	0.315/0.414/0.462/0.644/0.238	91.0/93.6/91.0/93.6/89.7	86.5/84.6/88.5/90.4/90.4
	Final model	0.964	0.373	91.4	90.3
Southwest region	Intermediate model	0.971/0.966/0.969/0.968/0.968	0.382/0.384/0.379/0.375/0.393	91.5/91.0/91.0/90.6/90.6	90.1/91.2/90.1/90.3/90.2
	Final model	0.966	0.393	90.6	91
Mid-south region	Intermediate model	0.965/0.965/0.953/0.968/0.965	0.397/0.391/0.470/0.418/0.451	89.1/88.9/88.6/89.2/90.4	88.4/88.7/87.4/89.2/90.3
	Final model	0.979	0.481	90.2	87.1
The whole study area	Intermediate model	0.949/0.954/0.951/0.912/0.942	0.419/0.431/0.448/0.372/0.327	86.8/88.2/87.8/83.2/84.4	85.9/88.0/87.6/83.0/83.9
	Final model	0.944	0.415	86	85.8

3.4. Likelihood of Forest Fire Occurrence

Figures 5 and 6 show that the areas with high probability of forest fires are concentrated in the Northeast and Mid-south regions as well as the south of East China region and the northwest of Northwest region. To compare the results of the national model and the regional model, we drew a map of the difference in the likelihood of forest fire occurrence calculated based on the whole study area model and the regional models (Figure 7). The map shows that the probability of the whole model was higher than those of the regional models in most areas of the Southwest region and North China region and in the centre of Northwest region and lower than those in most areas of the Northeast, Northwest, East China and Mid-south regions and in the north of Northwest region.

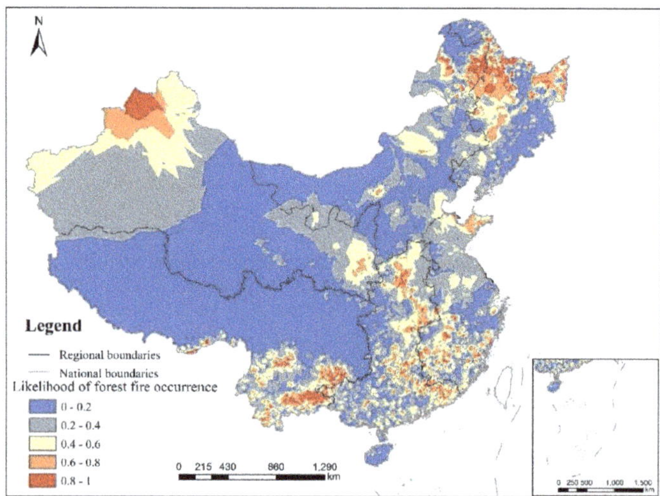

Figure 5. Map of the likelihood of forest fire occurrence in China obtained from the regional models.

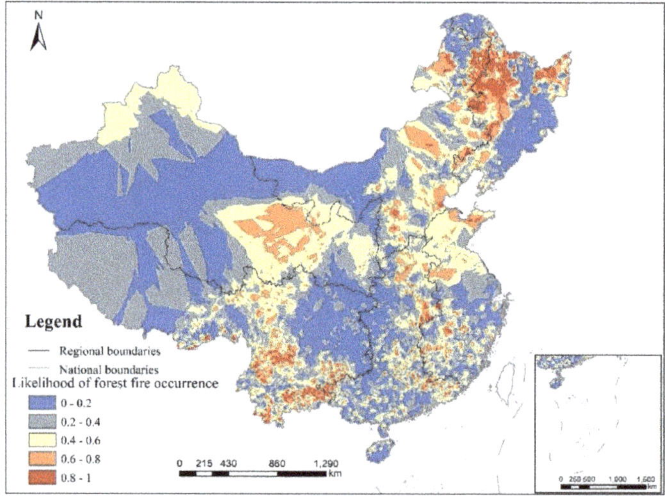

Figure 6. Map of the likelihood of forest fire occurrence in China obtained from the whole study area model.

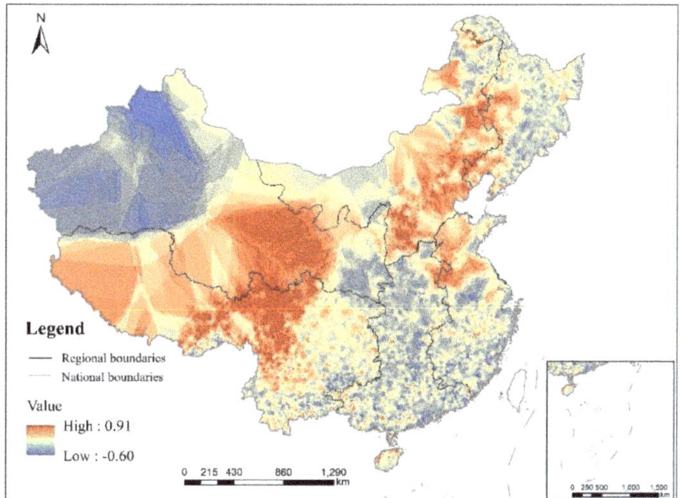

Figure 7. Map of the likelihood of forest fire occurrence obtained from the whole study area model minus that obtained from the regional model.

4. Discussion

4.1. Forest Fire Driving Factors and Their Influence

Previous studies have found regional and scale differences in forest fire factors [18,36,63]. This study also confirmed this point. We found that due to the differing geographical and social conditions in China from region to region, the forest fire driving factors vary in different regions, and the same variables also operate differently depending on the region and the scale of analysis, which illustrates the spatial applicability of forest fire research and the importance of formulating forest fire management systems based on regional characteristics. All final models included a smaller number of variables selected from the initial set. Previous studies have also shown that the simplified model is more stable. Previous studies have also noted that a parsimonious model would be more stable [28,49].

In this study, all final models included climate variables, which are considered the dominant factors affecting forest fires [64–66]. Among climate variables, daily average temperature was the forest fire driving factor in the most regions (Northeast, Southwest, Mid-south, East China and the whole study area). Previous studies [29,30] have shown thresholds and complex nonlinear relationships between temperature and forest fire occurrence probability, and our study confirms this point. The probability of forest fire occurrence initially increases or stabilizes at a higher value with the increase in temperature. When the temperature exceeds a certain threshold (12 °C in the Northeast region, 21 °C in the Southwest region and 20 °C in the Mid-south region and East China region), the probability shows a sharp downward trend. There may be two reasons for this situation. (1) Although high temperatures can increase plant evaporation, thereby reducing the moisture content of forest fire fuels [67], most parts of China experience a monsoon climate (Table 1), and rainfall and heat are synchronous. Therefore, high-temperature weather is often accompanied by high relative humidity levels, which have opposite effects on forest fires, so there were impact thresholds. (2) At high temperatures, forest fire prevention departments are vigilant, implementing strict fire prevention systems and limiting the occurrence of forest fires [68]. Relative humidity is also one of the main forest fire driving factors. In this study, relative humidity showed a similar influence trend in the respective regions, and it was negatively correlated with the occurrence probability of forest fires despite some moderate fluctuations. This is because high relative humidity increases the moisture content of combustible materials and reduces the possibility of fire [64]. It is noteworthy that the daily minimum relative humidity was also selected as the forest fire driving factor in the whole study area, which indicates that this variable operates at both regional and large scales. Air pressure affects the oxygen content and fuel ignition temperature, and a relatively lower pressure will lead to a lower oxygen content and higher ignition temperature, thus reducing the possibility of forest fire occurrence [69]. However, in the Southwest region, when the daily air pressure is higher than 860 hPa, the probability of forest fire occurrence shows a small decrease, which indicates that there is also an impact threshold of air pressure. The daily maximum wind speed is also one of the driving factors of forest fires in southwest China. The wind will increase evaporation capacity, and the higher the wind speed, the smaller the water content of forest combustibles; hence, the wind speed has a positive correlation with the occurrence probability of forest fires, which is consistent with the research results of Guo [18] in Fujian province of China. The soil moisture in the year of the fire directly affects the water content of forest combustibles [31], so this variable is negatively correlated with the occurrence probability of forest fire. Annual precipitation promotes the accumulation of plant fuels, thus having a positive impact on the occurrence probability of forest fires.

Among the topographic factors, elevation is a forest fire driving factor in the Northwest region and Southwest region, and it is negatively correlated with the occurrence probability of forest fires in both regions. We suspect that this is because the surface of these two areas fluctuates greatly; the elevation in most areas is 500~5000 m in the Northwest region and 500~6000 m in the Southwest region. As elevation increases, human activity decreases, and its impact on weather conditions, vegetation and soil moisture is also not conducive to forest fire occurrence [49,70,71]. Tian's research [32] also showed that forest fires mainly occurred at low elevations in China.

The vegetation variable (fractional vegetation cover) is a forest fire driving factor in all regions except the Northwest region, and its importance ranked first in four regions (Northeast region, Southwest region, East China region and the whole study area). Previous studies have also shown that vegetation cover has an important impact on forest fires [29,67]. Generally, the higher the vegetation coverage, the more fuel is available, so high vegetation coverage leads to a high forest fire rate. However, in this study, fractional vegetation cover showed a complicated influence trend: when the fractional vegetation cover is between 0.8 and 0.97, the occurrence probability of forest fires fluctuates at a higher value, and then it drops rapidly, reaching a minimum value when the fractional vegetation cover is approximately 0.98. We suspect that this is because in forest land with high vegetation coverage, canopy occlusion will lead to some small fires that are not easily detected by MODIS [67].

Compared with other variables, socioeconomic variables are the forest fire factors in few regions (Northwest region and Southwest region), with low degrees of importance (Figure 4). Distance from the road is negatively correlated with the probability of forest fires in the Northwest region because the forests close to the road are vulnerable to traffic accidents and human activities (i.e., smoking and picnics) [26,61]. GDP and population density show similar influence trends, and they have a positive impact on the occurrence probability of forest fires initially and then have a negative impact after exceeding a certain threshold (GDP of 200 RMB/km^2 and population density of 100 number/km^2). This may be because within a certain range, the increase in population density and GDP will increase human activity in forests, thereby promoting forest fire occurrence [51,72,73]. However, in economically prosperous and high-population-density areas, the forest coverage rate is low and there are few forest-related production activities conducted by humans, so the occurrence probability of forest fires decreases [18,29,32].

4.2. Implications for Forest Fire Prevention

There are differences in the forest fire driving factors (Figure 3 and Table 4) and the prediction results (Figures 5–7) between the regional models and the whole study area model. Therefore, it is necessary to study forest fire driving factors based on geographical regions, and regional differences should also be fully considered in forest fire management. Forest fire management departments should formulate forest fire prevention strategies according to the differences in forest fire driving factors and impact thresholds in different regions. E.g., in the Northwest region, elevation has the greatest impact on forest fire occurrence, and the probability of forest fires is higher in low-elevation areas. Therefore, the Northwest region forest fire management departments should strengthen forest fire monitoring in low-elevation areas, such as setting up more forest fire observation towers and forest fire brigades [30]. In the North China region, soil moisture has the greatest impact, so changes in soil moisture should be taken into account when developing a forest fire prevention strategy. In the Northeast region and East China region, when the daily average temperature reaches the impact threshold, the occurrence probability of forest fires reaches a maximum; hence, forest fire management departments should be more vigilant in the corresponding weather. In the Southwest region, there are 13 forest fire driving factors. These factors should be integrated into the local assessment index systems of forest fire hazard, and the influence of these factors should be considered comprehensively when judging the forest fire hazard. In the Mid-south region, forest fire management departments should pay attention to monitoring the daily minimum relative humidity.

The map of the likelihood of forest fire occurrence is also crucial to forest fire management [74]. Understanding the distribution of forest fire occurrence likelihood can help to determine the location and number of fire observation towers [28], contributing to more effective use of financial and human resources. Figure 5 shows that areas with a high probability of forest fires are concentrated in the Northeast and Mid-south regions as well as the south of East China region and the northwest of Northwest region; thus, more stringent forest fire prevention systems should be implemented in these areas.

4.3. Strengths and Limitations

Previous forest fire research has usually been based on eco-geographical areas or forest types [29,75,76]. However, these zoning methods ignore administrative boundaries, and forest fire management strategies are often formulated by administrative areas. Therefore, we chose a zoning method that takes administrative divisions into account, trying to provide a more practical reference for China's fire prevention department. Our research is based on geographical regions in China, a division method that considers both administrative divisions and natural conditions that has been used in some forestry analysis [77,78]. Each region is an aggregation of provinces with adjacent locations and similar topography, economy and climate. However, this zoning method has its shortcomings. First, if a province has complex topography and different climate and vegetation types (such as Tibet in the Southwest region), it must also be included in one region. We think that this may have led to far higher number of forest fire driving factors in the Southwest region than in the other regions. The second point is about the model. To reveal the nonlinear relationship and influence threshold between forest fire driving factors and forest fire occurrence probability, we used the random forest model, which has shown good prediction ability in previous studies on forest fire [18,30,31]. However, behaving as a "black box," this method cannot calculate regression coefficients or confidence intervals [63,79]. Based on these two points, in future research, we will try to use geographically weighted regression, a spatially explicit technique that would overcome the necessity of building predetermined regions, to analyse forest fire driving factors to address these limitations.

5. Conclusions

We used the random forest model to analyse the forest fire driving factors in different geographical regions in China for 2010 to 2018. The model predictive capability is good, with a prediction accuracy between 70.0% and 91.4%. Furthermore, we mapped the probability of forest fire occurrence in China based on the results of the model. In China, there are obvious regional differences in the types of forest fire driving factors and their impacts. Climate variables (especially temperature and humidity) have major impacts on forest fires occurrence, and the vegetation variable is secondary. Topographic variables and socioeconomic variables are only the forest fire driving factors in the Southwest and Northwest regions. There is a nonlinear relationship and influence threshold between forest fire driving factors and forest fire occurrence probability. High fire hazard areas are concentrated in the Northeast and Mid-south regions as well as the south of East China region and the northwest of Northwest region. This research will aid in providing a national-scale understanding of forest fire driving factors and fire hazard distribution in China and help policymakers to design fire management strategies and allocate resources reasonably to reduce potential fire hazards.

Author Contributions: Conceptualization, W.M.; Data curation, W.M. and Z.C.; Formal analysis, W.M.; Funding acquisition, Z.F.; Investigation, W.M.; Methodology, W.M.; Resources, F.W.; Supervision, Z.F.; Writing–original draft, W.M.; Writing–review & editing, W.M. and S.C. All authors have read and agreed to the published version of the manuscript.

Funding: This research was funded by the Fundamental Research Funds for the Central Universities (No. 2015ZCQ-LX-01) and the National Natural Science Foundation of China (No. U1710123).

Acknowledgments: We would like to acknowledge for the data support from National Earth System Science Data Center, National Science & Technology Infrastructure of China. (http://www.geodata.cn)

Conflicts of Interest: The authors declare no conflicts of interest.

Appendix A Appendix

Table A1. The results when identifying the significant variables in each intermediate model in six geographical regions and the whole study area.

Variable Type	Variable	Intermediate Models					Selected Frequency
		1	2	3	4	5	
				(a) Northeast region			
	Pre_year0						0
	Pre_year1						0
	Soil_mois0						0
	Soil_mois1	/	/	/	/	/	/
	Tem_avg	+	+	+	+	+	5
	GST_avg	/	/	/	/	/	/
Climatic	RH_avg						0
	RH_min						0
	Pre_daily						0
	Pres_avg						0
	SSD						0
	Win_avg						0
	Win_max						0
	DEM						0
Topographic	Aspect						0
	Slope						0
Vegetation	FVC	+	+	+	+	+	5
	Dis_road						0
Socioeconomic	Dis_sett						0
	Pop						0
	GDP						0
				(b) North China region			
	Pre_year0						0
	Pre_year1						0
	Soil_mois0	+		+	+	+	4
	Soil_mois1	/	/	/	/	/	/
	Tem_avg	+	+				2
	GST_avg	/	/	/	/	/	/
Climatic	RH_avg						0
	RH_min						0
	Pre_daily						0
	Pres_avg	+					1
	SSD						0
	Win_avg						0
	Win_max						0
	DEM	+					1
Topographic	Aspect						0
	Slope						0
Vegetation	FVC	+	+	+	+	+	5
	Dis_road						0
Socioeconomic	Dis_sett						0
	Pop	+					1
	GDP						0

Table A1. *Cont.*

Variable Type	Variable	Intermediate Models					Selected Frequency
		1	2	3	4	5	
(c) Northwest region							
Climatic	Pre_year0						0
	Pre_year1		+				1
	Soil_mois0						0
	Soil_mois1	/	/	/	/	/	/
	Tem_avg						0
	GST_avg	/	/	/	/	/	/
	RH_avg		+	+		+	3
	RH_min						0
	Pre_daily						0
	Pres_avg	+			+		2
	SSD						0
	Win_avg					+	1
	Win_max						0
	DEM	+	+	+		+	4
Topographic	Aspect					+	1
	Slope						0
Vegetation	FVC						0
	Dis_road			+	+	+	3
Socioeconomic	Dis_sett						0
	Pop						0
	GDP	+					1
(d) Southwest region							
Climatic	Pre_year0	+	+	+	+	+	5
	Pre_year1	+	+	+	+	+	5
	Soil_mois0	+	+	+	+	+	5
	Soil_mois1	+	+	+	+	+	5
	Tem_avg	+	+	+	+	+	5
	GST_avg	/	/	/	/	/	/
	RH_avg	+	+	+	+	+	5
	RH_min	+	+	+	+	+	5
	Pre_daily						0
	Pres_avg	+	+	+	+	+	5
	SSD						0
	Win_avg		+	+			2
	Win_max	+			+	+	3
	DEM	+	+	+	+	+	5
Topographic	Aspect						0
	Slope						0
Vegetation	FVC	+	+	+	+	+	5
	Dis_road						0
Socioeconomic	Dis_sett						0
	Pop	+	+	+	+	+	5
	GDP	+	+	+	+	+	5
(e) Mid-south region							
Climatic	Pre_year0			+	+		2
	Pre_year1				+		1
	Soil_mois0				+		1
	Soil_mois1	/	/	/	/	/	/
	Tem_avg	+	+	+	+	+	5
	GST_avg	/	/	/	/	/	/
	RH_avg				+		1
	RH_min	+	+	+	+	+	5
	Pre_daily				+		1
	Pres_avg				+		1
	SSD						0
	Win_avg						0
	Win_max						0
	DEM						0
Topographic	Aspect						0
	Slope						0
Vegetation	FVC	+	+	+	+	+	5
	Dis_road						0
Socioeconomic	Dis_sett						0
	Pop						0
	GDP				+		1

Table A1. *Cont.*

Variable Type	Variable	Intermediate Models					Selected Frequency
		1	2	3	4	5	
	(f) East China region						
Climatic	Pre_year0						0
	Pre_year1						0
	Soil_mois0						0
	Soil_mois1	/	/	/	/	/	/
	Tem_avg	+	+	+	+	+	5
	GST_avg	/	/	/	/	/	/
	RH_avg						0
	RH_min						0
	Pre_daily						0
	Pres_avg						0
	SSD						0
	Win_avg						0
	Win_max						0
Topographic	DEM						0
	Aspect						0
	Slope						0
Vegetation	FVC	+	+	+	+	+	5
Socioeconomic	Dis_road						0
	Dis_sett						0
	Pop						0
	GDP						0
	(g) The whole study area						
Climatic	Pre_year0						0
	Pre_year1						0
	Soil_mois0						0
	Soil_mois1	/	/	/	/	/	/
	Tem_avg						0
	GST_avg	/	/	/	/	/	/
	RH_avg						0
	RH_min	+	+	+	+	+	5
	Pre_daily						0
	Pres_avg						0
	SSD						0
	Win_avg						0
	Win_max						0
Topographic	DEM						0
	Aspect						0
	Slope						0
Vegetation	FVC	+	+	+	+	+	5
Socioeconomic	Dis_road						0
	Dis_sett						0
	Pop						0
	GDP						0

VIF (variance inflation factor) was used to measure the amount of multicollinearity in the explanatory variables. When VIF > 10, then collinearity in the explanatory variable exists and is excluded in the random forest model. "+" indicates that the variable was identified as being a forest fire driving factor in a given region, and "/" indicates that the variable is excluded due to multicollinearity. Pre_year0: annual precipitation in the year before the fire; Pre_year1: annual precipitation in the year of the fire; Soil_mois0: annual soil moisture in the year before the fire; Soil_mois1: annual soil moisture in the year of the fire; Tem_avg: daily average temperature; GST_avg: daily average ground surface temperature; RH_avg: daily average relative humidity; RH_min: daily minimum relative humidity; Pre_daily: daily precipitation; Pres_avg: daily average air pressure; SSD: sunshine hours; Win_avg: daily average wind speed; Win_max: daily maximum wind speed; DEM: elevation; FVC: fractional vegetation cover; Dis_road: the distance to road and railway; Dis_sett: the distance to settlement; Pop: population density; GDP: gross national product.

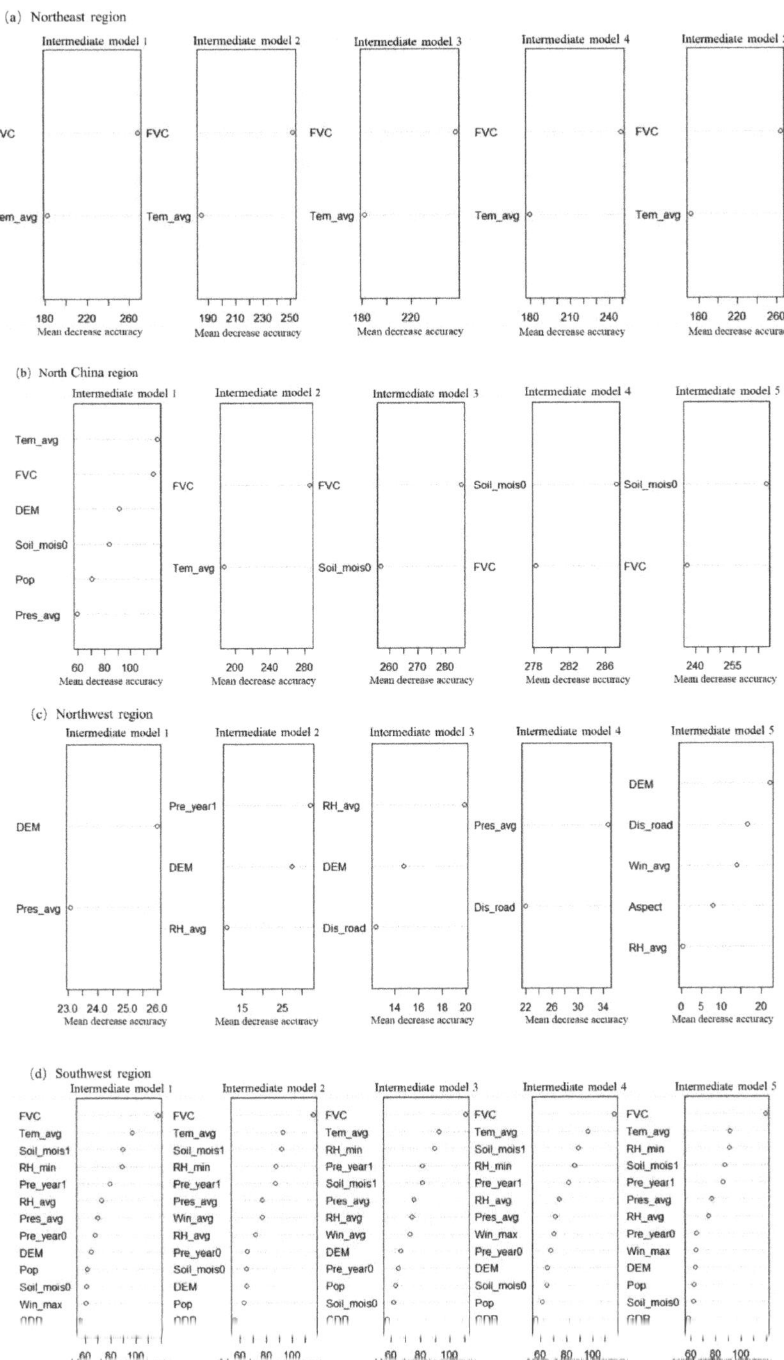

Figure A1. *Cont.*

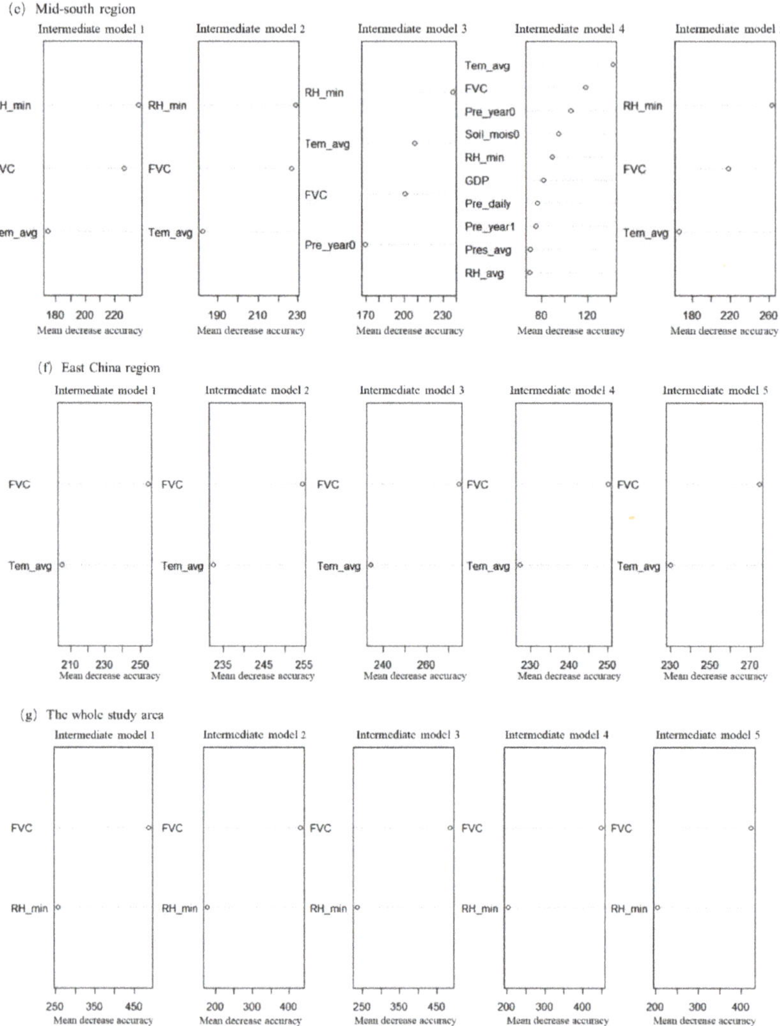

Figure A1. Importance rankings of the significant variables according to the mean decrease accuracy in each intermediate model in six geographical regions and the whole study area. The abbreviated variable names are the same as in Table A1.

References

1. Morales-Hidalgo, D.; Oswalt, S.N.; Somanathan, E. Status and trends in global primary forest, protected areas, and areas designated for conservation of biodiversity from the global forest resources assessment 2015. *For. Ecol. Manag.* **2015**, *352*, 68–77. [CrossRef]
2. Köhl, M.; Lasco, R.; Cifuentes, M.; Jonsson, Ö.; Korhonen, K.T.; Mundhenk, P.; Navar, J.D.J.; Stinson, G. Changes in forest production, biomass and carbon: Results from the 2015 un fao global forest resource assessment. *For. Ecol. Manag.* **2015**, *352*, 21–34. [CrossRef]
3. Keenan, R.J.; Reams, G.A.; Achard, F.; Freitas, J.V.D.; Grainger, A.; Lindquist, E. Dynamics of global forest area: Results from the fao global forest resources assessment 2015. *For. Ecol. Manag.* **2015**, *352*, 9–20. [CrossRef]

4. Bergeron, Y.; Gauthier, S.; Flannigan, M.; Kafka, V. Fire regimes at the transition between mixedwood and coniferous boreal forest in northwestern quebec. *Ecology* **2004**, *85*, 1916–1932. [CrossRef]

5. Piao, S.; Huang, M.; Zhuo, L.; Wang, X.; Ciais, P.; Canadell, J.G.; Kai, W.; Bastos, A.; Friedlingstein, P.; Houghton, R.A. Lower land-use emissions responsible for increased net land carbon sink during the slow warming period. *Nat. Geosci.* **2018**, *11*, 739–743. [CrossRef]

6. Podur, J.; Martell, D.L.; Csillag, F. Spatial patterns of lightning-caused forest fires in Ontario, 1976–1998. *Ecol. Model.* **2003**, *164*, 1–20. [CrossRef]

7. Bond, W.J.; Keeley, J.E. Fire as a global 'herbivore': The ecology and evolution of flammable ecosystems. *Trends Ecol. Evol.* **2005**, *20*, 387–394. [CrossRef]

8. Pastro, L.A.; Dickman, C.R.; Letnic, M. Burning for biodiversity or burning biodiversity? Prescribed burn vs. Wildfire impacts on plants, lizards, and mammals. *Ecol. Appl.* **2011**, *21*, 3238–3253. [CrossRef]

9. Thom, D.; Seidl, R. Natural disturbance impacts on ecosystem services and biodiversity in temperate and boreal forests. *Biol. Rev. Camb. Philos. Soc.* **2016**, *91*, 760–781. [CrossRef]

10. Westerling, A.; Bryant, B. Climate change and wildfire in California. *Clim. Chang.* **2008**, *87*, 231–249. [CrossRef]

11. Hering, A.S.; Bell, C.L.; Genton, M.G. Modeling spatio-temporal wildfire ignition point patterns. *Environ. Ecol. Stat.* **2009**, *16*, 225–250. [CrossRef]

12. Mckenzie, D.; Shankar, U.; Keane, R.E.; Stavros, E.N.; Heilman, W.E.; Fox, D.G.; Riebau, A.C. Smoke consequences of new wildfire regimes driven by climate change. *Earths Future* **2014**, *2*, 35–59. [CrossRef]

13. Shun, L.; Zhiwei, W.; Yu, L.; Hongshi, H. A review of fire controlling factors and their dynamics in boreal forest. *World For. Res.* **2017**, *30*, 41–45. [CrossRef]

14. Dimopoulou, M.; Giannikos, I. Towards an integrated framework for forest fire control. *Eur. J. Oper. Res.* **2004**, *152*, 476–486. [CrossRef]

15. Flannigan, M.D.; Krawchuk, M.A.; Groot, W.J.D.; Wotton, B.M.; Gowman, L.M. Implications of changing climate for global wildland fire. *Int. J. Wildland Fire* **2009**, *18*, 483–507. [CrossRef]

16. Moreno, M.V.; Chuvieco, E. Characterising fire regimes in Spain from fire statistics. *Int. J. Wildland Fire* **2013**, *22*, 296–305. [CrossRef]

17. Ganteaume, A.; Camia, A.; Jappiot, M.; San-Miguel-Ayanz, J.; Long-Fournel, M.; Lampin, C. A review of the main driving factors of forest fire ignition over Europe. *Environ. Manag.* **2013**, *51*, 651–662. [CrossRef]

18. Guo, F.; Wang, G.; Su, Z.; Liang, H.; Liu, A. What drives forest fire in Fujian, China? Evidence from logistic regression and random forests. *Int. J. Wildland Fire* **2016**, *25*, 505–519. [CrossRef]

19. Morgan, P.; Hardy, C.C.; Swetnam, T.W.; Rollins, M.G.; Long, D.G. Mapping fire regimes across time and space: Understanding coarse and fine-scale fire patterns. *Int. J. Wildland Fire* **2001**, *10*, 329–342. [CrossRef]

20. Rollins, M.G.; Morgan, P.; Swetnam, T. Landscape-scale controls over 20th century fire occurrence in two large rocky mountain (USA) wilderness areas. *Landsc. Ecol.* **2002**, *17*, 539–557. [CrossRef]

21. Sharples, J. An overview of mountain meteorological effects relevant to fire behaviour and bushfire risk. *Int. J. Wildland Fire* **2009**, *18*, 737–754. [CrossRef]

22. Minnichl, R.A.; Bahrez, C.J. Wildland fire and chaparral succession along the California-Baja California boundary. *Int. J. Wildland Fire* **1995**, *5*, 13–24. [CrossRef]

23. Pew, K.L.; Larsen, C.P.S. Gis analysis of spatial and temporal patterns of human-caused wildfires in the temperate rain forest of vancouver island, Canada. *For. Ecol. Manag.* **2001**, *140*, 1–18. [CrossRef]

24. Pausas, J.G.; Paula, S. Fuel shapes the fire—climate relationship: Evidence from mediterranean ecosystems. *Glob. Ecol. Biogeogr.* **2012**, *21*, 1074–1082. [CrossRef]

25. Maingi, J.K.; Henry, M.C. Factors influencing wildfire occurrence and distribution in eastern kentucky, USA. *Int. J. Wildland Fire* **2007**, *16*, 23–33. [CrossRef]

26. Cardille, J.A.; Ventura, S.J.; Turner, M.G. Environmental and social factors influencing wildfires in the upper midwest, united states. *Ecol. Appl.* **2001**, *11*, 111–127. [CrossRef]

27. Turco, M.; Llasat, M.C.; von Hardenberg, J.; Provenzale, A. Impact of climate variability on summer fires in a mediterranean environment (northeastern iberian peninsula). *Clim. Chang.* **2013**, *116*, 665–678. [CrossRef]

28. Catry, F.X.; Rego, F.C.; Bação, F.; Moreira, F. Modeling and mapping wildfire ignition risk in portugal. *Int. J. Wildland Fire* **2009**, *18*, 921–931. [CrossRef]

29. Guo, F.; Su, Z.; Wang, G.; Sun, L.; Tigabu, M.; Yang, X.; Hu, H. Understanding fire drivers and relative impacts in different chinese forest ecosystems. *Sci. Total Environ.* **2017**, *411*, 605–606. [CrossRef]

30. Su, Z.; Hu, H.; Wang, G.; Ma, Y.; Yang, X.; Guo, F. Using GIS and random forests to identify fire drivers in a forest city, Yichun, China. *Geomat. Nat. Hazards Risk* **2018**, *9*, 1207–1229. [CrossRef]

31. Oliveira, S.; Oehler, F.; San-Miguel-Ayanz, J.; Camia, A.; Pereira, J.M.C. Modeling spatial patterns of fire occurrence in mediterranean europe using multiple regression and random forest. *For. Ecol. Manag.* **2012**, *275*, 117–129. [CrossRef]

32. Tian, X.; Zhao, F.; Shu, L.; Wang, M. Distribution characteristics and the influence factors of forest fires in China. *For. Ecol. Manag.* **2013**, *310*, 460–467. [CrossRef]

33. Chang, Y.; Zhu, Z.; Bu, R.; Li, Y.; Hu, Y. Environmental controls on the characteristics of mean number of forest fires and mean forest area burned (1987–2007) in China. *For. Ecol. Manag.* **2015**, *356*, 13–21. [CrossRef]

34. Zhong, M.; Fan, W.; Liu, T.; Li, P. Statistical analysis on current status of China forest fire safety. *Fire Saf. J.* **2003**, *38*, 257–269. [CrossRef]

35. Aifeng, L.U. Study on the relationship among forest fire, temperature and precipitation and its spatial-temporal variability in China. *Agric. Sci. Technol.* **2011**, *12*, 1396–1400. [CrossRef]

36. Ying, L.; Han, J.; Du, Y.; Shen, Z. Forest fire characteristics in China: Spatial patterns and determinants with thresholds. *For. Ecol. Manag.* **2018**, *424*, 345–354. [CrossRef]

37. Huiling, L. *Based on Spatial and Non-Spatial Model and Influence Factors Analysis of the Space-Time Characteristics of Fujian Forest Fires*; Fujian Agriculture and Forestry University: Fuzhou, China, 2016.

38. Ma, W.; Feng, Z.; Cheng, Z.; Wang, F. Study on driving factors and distribution pattern of forest fires in shanxi province. *J. Cent. South Univ. For. Technol.* **2020**. Available online: https://kns.cnki.net/KCMS/detail/43.1470.S.20200115.1043.001.html (accessed on 5 March 2020). [CrossRef]

39. State Forestry Administ-Ration. *China Forest Resources Inventory Repor*; China Forestry Publishing House: Beijing, China, 2014; pp. 80–81, ISBN 978-7-5038-7424-6.

40. National Bureau of Statistics. China statistical Yearbook. 2018. Available online: http://www.stats.gov.cn/tjsj/ndsj/ (accessed on 10 July 2019).

41. Andela, N.; Morton, D.; Giglio, L.; Paugam, R.; Chen, Y.; Hantson, S.; Werf, G.; Randerson, J. The global fire atlas of individual fire size, duration, speed, and direction. *Earth Syst. Sci. Data Discuss.* **2018**, *11*, 1–28. [CrossRef]

42. Peng, S.; Ding, Y.; Liu, W.; Li, Z. 1 km monthly temperature and precipitation dataset for China from 1901 to 2017. *Earth Syst. Sci. Data* **2019**, *11*, 1931–1946. [CrossRef]

43. Zhang, Z.X.; Zhang, H.Y.; Zhou, D.W. Using GIS spatial analysis and logistic regression to predict the probabilities of human-caused grassland fires. *J. Arid Environ.* **2010**, *74*, 386–393. [CrossRef]

44. Yu, M. *The Research of Forest Fire Prediction Model in Fangshan District, Beijing and Sublot Fire Danger Rating Division*; Beijing Forestry University: Beijing, China, 2016.

45. Vilar, L.; Woolford, D.; Martell, D.; Martín, M. A model for predicting human-caused wildfire occurrence in the region of madrid, Spain. *Int. J. Wildland Fire* **2010**, *19*, 325–337. [CrossRef]

46. Prasad, V.K.; Badarinath, K.V.S.; Eaturu, A. Biophysical and anthropogenic controls of forest fires in the deccan plateau, india. *J. Environ. Manag.* **2008**, *86*, 1–13. [CrossRef] [PubMed]

47. Sturtevant, B.; Cleland, D. Human and biophysical factors influencing modern fire disturbance in northern wisconsin. *Int. J. Wildland Fire* **2007**, *16*, 398–413. [CrossRef]

48. Gitelson, A.A.; Kaufman, Y.J.; Stark, R.; Rundquist, D. Novel algorithms for remote estimation of vegetation fraction. *Remote Sens. Environ.* **2002**, *80*, 76–87. [CrossRef]

49. Leprieur, C.; Verstraete, M.M.; Pinty, B. Evaluation of the performance of various vegetation indices to retrieve vegetation cover from avhrr data. *Remote Sens. Rev.* **1994**, *10*, 265–284. [CrossRef]

50. Purevdorj, T.S.; Tateishi, R.; Ishiyama, T.; Honda, Y. Relationships between percent vegetation cover and vegetation indices. *Int. J. Remote Sens.* **1998**, *19*, 3519–3535. [CrossRef]

51. Xu, X. *China Quarterly Vegetation Index (ndvi) Spatial Distribution Data Set*; Data Registration and Publishing System of Resource and Environment Science Data Center of Chinese Academy of Science: Beijing, China, 2018. Available online: http://www.resdc.cn/10.12078/2018060603 (accessed on 10 January 2019). [CrossRef]

52. Rodrigues, M.; Riva, J. An insight into machine-learning algorithms to model human-caused wildfire occurrence. *Environ. Model. Softw.* **2014**, *57*, 192–201. [CrossRef]

53. Duro, D.C.; Franklin, S.E.; Dubé, M.G. Multi-scale object-based image analysis and feature selection of multi-sensor earth observation imagery using random forests. *Int. J. Remote Sens.* **2012**, *33*, 4502–4526. [CrossRef]

54. Marston, C.G.; Danson, F.M.; Armitage, R.P.; Giraudoux, P.; Pleydell, D.R.J.; Wang, Q.; Qui, J.; Craig, P.S. A random forest approach for predicting the presence of echinococcus multilocularis intermediate host ochotona spp. Presence in relation to landscape characteristics in western China. *Appl. Geogr.* **2014**, *55*, 176–183. [CrossRef]

55. Abdel-Rahman, E.M.; Ahmed, F.B.; Ismail, R. Random forest regression and spectral band selection for estimating sugarcane leaf nitrogen concentration using eo-1 hyperion hyperspectral data. *Int. J. Remote Sens.* **2013**, *34*, 712–728. [CrossRef]

56. Liang, H.; Lin, Y.; Yang, G.; Su, Z.; Wang, W.; Guo, F. Application of random forest algorithm on the forest fire prediction in Tahe area based on meteorological factors. *Sci. Silvae Sin.* **2016**, *52*, 89–98. [CrossRef]

57. Liaw, A.; Wiener, M. Classification and regression by randomforest. *R News* **2002**, *2*, 18–22.

58. Jiménez-Valverde, A. Insights into the area under the receiver operating characteristic curve (AUC) as a discrimination measure in species distribution modelling. *Glob. Ecol. Biogeogr.* **2012**, *21*, 498–507. [CrossRef]

59. Chang, Y.; Bu, R.; Chen, H.; Feng, Y.; Li, Y.; Hu, Y.; Wang, Z. Predicting fire occurrence patterns with logistic regression in heilongjiang province, China. *Landsc. Ecol.* **2013**, *28*, 1989–2004. [CrossRef]

60. Vilar, L.; Martín, M.; Martinez-Vega, J. Logistic regression models for human-caused wildfire risk estimation: Analysing the effect of the spatial accuracy in fire occurrence data. *Eur. J. For. Res.* **2011**, *130*, 983–996. [CrossRef]

61. Martínez, J.; Vega-Garcia, C.; Chuvieco, E. Human-caused wildfire risk rating for prevention planning in Spain. *J. Environ. Manag.* **2009**, *90*, 1241–1252. [CrossRef]

62. Vega-Garcia, C.; Woodard, P.M.; Titus, S.J.; Adamowicz, L.; Lee, B.S. A logit model for predicting the daily occurrence of human caused forest-fires. *Int. J. Wildland Fire* **1995**, *5*, 101–111. [CrossRef]

63. Prasad, A.M.; Iverson, L.R.; Liaw, A. Newer classification and regression tree techniques: Bagging and random forests for ecological prediction. *Ecosystems* **2006**, *9*, 181–199. [CrossRef]

64. Zumbrunnen, T.; Pezzatti, G.B.; Menéndez, P.; Bugmann, H.; Bürgi, M.; Conedera, M. Weather and human impacts on forest fires: 100 years of fire history in two climatic regions of switzerland. *For. Ecol. Manag.* **2011**, *261*, 2188–2199. [CrossRef]

65. Wotton, M.; Martell, D.; Logan, K. Climate change and people-caused forest fire occurrence in Ontario. *Clim. Chang.* **2003**, *60*, 275–295. [CrossRef]

66. Varela, V.; Vlachogiannis, D.; Sfetsos, A.; Karozis, S.; Politi, N.; Giroud, F. Projection of forest fire danger due to climate change in the french mediterranean region. *Sustainability* **2019**, *11*, 4284. [CrossRef]

67. Chuvieco, E.; Cocero, D.; Riaño, D.; Martin, P.; Martínez-Vega, J.; de la Riva, J.; Pérez, F. Combining ndvi and surface temperature for the estimation of live fuel moisture content in forest fire danger rating. *Remote Sens. Environ.* **2004**, *92*, 322–331. [CrossRef]

68. Hu, T.; Zhou, G. Drivers of lightning- and human-caused fire regimes in the great xing'an mountains. *For. Ecol. Manag.* **2014**, *329*, 49–58. [CrossRef]

69. Song, Z. *Principle and Forecast of Forest Fire*, 1st ed.; China Meteorological Press: Beijing, China, 1991; pp. 56–57, ISBN 7502905820.

70. Sebastian, A.; Salvador, R.; Gonzalo Jimenez, J.; San-Miguel-Ayanz, J. Integration of socio-economic and environmental variables for modelling long-term fire danger in southern europe. *Eur. J. For. Res.* **2008**, *127*, 149–163. [CrossRef]

71. González, J.R.; Palahí, M.; Trasobares, A.; Pukkala, T. A fire probability model for forest stands in Catalonia (north-east Spain). *Ann. For. Sci.* **2006**, *63*, 169–176. [CrossRef]

72. Syphard, A.; Radeloff, V.; Keeley, J.; Hawbaker, T.; Clayton, M.; Stewart, S.; Hammer, R. Human influence on California fire regimes. *Ecol. Appl. Publ. Ecol. Soc. Am.* **2007**, *17*, 1388–1402. [CrossRef] [PubMed]

73. Pereira, M.G.; Malamud, B.D.; Trigo, R.M.; Alves, P.I. The history and characteristics of the 1980–2005 Portuguese rural fire database. *Nat. Hazards Earth Syst. Sci.* **2011**, *11*, 3343–3358. [CrossRef]

74. Saglam, B.; Bilgili, E.; Dincdurmaz, B.; Kadiogulari, A.I.; Kücük, Ö. Spatio-temporal analysis of forest fire risk and danger using landsat imagery. *Sensors* **2008**, *8*, 3970–3987. [CrossRef]

75. Tian, X.; Shu, L.; Zhao, F.; Wang, M. Dynamic characteristics of forest fires in the main ecological geographic districts of China. *Sci. Silvae Sin.* **2015**, *51*, 71–77. [CrossRef]

76. Wu, Z.; He, H.S.; Keane, R.E.; Zhu, Z.; Wang, Y.; Shan, Y. Current and future patterns of forest fire occurrence in China. *Int. J. Wildland Fire* **2020**, *29*, 104–119. [CrossRef]

77. Lu, J.; Feng, Z.; Zhu, Y. Estimation of forest biomass and carbon storage in China based on forest resources inventory data. *Forests* **2019**, *10*, 650. [CrossRef]
78. Qiu, Z.; Feng, Z.; Song, Y.; Li, M.; Zhang, P. Carbon sequestration potential of forest vegetation in China from 2003 to 2050: Predicting forest vegetation growth based on climate and the environment. *J. Clean. Prod.* **2020**, *252*, 119715. [CrossRef]
79. Cutler, D.R.; Edwards, T.C., Jr.; Beard, K.H.; Cutler, A.; Hess, K.T.; Gibson, J.; Lawler, J.J. Random forests for classification in ecology. *Ecology* **2007**, *88*, 2783–2792. [CrossRef] [PubMed]

Article

Forest Fire Probability Mapping in Eastern Serbia: Logistic Regression versus Random Forest Method

Slobodan Milanović [1,2,*], Nenad Marković [3], Dragan Pamučar [4], Ljubomir Gigović [4], Pavle Kostić [5] and Sladjan D. Milanović [6]

1 Chair of Forest Protection, University of Belgrade Faculty of Forestry, 11030 Belgrade, Serbia
2 Department of Forest Protection and Wildlife Management, Faculty of Forestry and Wood Technology, Mendel University in Brno, 61300 Brno, Czech Republic
3 State Enterprise "Srbijašume", 11000 Belgrade, Serbia; marnen@gmail.com
4 Military Academy, University of Defence in Belgrade, 11000 Belgrade, Serbia; dragan.pamucar@gmail.com (D.P.); gigoviclj@gmail.com (L.G.)
5 Center for Predictive Analytics, 11000 Belgrade, Serbia; pavle.kostic@gmail.com
6 Department for Biomedical Engineering and Biophysics, University of Belgrade Institute for Medical Research, 11000 Belgrade, Serbia; sladjan.milanovic@imi.bg.ac.rs
* Correspondence: slobodan.milanovic@sfb.bg.ac.rs; Tel.: +381-63-888-6119

Abstract: Forest fire risk has increased globally during the previous decades. The Mediterranean region is traditionally the most at risk in Europe, but continental countries like Serbia have experienced significant economic and ecological losses due to forest fires. To prevent damage to forests and infrastructure, alongside other societal losses, it is necessary to create an effective protection system against fire, which minimizes the harmful effects. Forest fire probability mapping, as one of the basic tools in risk management, allows the allocation of resources for fire suppression, within a fire season, from zones with a lower risk to those under higher threat. Logistic regression (LR) has been used as a standard procedure in forest fire probability mapping, but in the last decade, machine learning methods such as fandom forest (RF) have become more frequent. The main goals in this study were to (i) determine the main explanatory variables for forest fire occurrence for both models, LR and RF, and (ii) map the probability of forest fire occurrence in Eastern Serbia based on LR and RF. The most important variable was drought code, followed by different anthropogenic features depending on the type of the model. The RF models demonstrated better overall predictive ability than LR models. The map produced may increase firefighting efficiency due to the early detection of forest fire and enable resources to be allocated in the eastern part of Serbia, which covers more than one-third of the country's area.

Keywords: occurrence of forest fire; machine learning; variable importance; prediction accuracy

1. Introduction

Forest fires, as global phenomena, present numerous forms of threats to many countries around the world, from Canada and Siberia through to Indonesia, Australia, Africa, and the Amazonia. Although statistics show that the frequency of fires, alongside the total burnt area, have been decreasing from year to year globally [1], several regions will experience larger and more intense fires [2–4]. The Mediterranean region is traditionally the most at risk in Europe [5], but in recent years, forest fires have also become an important issue in Central Europe [6]. Among other European countries, Serbia experienced increased fire activity during the last two decades [7].

The increasing risk and associated damage caused by fires are usually linked to climate changes [8]. Considering the current climate scenario [9], which predicts an average temperature rise of 4–6 °C by the end of this century [10,11], and a decrease in total rainfall with an uneven distribution throughout the year in the south of Europe featuring long

67

periods of drought in summer, an increased risk of forest fire is expected in Europe [5]. The greatest changes in Europe are expected in the transition between the Mediterranean and Euro-Siberian regions [12], where the Balkan Peninsula is situated. The most dominant oak and beech forest may be replaced by evergreen Mediterranean vegetation [13,14], which is more prone to forest fire. Therefore, a further increase in forest fire risk can be expected due to changes in the fuel type.

Aside from the direct damage in terms of wood loss, which is obvious and easy to quantify, many other risks may appear following forest fires, due to the slow process of regeneration, especially in conifer forests [15]. Secondary pests, such as bark (Coleoptera: Curculionidae: Scolytinae) and ambrosia beetles (Coleoptera: Buprestidae) and diseases, that attack physiologically weakened trees may reach an outbreak population level and affect a much larger area than forest fire alone [16]. To minimize the harmful effects of forest fires, it is necessary to create an effective protection system against fire. According to San-Miguel-Ayanz [17], forest fire risk is defined as the probability of a fire happening and its consequences. Therefore, any increase in fire probability or its consequences directly results in an increase in fire risk. Our research has focused on forest fire occurrence probability mapping as a component of the future system for forest fire risk assessment. Forest fire probability mapping is a basic tool in risk management [18,19] and allows the allocation of resources for fire suppression, within a fire season, from zones with a lower risk to those under higher threat. Also, in the long run, mapping fire occurrence probability is a very valuable tool in forest management planning that improves forest resilience to fire through the implementation of various types of fire preventive silvicultural measures [20–23].

Among traditional methods, logistic regression (LR) is the most common when dealing with binary outcomes, like the presence or absence of fire [24]. Conversely, machine learning (ML) methods, as a form of artificial intelligence, are widely used in wildfire science and management, with more than 300 articles published on this topic since the 1990s [25]. Random forest (RF) belongs to the decision trees branch of the same group of ML methods [26]. Our study had the following objectives: (1) to map the probability of forest fire occurrence in Eastern Serbia based on LR and RF; (2) to determine the main explanatory variables for forest fire occurrence for both models.

2. Materials and Methods

2.1. Study Area

The study area is in the eastern and southern parts of Serbia (Figure 1). It has a total land area of 30,235.5 km^2. Broadleaved, conifer, and mixed forest cover 12,587.5, 170.7, 340.6 km^2, respectively. Natural grasslands, transitional woodland-shrub, and sparsely vegetated areas cover 925.7, 2949.5, and 48.0 km^2, respectively, while other areas represent agricultural, urban, or other non-wood areas. The elevation ranges from 28 to 2169 m. It has been observed that durations of extremely hot weather last longer and periods of extremely cold weather are shortened compared to the reference period of 1960–1990 [27]. These trends of extreme temperature conditions are most pronounced in the summer season [27]. A decrease in spring precipitation has been detected over the central and eastern parts of Serbia [28]. The annual quantity of rainfall is often insufficient, and droughts are evident in eastern and south-eastern parts of Serbia [28]. Scenarios where the monthly quantity of precipitation falls in only a few days of the month are expected to become more frequent, which will lead to more extreme weather events such as floods and droughts [29]. Serbia has two peaks in its fire season [30]. The first peak occurs in March, which is associated with 25% of the annual burnt area, while the second and largest peak appears in August [31]. In August 2012, more than 40 large fires were recorded, including two fires that were more than 1000 ha in size [32]. More than 16,000 ha of forest land were destroyed in 2007, with the value of the wood lost was estimated at 40 million euros [33].

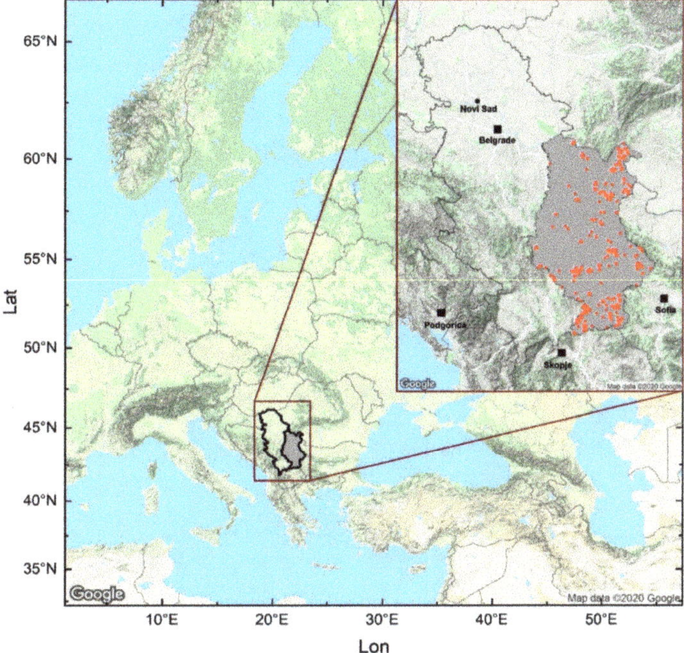

Figure 1. The geographic position of study area, Eastern Serbia, is marked in gray (latitudes 42.27–44.82° N and longitudes 20.90–23.01° E) with the layer of forest fire hotspots obtained by NASA's Fire Information for Resource Management System (FIRMS) (MODIS fire hotspot) for the period of 2001–2018.

2.2. Data Preparation

2.2.1. Dependent Variable

Data regarding the fire location were obtained from NASA's Fire Information for Resource Management System (FIRMS), which distributes near real-time fire products from Moderate Resolution Imaging Spectroradiometer (MODIS) from the Terra and Aqua platforms with a spatial resolution of 1 km (https://firms.modaps.eosdis.nasa.gov). Each position of a MODIS-identified active fire represents the center of a 1×1 km pixel that is labeled by the algorithm as containing one or more fires inside the pixel. Only fire pixels with a confidence higher than 80% for the period from January 2001 to December 2018 were considered for further analysis, because in certain cases the product underestimates the occurrence of fire [34].

2.2.2. Independent Variables

Independent variables were grouped into four categories: topography, vegetation, anthropogenic factors, and climate. Specific variables within categories were selected based on previous studies on forest fire occurrence [24,34–39]. Detailed information on preselected variables is presented in Table 1.

Table 1. Independent variables considered for forest fire occurrence models with codes, units, source and calculated variable inflation factors (VIFs).

Variable	Code	Units	Source	VIF
Vegetation				
Broad-leaved forest	BF	m^2		4.121
Coniferous forest	CF	m^2		1.160
Mixed forest	MF	m^2		1.224
Natural grasslands	NG	m^2	CORINE 2012	2.114
Transitional woodland-shrub	TWS	m^2		2.482
Sparsely vegetated area	SVA	m^2		1.066
[1] Total forested area	TFA	m^2		
Anthropogenic				
Distance to Municipality	DisM	m		1.576
Distance to Road	DisRo	m	OpenStreetMap	1.117
Distance to Rail	DisRa	m		1.224
Population Density	PopD	N/km^2	CIESIN	1.147
Distance to Arable Land	DisAL	m	CORINE 2012	1.012
Distance to Agricultural Land	DisAgL	m		1.619
Topographic				
Distance to Water	DisW	m	OpenStreetMap	1.804
Slope Degree Classes *	SD.C			1.606
Aspect Classes *	A.C4		DEM	2.132
[2] Elevation Classes*	E.C2			1.009
Climatic				
Drought code	DC		RHMS	1.365

[1] excluded from further analysis due to high Spearman's rho correlation coefficient with the distance to agricultural land (DAgL). [2] included as matching criterion in propensity-score matching and therefore excluded in further analyses * categorical explanatory variable.

Topographic parameters, elevation, slope, and aspect, were derived from the digital elevation model (DEM) with a precision of 3 arcsec previously downloaded from the United States Geological Survey (http://landsat.usgs.gov//landsatcollections.php). Average values of elevation, slope, and dominant aspect were calculated for each polygon in a 1×1 km grid using ArcGIS software (ESRI, Redlands, CA, USA). Obtained values of the slope and dominant aspect were divided into classes according to Carmo et al. [35], while elevations were divided into classes of 200 m (Figure 2).

Vegetation and land cover data were obtained from the CORINE 2012 data set (https://land.copernicus.eu/pan-european/corine-land-cover/clc-2012). The vector layer was intersected with polygon grid data. Objects in this vector layer were filtered for the following CORINE 2012 land cover classes (CLCs): broad-leaved forest (BF), coniferous forest (CF), mixed forest (MF), natural grasslands (NG), transitional woodland-shrub (TWS), and sparsely vegetated areas (SVA) (Figure 2). Intersecting the polygon grid data with the polygon CLC layer filtered this way, a new polygon layer with a table of attributes containing a polygon grid Object ID, CLC class ID with its description, and an area of CLC class that falls into the respective grid polygon were generated.

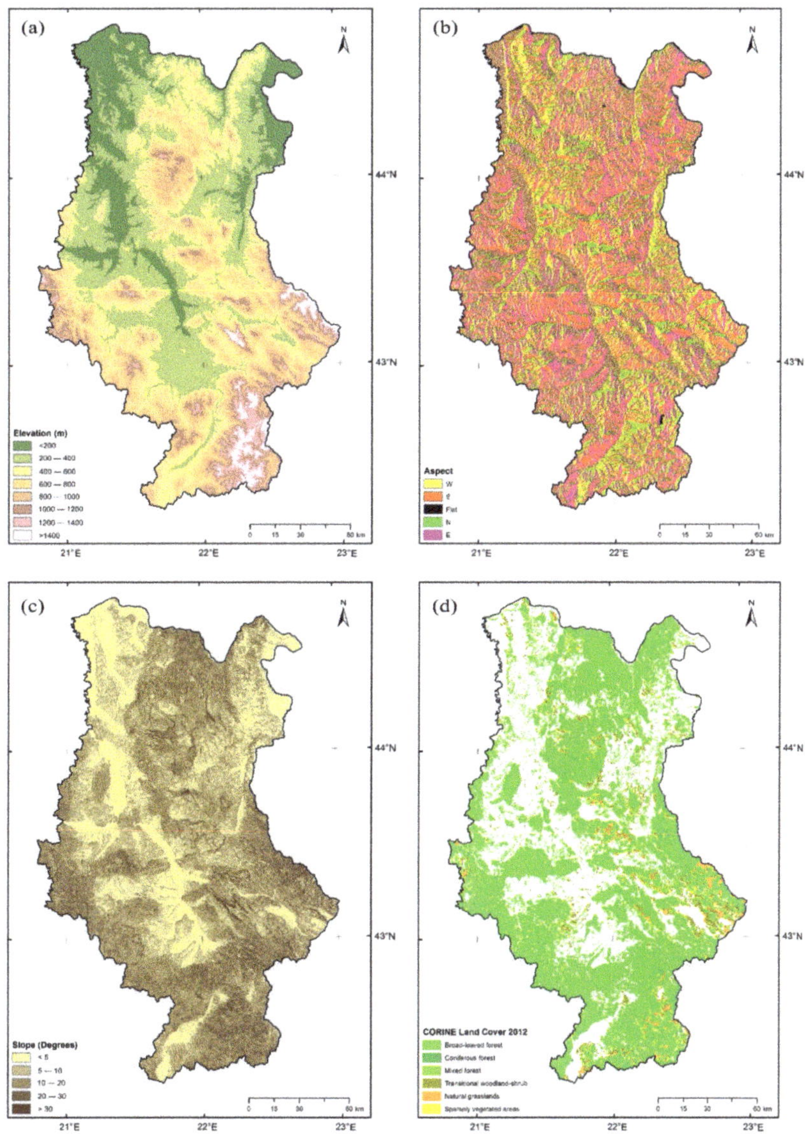

Figure 2. Categorical predictors related to forest fire in Eastern Serbia. (**a**) Classes of elevation; (**b**) aspect classes; (**c**) slope categories; (**d**) land cover categories obtained by CORINE 2012.

Data on drought code (DC), as a component of the Canadian Forest Fire Weather Index [40], were obtained from the Republic Hydrometeorological Service of Serbia (http://www.meteoalarm.rs/eng/fwi_osm.php) as tables with coordinates of meteorological stations with values on the drought code for each day in the observed period (2012–2018). Each of these tables was then converted into a spatial point layer with points representing meteorological stations and a respective table of attributes with values of the drought code. The layer was then used for interpolation using the ordinary kriging method [41], resulting in raster layers whose pixels represent values of drought code. This way, we were able to calculate zonal statistics of drought code for each day and each polygon of a grid layer for raster layers that represent the observed period. Finally, average values for each cell in the grid cells were calculated based on all of the drought code data for the observed period.

Layers of roads, populated places, railroads, and agricultural land were used for the calculation of the distance from each object of a 1 × 1 km grid to the nearest object of the respective layer, so the attribute table of the generated point and polygon grid layer was extended with distances to the nearest populated place, road, railway and agricultural land (https://www.openstreetmap.org).

Population data were obtained from a raster dataset available for download in Geo-TIFF format at the Center for International Earth Science Information Network (CIESIN), Columbia University (http://dx.doi.org/10.7927/H4639MPP). The sum of the number of people per polygon grid was also calculated by a zonal statistic tool.

At the end of the GIS portion of the analysis, Boolean values (Yes/No) were assigned to the elements of the grid by using the Spatial Join tool. Values of Yes were assigned to the elements for which the historical wildfire(s) occurred, and values of No were assigned to those for which it did not.

Each cell with at least one historical record of a fire event from 2001 to 2018 was classified as a fire cell and coded as one (1). In total, 429 cells were selected as a fire cell for Eastern Serbia. The use of binary LR alongside RF classification assumes that the predicted variable is dichotomous. Therefore, necessary non-fire cells were sampled based on the propensity score matching (PSM) method [42] and coded as zero (0). In general, a propensity score analysis reduces the effect of confounding in observational (nonrandomized) studies and can be used for matching, stratification, inverse probability of treatment weighting, and co-variate adjustment, all based on the propensity score [43]. Thus, PSM forms matched sets with equal ratio of treated and untreated subjects [44], in our case, of fire and non-fire cells, who share a similar value of propensity score. Matching without replacement [45] was applied for the selection of non-fire cells in this study. Namely, non-fire cells can be selected only once as a match for a given fire cell. Among greedy and optimal matching procedures of PSM, the latter was used in this study. Both approaches choose the same sets of controls for the overall matched samples, but optimal matching does a better job of minimizing the difference in propensity score value within each pair [46]. PSM was conducted in the R package 'MatchIt' v 3.02 [47].

As it was important to avoid the selection of non-fire cells in the vicinity of the fire cells, because of the similar environmental conditions, we tested distances among created pairs of cells. The distance between pairs of fire and non-fire cells varied from 2.2 to 268.1 km, with an average of 134.2 km. Additionally, to validate the created models, whole paired samples were randomly divided into two equal subsamples, training and validation, with the same number of fire and non-fire cells [24,39,48,49].

One of the basic assumptions that must be met before applying LR is the absence of high correlations (multicollinearity) among the explanatory variables (predictors) included in the model. Explanatory variables were checked for multicollinearity by the variance inflation factor (VIF) and Spearman's rho correlation [50]. Only variables with VIF $\leq$ 10 and a Spearman's rho coefficient lower than 0.7 [51] were considered for model building.

2.2.3. Forest Fire Occurrence Frequency across Categorical Predictors

Forest fire distribution across categorical predictors was analyzed for all events in the 15 years. Observed versus expected frequencies were analyzed and compared. Observed frequencies represented the number of forest fires that occurred within the explanatory variable category, while the expected frequencies were represented by the surface covered by the respective variable category in the study area [24]. Comparisons between observed and expected frequencies were based on chi-square statistics [52] at a significance level of 0.001. Mean values were calculated for each class of categorical predictors, and histograms were created accordingly.

2.3. Modeling Procedures

Based on collected and generated data, the model of probabilities for forest fire occurrences was generated for each cell of the polygon grid for the territory of Eastern Serbia.

2.3.1. LR Models

The algorithm obtained by the LR calculates the conditional probability of the fire event occurring from one or more input variables [53]. Also, LR can be used for estimating the contribution and significance of each variable by the Wald test [54], and afterwards can select a combination of variables that can be introduced in more complex models [55]. To estimate forest fire occurrence probability in Eastern Serbia by LR, the following equation was applied:

$$Pi = \frac{1}{1 + e^{-\ (\alpha + \beta 1 Xi1 + \beta 2 Xi2 + \cdots + \beta p Xip)}} \tag{1}$$

with "Pi" as the probability of the forest fire occurrence, "α" as a constant, "β" as a coefficient from regression, and "X" as original values of variables.

2.3.2. RF Models

This method uses a large number of decision trees, which produce their predictions and combine them into a single, more accurate, prediction [56]. RF has gained popularity in recent years, as it has been proven to perform better than LR according to several studies [39,57,58]. Specifically, it considerably outperforms LR in the accuracy measured, as well as the Brier score and area under curve (AUC) [59–61], on large and diverse datasets. Thus, considering the similarity of forest fire probability prediction to other risk-assessment applications, we considered RF as a strongly supported model of choice worth investigation. Similar to LR, the RF method is also good at handling both categorical and continuous types of explanatory variables.

The method proposed by Ye et al. and Genuer et al. [37,62] was applied for the variable selection. Each variable was included in the model $N - 1$ times, where N represents the number of variables considered as a predictor. RF models were run 16 times and an additional variable was excluded in each iteration. The variable importance obtained from each iteration was used for the calculation of relative variable importance, which represents an average value for a specific variable ranked from 0 to 1. All variables were scored from 1 to 16, according to the average importance. The variable with the highest average importance was scored as 1, and the variable with the lowest importance was scored as 16. Then, a RF model with each of the 15 highest ranked variables was generated for comparison with LR models. RF analyses were conducted by the default value of *mtry* (4), which represents the number of variables at each split and the 100 trees in the forest (ntree).

2.4. Model Validation

Model evaluation was conducted by a receiver operating curve (ROC) analysis, calculating the proportion of fire cells, classified as fire (sensitivity), and non-fire cells, classified as non-fire (specificity), for the obtained models. A ROC curve plots the values that represent sensitivity versus values that represent 1—specificity for the range of possible cut points [63]. AUC values between 0.5–0.7 indicate poor precision, values between 0.7–0.8 indicate acceptable precision, values between 0.8–0.9 indicate excellent precision, and values higher than 0.9 indicate outstanding model precision [63]. Additional evaluation of the models was conducted by 2 × 2 classification tables based on the overall accuracy. Thus, the tables are a result of the cross classification of the dichotomous outcome variable, whose values are derived from the predicted and observed probabilities. Also, to compare the predictive capacity of the models produced, the distribution of fire cells of the validation group across classes of forest fire occurrence probability was analyzed. Due to the limited size of data, the k-fold cross-validation (CV) method was also applied to compare obtained models. Analysis was conducted in the R package 'Caret' v 6.0–86 [64] using "glm" and "rf" methods for LR and RF, respectively. ROC analysis, accuracy, and Kappa values obtained from CV were used to compare models, with k-value set as 10.

To designate each cell as fire or non-fire, an optimal cutoff point must first be defined. For the models selected by the LR and RF procedures, based on the ROC analysis and prediction accuracy obtained under the training subsets, the optimal cutoff point was determined by the sensitivity equals specificity method [65] using the easyROC web-tool [66]. Then, the estimated probability for each cell was compared to the optimal cutoff point. If the estimated probability for the particular cell exceeded the optimal cutoff point, then that cell was designated as a fire cell. Conversely, if the estimated probability was lower than the optimal cutoff point, a particular cell was designated as a non-fire cell. Cutoff values were applied within selected models to both training and validation subsets.

2.5. Variable Importance Analysis

The quantification of variable importance is used for variable selection in many applied studies. To assess the importance of individual variables selected by the LR procedure, Wald statistics were applied to the models [67]. Variable importance in RF classification could be assessed by the permutation importance indices [68] or by the Gini impurity function for a classification problem [69]. For RF, the variable importance measures are summed across all trees in the forest and scaled in the same manner so that the most important variable has a value of 1.

2.6. Mapping Forest Fire Occurrence Probability

Forest fire occurrence probability calculated by the LR or RF models for fire cells and non-fire cells was used to map the forest fire occurrence probability in Eastern Serbia in ArcGIS 10.2. According to Nhongo et al. [34], the produced map was classified into five categories: very low ($0.01-0.20$), low ($0.21-0.40$), medium ($0.41-0.60$), high ($0.61-0.80$), and very high ($0.81-1.00$).

All statistical analyses related to LR and RF were conducted using the software Statistica 13 (TIBCO Software Inc., Palo Alto, CA, USA).

3. Results

3.1. Forest Fire Occurrence Frequency

Forest fire distributions across the categorical explanatory variables are displayed in Figure 3. An almost normal distribution of forest fires was recorded across elevation classes, with higher frequencies at elevations between 200 and 1000 (with a less pronounced peak in the class of $600-800$ m), and with lower frequencies below and above this range (Figure 3a). The frequencies of forest fire across elevation classes highly correspond to surface area covered by the respective classes ($\chi^2 = 39.83$, $p < 0.001$). This connection was not significant between forest fire occurrence and aspect classes ($\chi^2 = 1.11$, $p < 0.774$). Regarding aspect, the highest frequencies of fires occurred on southern aspects, then almost equally eastern and western aspects, while the lowest frequency was recorded on northern aspects (Figure 3b). The highest frequency of forest fire was recorded on the slopes with an inclination between $10°$ and $19.99°$, then between $5°$ and $9.99°$, while the lowest frequency was recorded at the lowest inclination ($0-4.99°$). No fire was recorded at the highest inclination of over $30°$ (Figure 3c). Forest fires were influenced by slope ($\chi^2 = 47.54$, $p < 0.001$). Also, land cover type had a significant influence on forest fire occurrence ($\chi^2 = 107.89$, $p < 0.001$). More than 55% of forest fires occurred in broad-leaved forests, and almost 29% and 14% occurred in the transitional woodland-shrubs and natural grasslands respectively, while less than 2% of forest fires occurred in the mixed forest and coniferous forests. No forest fire was recorded in the sparsely vegetated area (Figure 3d).

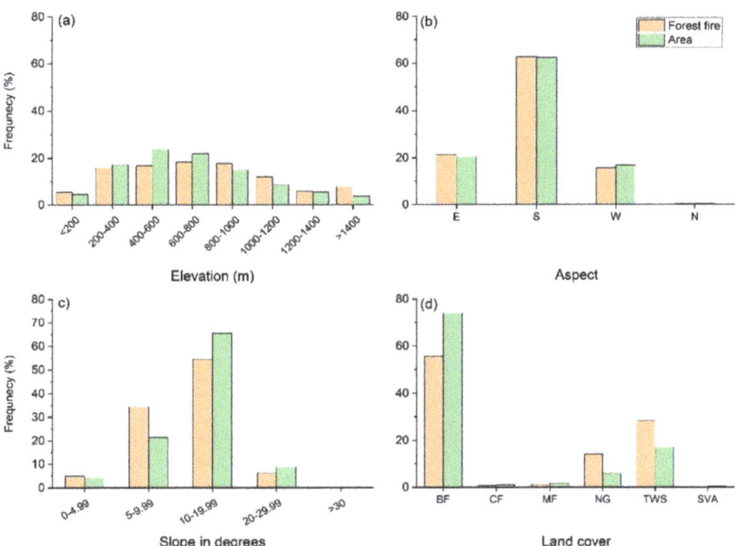

Figure 3. Frequencies of forest fires and areas covered by categorical variables: (**a**) elevation classes, (**b**) aspect classes (exposure), (**c**) slope degree classes, and (**d**) vegetation classes obtained from CORINE land cover 2012 that are present in Eastern Serbia: BF: broad-leaved forest; CF: coniferous forest; MF: mixed forest; NG: natural grasslands; TWS: transitional woodland-shrubs, and SVA: sparsely vegetated areas.

3.2. Models of Forest Fire Occurrence

In the first step, the total forested area (TFA) variable was excluded for further analysis due to the high correlation ($R = 0.84$) with distance agricultural land (DAgL), as displayed in Figure S1 (see the Supplementary Materials). The remaining 16 explanatory variables met the conditions of VIF $\leq$ 10 and were considered for the future models. After several iterations, the final models were created by LR and RF procedure for 15 variables. The highest impact on fire probability had a drought code, followed by distance to municipality, distance to water, distance to railway, and distance to arable land, while a relative contribution of mixed forest had the lowest impact in the RF model. Also, in the LR models, the highest impact on fire probability had a drought code, while a relative contribution of coniferous forest had the lowest impact on forest fire occurrence (Table 2).

The overall prediction accuracy of the LR and RF models, calculated by using an optimal cut-off value, was 86.7 and 87.7%, respectively (Table 3). Moreover, the LR and RF models display AUC values of 94.2% and 94.5%, respectively, affirming a high predictive power (Figure 4b).

Table 2. Explanatory variable importance evaluation based on Wald statistics for linear regression (LR) and Gini impurity for random forest (RF) models.

LR		RF	
Variable	**Wald**	**Variable**	**Gini Impurity**
Drought Code	44.968 ***	Drought Code	1
Distance to Rail	36.085 ***	Distance to Municipality	0.697
Distance to Agricultural Land	19.407 ***	Distance to Water	0.688
Distance to Water	18.851 ***	Distance to Rail	0.658
Natural Grasslands	17.136 ***	Distance to Arable Land	0.544
Transitional Woodland-Shrub	13.054 ***	Broad-Leaved Forest	0.503
Distance to Arable Land	7.080 **	Distance to Agricultural Land	0.461
Broad-Leaved Forest	4.182 *	Transitional Woodland-Shrub	0.365
Population Density	3.944 *	Natural Grasslands	0.324
Distance to Municipality	3.694 ^{ns}	Population Density	0.283
Slope Degree Classes	3.226 ^{ns}	Distance to Road	0.272
Distance to Road	1.914 ^{ns}	Slope Degree Classes	0.204
Aspect Classes	1.868 ^{ns}	Aspect Classes	0.185
Mixed Forest	1.313 ^{ns}	Coniferous Forest	0.093
Coniferous Forest	0.034 ^{ns}	Mixed Forest	0.085

^{ns} not-significant, * significant at $p < 0.05$, ** significant at $p < 0.01$, *** significant at $p < 0.001$.

Table 3. Classification tables for the training and validation sets of data based on LR and RF models, with applied cut off values, according to the sensitivity equals specificity method.

Model	Cutoff			**Predicted**					
				Training			**Validation**		
				0	**1**	**% Correct**	**0**	**1**	**% Correct**
LR	0.483	Observed	0	186	28	86.9	187	28	87.0
			1	29	185	86.4	29	186	86.5
		Overall %				86.7			86.7
RF	0.460	Observed	0	196	18	91.6	192	23	89.3
			1	19	195	91.1	30	185	86.0
		Overall %				91.4			87.7

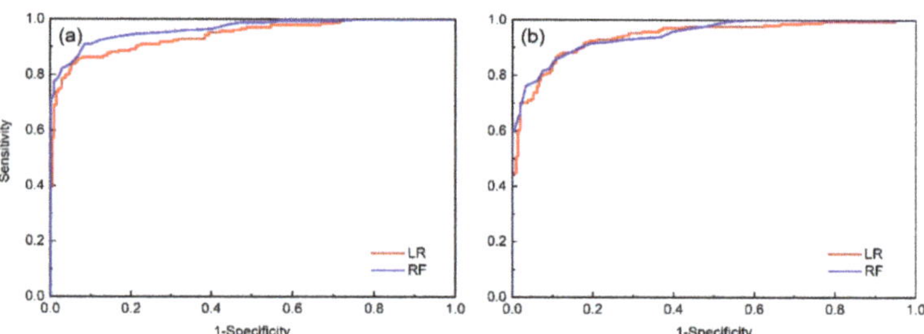

Figure 4. Receiver operating curve (ROC) for LR and RF models: (**a**) training data set, (**b**) validation data set.

3.3. Relative Importance of Variables

In the LR models, drought code was the most important variable for fire occurrences, followed by the distance to rail and agricultural land. Conversely, the lowest importance of the forest fire occurrence was associated with the contribution of coniferous and mixed forest, and also aspect classes (Table 2). In the RF models, drought code was the most important variable followed by the distances to municipality, water, rail, and arable land. Similarly, to the LR models, contribution of mixed and coniferous forest as well as aspect classes had the lowest impact on the RF models (Table 2). Interestingly, distance to municipality was recognized by RF as very important model variable, while in the LR models this variable did not have a statistically significant effect on model performance.

3.4. Spatial Modeling of Probability of Fire Occurrence

Zones with a very high probability of forest fire occurrence were situated in the southeastern part of the study area in both models and vary from 19.7% (LR) to 18.9% (RF) of the forested area. Zones with a very low probability of forest fire occurrence were situated in the northwestern part of the study area and range from 21.1% (LR) to 22.2% (RF) of the forested area (Figure 5).

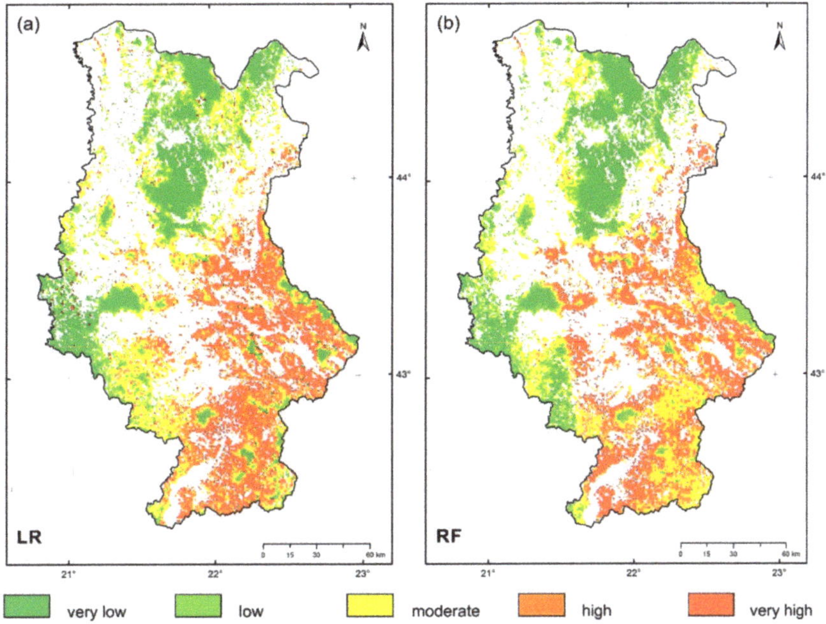

Figure 5. Maps of forest fire probability based on (**a**) LR models, (**b**) RF models.

3.5. Model Validation

LR models performed better in the very low fire distribution class, compared to the RF models, identifying lower forest fire incidence in the validation set of data. On the other hand, RF models performed better in high and very high classes than LR models, identifying a higher number of forest fire events in the same set of data (Table 4). If the two lowest classes (low and very low) classes are compared among models, then there is a very slight difference between the LR and RF models (e.g., 23.3% vs. 23.7%, respectively). On the other hand, the RF models classified 58.6% of forest fire events in the two higher classes (high and very high), while the LR models classified 54.0% in the same classes of the validation set of data. The overall prediction accuracy of the LR and RF models, based on 10-fold cross validation, was 86.5 (kappa = 0.7296) and 91.7% (kappa = 0.8345), respectively.

Moreover, the LR and RF models display AUC values of 92.4% and 97.5%, respectively, after 10-fold cross validation.

Table 4. Forest fire distribution (%) across different probability classes based on LR and RF models for the validating set of data.

Forest Fire Probability Percentile	Forest Fire Probability Class	LR	RF
0–20	Very low	4.7	8.4
20–40	Low	18.6	15.3
40–60	Moderate	22.8	17.7
60–80	High	21.4	22.3
80–100	Very high	32.6	36.3

4. Discussion

There has been a huge gap in forest fire risk assessment in Serbia without any system for forest fire risk assessment at the national or even regional scale [70]. Within this study, the first step has been made on that course by producing maps with forest fire occurrence probability that cover more than one-third of the state territory. To map the forest fire probability in Eastern Serbia, the area that has been the most affected by forest fires in the past [32], two different methods were applied, LR and RF. Both methods have been widely used in forest fire risk assessment within the past few decades, with the dominance of LR at the beginning of this century [24,35,50,55,71], while the RF method has prevailed since the last decade [25,37,58,60,70]. The RF models had slightly better performance than the LR models in the training and validation sets. Better performance of the RF models over the LR models in fire occurrence prediction is consistent with similar studies in our region [72,73] and other regions [39,58]. The LR models shift performance from outstanding in training data sets to excellent in validation data sets, while in the RF procedure this shift is less pronounced according to applied model classification [63]. Based on the results obtained within this study and literature survey, we can strongly recommend the use of the RF method for the forest fire occurrence mapping of the entire territory of Serbia.

The propensity-score matching method was used to select non-fire events for known fire events. This method was adopted from the field of medical sciences [74–77], like many others, and it was used for the first time in natural hazard risk assessment by Hudson et al. [78] for evaluating the effectiveness of flood damage mitigation measures. We used this method for the first time in forest fire risk assessment to pair historical fire event data obtained from the NASA FIRMS with non-fire events using elevation as matching criteria.

Elevation had a significant and positive effect on forest fire occurrence probability, with higher frequencies observed between 200 and 1000 m. The fire season is shorter at higher elevations due to snow melting later, leading to a lower fire frequency [79]. Also, a higher relative humidity due to a decrease in temperature of 1 °C per 100 m rise in elevation reduces chances for fuel ignition [80]. However, the rapid decrease in fire frequency at altitudes over 1000 m, observed by Ramón González et al. [71], was not recorded in our study. It is more likely that climate changes will influence a positive effect of the elevation on forest fire frequency in the future, as has already been shown by Schwartz et al. [81]. Southern, intermediate slopes between 5° and 20° experienced increased fire activity in our study due to the lower moisture of combustible material being exposed to the sun radiation more than other aspects with milder and steeper slopes [34,81,82]. All topographical features strongly affect vegetation and its burn ability [83]. Forest fire frequency was much higher in the broad-leaved forests than in the coniferous forests. The negative effect of the distance to water on forest fire occurrence probability may be linked to the NASA sensor's ability to detect only larger forest fires, while a smaller fire is suppressed easily when it is in the vicinity of the water bodies and the early phase of development are thereby not detected. Conifers cover less than 1% of the study area and were scattered in the higher mountains within the bigger broad-leaved forest complex and therefore were lesser exposed to anthropogenic activity, explaining the negative effect of its proportion on fire activity. Transitional woodland-shrubs are usually situated in the zone between forests and agricultural or arable land. Both types of land cover are connected to spring and autumn burning activity, and therefore transitional woodland-shrubs had increased fire frequency. Forests near human settlements and infrastructures that are densely populated are more prone to small fires due to negligence and/or accidental ignitions [84], while distant locations are prone to often larger, but less frequent, forest fires [85]. Therefore, an expected decrease in the population density in rural areas in the southern part of the study area [86] may result in lower fire activity in the future. Conversely, the expansion of urbanized areas and road cover with an associated increase in population density due to migration from south to north can be expected to lead to higher fire activity in the northern part of the study area.

Zones with the highest probability for forest fire occurrence are located in the southeastern part of the study area in all models, which correspond to the more pronounced drought periods during summer [27,28]. A higher drought code (higher temperature and lack of precipitation) leads to the lower moisture of fuel and makes an area more susceptible to ignition. It is well known that higher temperatures [87] reduce fuel moisture, making the fuels highly susceptible to ignition. Additionally, a study by Chang et al. [88] described low precipitation as a determinant factor for ignition. Drought code was the most important variable, followed by anthropogenic features, in both the LR and RF models. These results were consistent with other studies on determinant factors for the occurrence of wildfire where climatic and anthropogenic predictors had a higher influence on the fire occurrence probability [70,89–92]. All those models were efficiently applied at a smaller scale (such as national parks or protected areas), while our models showed similar efficacy at a larger scale. The produced maps can be used by firefighting services for strategic and operative planning. Defined zones with higher forest fire occurrence probability, in the southeast of the study area, should be intensively monitored during the fire season, especially during the second peak in August [30], when the largest forest fires can be expected [31]. Intense monitoring allows early detection of forest fires and leads to rapid response. All of these measures, along with enhanced equipment for fire suppression, can significantly decrease the burnt area in Eastern Serbia. Also, silvicultural measures that reduce fire risk [23] can be applied under zones with a higher forest fire occurrence probability according to the created maps. Thus, in the fire prone zone, less flammable tree species [93] should be selected for afforestation. Additionally, to prevent the transfer of ground fire to the crown, the introduction of silvicultural measures, such as pruning of the lower branches, should be obligatory in order to reduce fire hazard in the vulnerable zones. Other fuel reduction treatments, such as thinning, prescribed burning, and fuel breaks, can be useful tools to achieve these objectives at the landscape level [20,21].

5. Conclusions

The overall accuracy of the RF models was higher than those of the LR models. Both types of model identified drought code and anthropogenic features as the most important forest fire predictors. The models displayed a very high predictive ability, but the RF models were slightly more efficient and could be recommended for forest fire occurrence mapping in the eastern part of Serbia. The obtained maps could improve the efficacy of forest fire suppression in the study area in several ways. First, the fire probability map could be used for position optimization of the devices used in the early detection of forest fires. Also, firefighting resource allocation could be planned and applied in a manner consistent with the fire frequency. Finally, forest management planning and silvicultural measures should be adapted in terms of the forest fire risk reduction, based on the obtained maps.

Supplementary Materials: The following are available online at https://www.mdpi.com/1999-4907/12/1/5/s1: Figure S1: Correlation plot for all preselected variables based on Spearman's rho coefficient; Table S1: Tables of contingency for training and validation sets of data for the tested random forest models.

Author Contributions: Conceptualization: S.M.; methodology: S.M. and S.D.M.; validation: D.P., L.G., and P.K.; formal analysis of Logistic Rregression: S.D.M., and D.P.; formal analysis of Random Forest: S.M., and P.K. investigation: S.M. and S.D.M.; resources: N.M.; data curation: N.M. and L.G.; writing—original draft preparation: S.M.; writing—review and editing: N.M., D.P., and L.G.; visualization: S.D.M. and N.M.; supervision: S.M. and P.K.; project administration: S.D.M.; funding acquisition: S.M. and S.D.M. All authors have read and agreed to the published version of the manuscript.

Funding: This research was funded by the Ministry of Agriculture, Forestry and Water Management of the Republic of Serbia-Forest Directorate (contract: 401-00-1713/2019-10) and by the Ministry of Education, Science and Technological Development of the Republic of Serbia (contract: 451-03-68/2020-14/200015).

Data Availability Statement: The data presented in this study are available on request from the corresponding author.

Conflicts of Interest: The authors declare no conflict of interest.

References

1. Doerr, S.H.; Santín, C. Global trends in wildfire and its impacts: Perceptions versus realities in a changing world. *Philos. Trans. R. Soc. Lond. B Biol. Sci.* **2016**, *371*, 20150345. [CrossRef] [PubMed]
2. Flannigan, M.D.; Krawchuk, M.A.; de Groot, W.J.; Wotton, B.M.; Gowman, L.M. Implications of changing climate for global wildland fire. *Int. J. Wildl. Fire* **2009**, *18*, 483–507. [CrossRef]
3. Flannigan, M.; Cantin, A.S.; de Groot, W.J.; Wotton, M.; Newbery, A.; Gowman, L.M. Global wildland fire season severity in the 21st century. *Ecol. Manag.* **2013**, *294*, 54–61. [CrossRef]
4. Moritz, M.A.; Batllori, E.; Bradstock, R.A.; Gill, A.M.; Handmer, J.; Hessburg, P.F.; Leonard, J.; McCaffrey, S.; Odion, D.C.; Schoennagel, T.; et al. Learning to coexist with wildfire. *Nature* **2014**, *515*, 58–66. [CrossRef]
5. Turco, M.; Rosa-Cánovas, J.J.; Bedia, J.; Jerez, S.; Montávez, J.P.; Llasat, M.C.; Provenzale, A. Exacerbated fires in Mediterranean Europe due to anthropogenic warming projected with non-stationary climate-fire models. *Nat. Commun.* **2018**, *9*, 1–9. [CrossRef]
6. Feurdean, A.; Vannière, B.; Finsinger, W.; Warren, D.; Connor, S.C.; Forrest, M.; Liakka, J.; Panait, A.; Werner, C.; Andrič, M.; et al. Fire hazard modulation by long-term dynamics in land cover and dominant forest type in eastern and central Europe. *Biogeosciences* **2020**, *17*, 1213–1230. [CrossRef]
7. Costa, H.; de Rigo, D.; Libertà, G.; Durrant, T.; San-Miguel-Ayanz, J. *European Wildfire Danger and Vulnerability in a Changing Climate: Towards Integrating Risk Dimensions*; Publications Office of the European Union: Luxembourg, 2020; ISBN 978-92-76-16898-0.
8. Moritz, M.A.; Parisien, M.-A.; Batllori, E.; Krawchuk, M.A.; van Dorn, J.; Ganz, D.J.; Hayhoe, K. Climate change and disruptions to global fire activity. *Ecosphere* **2012**, *3*, art49. [CrossRef]
9. Räisänen, J.; Hansson, U.; Ullerstig, A.; Döscher, R.; Graham, L.P.; Jones, C.; Meier, H.E.M.; Samuelsson, P.; Willén, U. European climate in the late twenty-first century: Regional simulations with two driving global models and two forcing scenarios. *Clim. Dyn.* **2004**, *22*, 13–31. [CrossRef]
10. Schär, C.; Vidale, P.L.; Lüthi, D.; Frei, C.; Häberli, C.; Liniger, M.A.; Appenzeller, C. The role of increasing temperature variability in European summer heatwaves. *Nature* **2004**, *427*, 332–336. [CrossRef]
11. Fronzek, S.; Carter, T.R.; Jylhä, K. Representing two centuries of past and future climate for assessing risks to biodiversity in Europe. *Glob. Ecol. Biogeogr.* **2012**, *21*, 19–35. [CrossRef]
12. Thuiller, W.; Lavorel, S.; Araújo, M.B.; Sykes, M.T.; Prentice, I.C. Climate change threats to plant diversity in Europe. *Proc. Natl. Acad. Sci. USA* **2005**, *102*, 8245–8250. [CrossRef] [PubMed]
13. Brewer, S.; Cheddadi, R.; de Beaulieu, J.L.; Reille, M. The spread of deciduous Quercus throughout Europe since the last glacial period. *Ecol. Manag.* **2002**, *156*, 27–48. [CrossRef]
14. Hernández, L.; Sánchez de Dios, R.; Montes, F.; Sainz-Ollero, H.; Cañellas, I. Exploring range shifts of contrasting tree species across a bioclimatic transition zone. *Eur. J. Res.* **2017**, *136*, 481–492. [CrossRef]
15. Brovkina, O.; Stojanović, M.; Milanović, S.; Latypov, I.; Marković, N.; Cienciala, E. Monitoring of post-fire forest scars in Serbia based on satellite Sentinel-2 data. *Geomat. Nat. Hazards Risk* **2020**, *11*, 2315–2339. [CrossRef]
16. Krawchuk, M.A.; Meigs, G.W.; Cartwright, J.M.; Coop, J.D.; Davis, R.; Holz, A.; Kolden, C.; Meddens, A.J.H. Disturbance refugia within mosaics of forest fire, drought, and insect outbreaks. *Front. Ecol. Env.* **2020**, *18*, 235–244. [CrossRef]
17. San-Miguel-Ayanz, J. Methodologies for the Evaluation of Forest Fire Risk: From Long-Term (Static) to Dynamic Indices. *Corso Cult. Ecol.* **2002**, *117*, 117.
18. San-Miguel-Ayanz, J.; Carlson, J.D.; Alexander, M.; Tolhurst, K.; Morgan, G.; Sneeuwjagt, R.; Dudley, M. *Current Methods to Assess Fire Danger Potential*; World Scientific: Singapore, 2003; pp. 21–61.
19. Mohammadi, F.; Bavaghar, M.P.; Shabanian, N. Forest Fire Risk Zone Modeling Using Logistic Regression and GIS: An Iranian Case Study. *Small-Scale* **2014**, *13*, 117–125. [CrossRef]
20. Agee, J.K.; Bahro, B.; Finney, M.A.; Omi, P.N.; Sapsis, D.B.; Skinner, C.N.; van Wagtendonk, J.W.; Phillip-Weatherspoon, C. The use of shaded fuelbreaks in landscape fire management. *Ecol. Manag.* **2000**, *127*, 55–66. [CrossRef]

21. Agee, J.K.; Skinner, C.N. Basic principles of forest fuel reduction treatments. *Ecol. Manag.* **2005**, *211*, 83–96. [CrossRef]
22. Fernandes, P.M. Fire-smart management of forest landscapes in the Mediterranean basin under global change. *Landsc. Urban. Plan.* **2013**, *110*, 175–182. [CrossRef]
23. Khabarov, N.; Krasovskii, A.; Obersteiner, M.; Swart, R.; Dosio, A.; San-Miguel-Ayanz, J.; Durrant, T.; Camia, A.; Migliavacca, M. Forest fires and adaptation options in Europe. *Reg. Env. Chang.* **2016**, *16*, 21–30. [CrossRef]
24. Catry, F.X.; Rego, F.C.; Bação, F.L.; Moreira, F. Modeling and mapping wildfire ignition risk in Portugal. *Int. J. Wildl. Fire* **2009**, *18*, 921–931. [CrossRef]
25. Jain, P.; Coogan, S.C.P.; Subramanian, S.G.; Crowley, M.; Taylor, S.; Flannigan, M.D. A review of machine learning applications in wildfire science and management. *arXiv* **2020**, arXiv:2003.00646v1. [CrossRef]
26. Breiman, L. Statistical modeling: The two cultures. *Stat. Sci.* **2001**, *16*, 199–215. [CrossRef]
27. Malinovic-Milicevic, S.; Radovanovic, M.M.; Stanojevic, G.; Milovanovic, B. Recent changes in Serbian climate extreme indices from 1961 to 2010. *Appl. Clim.* **2016**, *124*, 1089–1098. [CrossRef]
28. Luković, J.; Bajat, B.; Blagojević, D.; Kilibarda, M. Spatial pattern of recent rainfall trends in Serbia (1961–2009). *Reg. Env. Chang.* **2014**, *14*, 1789–1799. [CrossRef]
29. Mimić, G.; Mihailović, D.T.; Kapor, D. Complexity analysis of the air temperature and the precipitation time series in Serbia. *Appl. Clim.* **2017**, *127*, 891–898. [CrossRef]
30. Milomir, V. *Forest Fire: Manual for Forest Engineers and Technicians*; Level of Thesis; Faculty of Forestry University of Belgrade: Belgrade, Serbia, 1992.
31. San-Miguel-Ayanz, J.; Durrant, T.; Boca, R.; Libertà, G.; Branco, A.; de Rigo, D.; Ferrari, D.; Maianti, P.; Artes, T.; Oom, D.; et al. *Forest Fires in Europe, Middle East and North Africa 2018*; Publications Office of the European Union: Luxembourg, 2019; ISBN 978-92-76-12591-4.
32. Schmuck, G.; San-Miguel-Ayanz, J.; Camia, A.; Durrant, T.; Boca, R.; Libertà, G.; Schulte, E. Forest fires in Europe Middle East and North Africa 2012. *Sci. Tech. Res. Ser.* **2013**, 10–30.
33. Goldstein, E. *Serbia's Potential For. Sustainable Growth And Shared Prosperity Systematic Country Diagnostic Report*; World Bank: Washington, DC, USA, 2015.
34. Nhongo, E.J.S.; Fontana, D.C.; Guasselli, L.A.; Bremm, C. Probabilistic modelling of wildfire occurrence based on logistic regression, Niassa Reserve, Mozambique. *Geomat. Nat. Hazards Risk* **2019**, *10*, 1772–1792. [CrossRef]
35. Carmo, M.; Moreira, F.; Casimiro, P.; Vaz, P. Land use and topography influences on wildfire occurrence in northern Portugal. *Landsc. Urban. Plan.* **2011**, *100*, 169–176. [CrossRef]
36. Konkathi, P.; Shetty, A.; Kolluru, V.; Yathish, P.; Pruthviraj, U. Static Fire Risk Index for the Forest Resources of Karnataka. In Proceedings of the IGARSS 2019-2019 IEEE International Geoscience and Remote Sensing Symposium, Yokohama, Japan, 28 July–2 August 2019; pp. 6716–6719.
37. Ye, T.; Wang, Y.; Guo, Z.; Li, Y. Factor contribution to fire occurrence, size, & burn probability in a subtropical coniferous forest in East China. *PLoS ONE* **2017**, *12*. [CrossRef]
38. Jaafari, A.; Gholami, D.M.; Zenner, E.K. A Bayesian modeling of wildfire probability in the Zagros Mountains, Iran. *Ecol. Inf.* **2017**, *39*, 32–44. [CrossRef]
39. Guo, F.; Su, Z.; Wang, G.; Sun, L.; Lin, F.; Liu, A. Wildfire ignition in the forests of southeast China: Identifying drivers and spatial distribution to predict wildfire likelihood. *Appl. Geogr.* **2016**, *66*, 12–21. [CrossRef]
40. van Wagner, C.E.; Forest, P.; Station, E.; Ontario, C.R.; Francais, R.U.E.; Davis, H.J. *Development and Structure of the Canadian Forest Fire Weather Index System*; Canadian Forestry Service: Ottawa, ON, Canada, 1987.
41. Oliver, M.A.; Webster, R. Kriging: A method of interpolation for geographical information systems. *Int. J. Geogr. Inf. Syst.* **1990**, *4*, 313–332. [CrossRef]
42. Rosenbaum, P.R.; Rubin, D.B. The central role of the propensity score in observational studies for causal effects. *Biometrika* **1983**, *70*, 41–55. [CrossRef]
43. Austin, P.C. An introduction to propensity score methods for reducing the effects of confounding in observational studies. *Multivar. Behav. Res.* **2011**, *46*, 399–424. [CrossRef]
44. Rosenbaum, P.R.; Rubin, D.B. Constructing a control group using multivariate matched sampling methods that incorporate the propensity score. *Am. Stat.* **1985**, *39*, 33–38. [CrossRef]
45. Rosenbaum, P.R. *Observational Studies*; Springer: Berlin/Heidelberg, Germany, 2002; pp. 1–17.
46. Gu, X.S.; Rosenbaum, P.R. Comparison of Multivariate Matching Methods: Structures, Distances, and Algorithms. *J. Comput. Graph. Stat.* **1993**, *2*, 405–420. [CrossRef]
47. Ho, D.E.; Imai, K.; King, G.; Stuart, E.A. Matching as nonparametric preprocessing for reducing model dependence in parametric causal inference. *Polit. Anal.* **2007**, *15*, 199–236. [CrossRef]
48. Chuvieco, E.; González, I.; Verdú, F.; Aguado, I.; Yebra, M. Prediction of fire occurrence from live fuel moisture content measurements in a Mediterranean ecosystem. *Int. J. Wildl. Fire* **2009**, *18*, 430. [CrossRef]
49. Vilar del Hoyo, L.; Isabel, M.P.M.; Vega, F.J.M. Logistic regression models for human-caused wildfire risk estimation: Analysing the effect of the spatial accuracy in fire occurrence data. *Eur. J. Res.* **2011**, *130*, 983–996. [CrossRef]
50. Midi, H.; Sarkar, S.K.; Rana, S. Collinearity diagnostics of binary logistic regression model. *J. Interdiscip. Math.* **2010**, *13*, 253–267. [CrossRef]

51. Dormann, C.F.; Elith, J.; Bacher, S.; Buchmann, C.; Carl, G.; Carré, G.; Marquéz, J.R.G.; Gruber, B.; Lafourcade, B.; Leitão, P.J.; et al. Collinearity: A review of methods to deal with it and a simulation study evaluating their performance. *Ecography* **2013**, *36*, 27–46. [CrossRef]

52. Sokal, R.; Rohlf, F. Biometry: The principles and practice of statistics in biological research. *J. R. Stat. Soc. Ser. A* **2012**, *133*. [CrossRef]

53. Bisquert, M.M.; Sánchez, J.M.; Caselles, V. Fire danger estimation from MODIS Enhanced Vegetation Index data: Application to Galicia region (north-west Spain). *Int. J. Wildl. Fire* **2011**, *20*, 465–473. [CrossRef]

54. Martínez, J.; Vega-Garcia, C.; Chuvieco, E. Human-caused wildfire risk rating for prevention planning in Spain. *J. Environ. Manag.* **2009**, *90*, 1241–1252. [CrossRef] [PubMed]

55. Bisquert, M.; Caselles, E.; Snchez, J.M.; Caselles, V. Application of artificial neural networks and logistic regression to the prediction of forest fire danger in Galicia using MODIS data. *Int. J. Wildl. Fire* **2012**, *21*, 1025–1029. [CrossRef]

56. Breiman, L. Random forests. *Mach. Learn.* **2001**, *45*, 5–32. [CrossRef]

57. Guo, F.; Wang, G.; Su, Z.; Liang, H.; Wang, W.; Lin, F.; Liu, A. What drives forest fire in Fujian, China? Evidence from logistic regression and Random Forests. *Int. J. Wildl. Fire* **2016**, *25*, 505–519. [CrossRef]

58. Chen, M.-M.; Chen, M.-C. Modeling Road Accident Severity with Comparisons of Logistic Regression, Decision Tree and Random Forest. *Information* **2020**, *11*, 270. [CrossRef]

59. Caruana, R.; Niculescu-Mizil, A. An Empirical Comparison of Supervised Learning Algorithms. In Proceedings of the ACM International Conference Proceeding Series; ACM Press: New York, NY, USA, 2006; Volume 148, pp. 161–168.

60. Couronné, R.; Probst, P.; Boulesteix, A.L. Random forest versus logistic regression: A large-scale benchmark experiment. *BMC Bioinform.* **2018**, *19*, 1–14. [CrossRef] [PubMed]

61. Kaitlin, K.; Smith, T.; Sadler, B. Random Forest vs. Logistic Regression: Binary Classification for Heterogeneous Datasets. *SMU Data Sci. Rev.* **2018**, *1*, 9.

62. Genuer, R.; Poggi, J.M.; Tuleau-Malot, C. Variable selection using random forests. *Pattern Recognit. Lett.* **2010**, *31*, 2225–2236. [CrossRef]

63. Hosmer, D.W.; Lemeshow, S.; Sturdivant, R.X. *Applied Logistic Regression*, 3rd ed.; Wiley: Hoboken, NJ, USA, 2013; ISBN 9781118548387.

64. Kuhn, M. Building Predictive Models in R Using the caret Package. *J. Stat. Softw.* **2008**, *28*. [CrossRef]

65. López-Ratón, M.; Rodríguez-Álvarez, M.X.; Suárez, C.C.; Sampedro, F.G. OptimalCutpoints: An R Package for Selecting Optimal Cutpoints in Diagnostic Tests. *J. Stat. Softw.* **2014**, *61*, 1–36. [CrossRef]

66. Goksuluk, D.; Korkmaz, S.; Zararsiz, G.; Karaagaoglu, A.E. EasyROC: An. interactive web-tool for roc curve analysis using r language environment. *R J.* **2016**, *8*, 2. [CrossRef]

67. Martínez-Fernández, J.; Chuvieco, E.; Koutsias, N. Modelling long-term fire occurrence factors in Spain by accounting for local variations with geographically weighted regression. *Nat. Hazards Earth Syst. Sci.* **2013**, *13*, 311–327. [CrossRef]

68. Breiman, L.; Friedman, J.; Stone, C.J.; Olshen, R.A. *Classification and Regression Trees*; CRC Press: Boca Raton, FL, USA, 1984; ISBN 0412048418.

69. Nembrini, S.; König, I.R.; Wright, M.N. The revival of the Gini importance? *Bioinformatics* **2018**, *34*, 3711–3718. [CrossRef]

70. Gigović, L.; Pourghasemi, H.R.; Drobnjak, S.; Bai, S. Testing a new ensemble model based on SVM and random forest in forest fire susceptibility assessment and its mapping in Serbia's Tara National Park. *Forests* **2019**, *10*. [CrossRef]

71. Ramón González, J.; Palahí, M.; Trasobares, A.; Pukkala, T. A fire probability model for forest stands in Catalonia (north-east Spain). *Ann. Sci.* **2006**, *63*, 169–176. [CrossRef]

72. Mallinis, G.; Petrila, M.; Mitsopoulos, I.; Lorenţ, A.; Neagu, Ş.; Apostol, B.; Gancz, V.; Popa, I.; Goldammer, J.G. Geospatial Patterns and Drivers of Forest Fire Occurrence in Romania. *Appl. Spat. Anal. Policy* **2019**, *12*, 773–795. [CrossRef]

73. Šturm, T.; Podobnikar, T. A probability model for long-term forest fire occurrence in the Karst forest management area of Slovenia. *Int. J. Wildl. Fire* **2017**, *26*, 399. [CrossRef]

74. Pourhoseingholi, M.A.; Baghestani, A.R.; Vahedi, M. How to control confounding effects by statistical analysis. *Gastroenterol. Hepatol. Bed Bench* **2012**, *5*, 79–83. [CrossRef] [PubMed]

75. Li, L.; Greene, T. A Weighting Analogue to Pair Matching in Propensity Score Analysis. *Int. J. Biostat.* **2013**, *9*, 215–234. [CrossRef]

76. Deb, S.; Austin, P.C.; Tu, J.V.; Ko, D.T.; Mazer, C.D.; Kiss, A.; Fremes, S.E. A Review of Propensity-Score Methods and Their Use in Cardiovascular Research. *Can. J. Cardiol.* **2016**, *32*, 259–265. [CrossRef]

77. Yan, H.; Karmur, B.S.; Kulkarni, A.V. Comparing Effects of Treatment: Controlling for Confounding. *Clin. Neurosurg.* **2020**, *86*, 325–331. [CrossRef]

78. Hudson, P.; Botzen, W.J.W.; Kreibich, H.; Bubeck, P.H.; Aerts, J.C.J. Evaluating the effectiveness of flood damage mitigation measures by the application of propensity score matching. *Nat. Hazards Earth Syst. Sci.* **2014**, *14*, 1731–1747. [CrossRef]

79. Heyerdahl, E.K.; Brubaker, L.B.; Agee, J.K. Spatial controls of historical fire regimes: A multiscale example from the interior west, USA. *Ecology* **2001**, *82*, 660–678. [CrossRef]

80. Rogeau, M.P.; Armstrong, G.W. Quantifying the effect of elevation and aspect on fire return intervals in the Canadian Rocky Mountains. *Ecol. Manag.* **2017**, *384*, 248–261. [CrossRef]

81. Schwartz, M.W.; Butt, N.; Dolanc, C.R.; Holguin, A.; Moritz, M.A.; North, M.P.; Safford, H.D.; Stephenson, N.L.; Thorne, J.H.; van Mantgem, P.J. Increasing elevation of fire in the Sierra Nevada and implications for forest change. *Ecosphere* **2015**, *6*, art121. [CrossRef]
82. Everett, R.L.; Schellhaas, R.; Keenum, D.; Spurbeck, D.; Ohlson, P. Fire history in the ponderosa pine/Douglas-fir forests on the east slope of the Washington Cascades. *Ecol. Manag.* **2000**, *129*, 207–225. [CrossRef]
83. Castro, R.; Chuvieco, E. Modeling forest fire danger from geographic information systems. *Geocarto Int.* **1998**, *13*, 15–23. [CrossRef]
84. Curt, T.; Fréjaville, T.; Lahaye, S. Modelling the spatial patterns of ignition causes and fire regime features in southern France: Implications for fire prevention policy. *Int. J. Wildl. Fire* **2016**, *25*, 785–796. [CrossRef]
85. Zumbrunnen, T.; Pezzatti, G.B.; Menéndez, P.; Bugmann, H.; Bürgi, M.; Conedera, M. Weather and human impacts on forest fires: 100 years of fire history in two climatic regions of Switzerland. *Ecol. Manag.* **2011**, *261*, 2188–2199. [CrossRef]
86. Petrić, J.; Maričić, T.; Basarić, J. The population conundrums and some implications for urban development in Serbia. *Spatium* **2012**, *314*, 7–14. [CrossRef]
87. Fried, J.S.; Gilless, J.K.; Riley, W.J.; Moody, T.J.; Simon de Blas, C.; Hayhoe, K.; Moritz, M.; Stephens, S.; Torn, M. Predicting the effect of climate change on wildfire behavior and initial attack success. *Clim. Chang.* **2008**, *87*, 251–264. [CrossRef]
88. Chang, Y.; He, H.S.; Hu, Y.; Bu, R.; Li, X. Historic and current fire regimes in the Great Xing'an Mountains, northeastern China: Implications for long-term forest management. *Ecol. Manag.* **2008**, *254*, 445–453. [CrossRef]
89. Sadori, L.; Masi, A.; Ricotta, C. Climate-driven past fires in central Sicily. *Plant Biosyst. Int. J. Deal. All Asp. Plant Biol.* **2015**, *149*, 166–173. [CrossRef]
90. Kalabokidis, K.; Palaiologou, P.; Gerasopoulos, E.; Giannakopoulos, C.; Kostopoulou, E.; Zerefos, C. Effect of Climate Change Projections on Forest Fire Behavior and Values-at-Risk in Southwestern Greece. *Forests* **2015**, *6*, 2214–2240. [CrossRef]
91. Varela, V.; Vlachogiannis, D.; Sfetsos, A.; Karozis, S.; Politi, N.; Giroud, F. Projection of Forest Fire Danger due to Climate Change in the French Mediterranean Region. *Sustainability* **2019**, *11*, 4284. [CrossRef]
92. Pham, B.T.; Jaafari, A.; Avand, M.; Al-Ansari, N.; Du, T.D.; Hai-Yen, H.P.; van Phong, T.; Nguyen, D.H.; van Le, H.; Mafi-Gholami, D.; et al. Performance evaluation of machine learning methods for forest fire modeling and prediction. *Symmetry* **2020**, *12*, 1022. [CrossRef]
93. Xanthopoulos, G.; Calfapietra, C.; Fernandes, P. Fire Hazard and Flammability of European Forest Types. In *Post-Fire Management and Restoration of Southern European Forests*; Moreira, F., Arianoutsou, M., Corona, P., de las Heras, J., Eds.; Springer: Berlin/Heidelberg, Germany, 2012; Volume 24, pp. 79–92. ISBN 978-94-007-2207-1.

 *forests* 

Article

Impacts of Climate Change on Wildfires in Central Asia

Xuezheng Zong [1], Xiaorui Tian [1,*] and Yunhe Yin [2]

[1] Research Institute of Forest Ecology, Environment and Protection, Chinese Academy of Forestry, Key Laboratory of Forest Protection of National Forestry and Grassland Administration, Beijing 100091, China; zongxuezheng1995@163.com

[2] Key Laboratory of Land Surface Pattern and Simulation, Institute of Geographic Sciences and Natural Resources Research, Chinese Academy of Sciences, Beijing 100101, China; yinyh@igsnrr.ac.cn

* Correspondence: tianxr@caf.ac.cn; Tel.: +86-10-6288-9849

Received: 18 June 2020; Accepted: 24 July 2020; Published: 25 July 2020

Abstract: This study analyzed fire weather and fire regimes in Central Asia from 2001–2015 and projected the impacts of climate change on fire weather in the 2030s (2021–2050) and 2080s (2071–2099), which would be helpful for improving wildfire management and adapting to future climate change in the region. The study area included five countries: Kazakhstan, Kyrgyzstan, Tajikistan, Uzbekistan, and Turkmenistan. The study area could be divided into four subregions based on vegetation type: shrub (R1), grassland (R2), mountain forest (R3), and rare vegetation area (R4). We used the modified Nesterov index (MNI) to indicate the fire weather of the region. The fire season for each vegetation zone was determined with the daily MNI and burned areas. We used the HadGEM2-ES global climate model with four scenarios (RCP2.6, RCP4.5, RCP6.0, and RCP8.5) to project the future weather and fire weather of Central Asia. The results showed that the fire season for shrub areas (R1) was from 1 April to 30 November, for grassland (R2) was from 1 March to 30 November, and for mountain forest (R3) was from 1 April to 30 October. The daily burned areas of R1 and R2 mainly occurred in the period from June–August, while that of R3 mainly occurred in the April–June and August–October periods. Compared with the baseline (1971–2000), the mean daily maximum temperature and precipitation, in the fire seasons of study area, will increase by 14%–23% and 7%–15% in the 2030s, and 21%–37% and 11%–21% in the 2080s, respectively. The mean MNI will increase by 33%–68% in the 2030s and 63%–146% in the 2080s. The potential burned areas of will increase by 2%–8% in the 2030s and 3%–13% in the 2080s. Wildfire management needs to improve to adapt to increasing fire danger in the future.

Keywords: climate change; fire weather; MNI; fire season

1. Introduction

Wildfires are a dominant disturbance in most forests and are strongly influenced by climate [1]. Climate warming has recently caused changes in the fire regime in the Northern Hemisphere [2], which has experienced extreme wildfire seasons and fire frequency increases in forests. Notably, high-intensity fires have occurred in summer in some regions. During the summer of 2010, climate warming caused several hundred wildfires and burned areas of approximately 5 million ha in Russia [3]. In the summer of 2017, British Columbia, Canada, experienced the worst wildfire, which caused a burned area of 1.2 million ha [4]. In addition, in the boreal forest of North America, climate warming has led to greater and more severe wildfire activity, increased fire frequency and fire sizes, and longer fire seasons [5]. The large-scale wildfires in the United States in 2019 and Australia from 2019–2020 attracted the attention of global society. Central Asia is located in the arid and semiarid zone, which includes

Kazakhstan, Kyrgyzstan, Tajikistan, Uzbekistan, and Turkmenistan [6]. Wildfires caused great losses in forest resources and properties in the region. The region is an important region for global biodiversity and has important species, such as snow leopards and brown bears [7]. Currently, Central Asian countries are in a critical period of economic and social transformation and are important components of "the Belt and Road". The countries in the region have made great efforts to increase the forest coverage rate and enhance biodiversity [8]. It is important to protect the existing forest resources to improve the ecological environment in Central Asia, which would be helpful for improving the living environment of local residents and promoting economic development.

Climate change will increase fire danger, and potential wildfires will increase significantly around the world in the future [9,10]. Additionally, climate change has increased wildfires in northern Europe and Greenland [11–13]. The fifth assessment report of the IPCC indicated that future wildfire risk would increase and that the fire season would become longer in southern Europe [14]. Compared to the present scenario, the annual burned area was projected to increase three to five times under the A2 scenario by 2100 [14]. The frequency of fires will increase to 25% by 2030 and 75% by the end of the 21st century in Canada, under the Canadian Climate Centre GCM (general circulation models) scenarios, and fire occurrence will increase by 140% under the Hadley Centre GCM scenario [15]. The higher number of wildfires resulted in an increased budget for fire management. In Russia, increasing wildfire frequency is expected to cause the total cost of forest fire management to increase by 211,114 and 248,956 thousand rubles (approximately 286.6–340 million US$) by the end of the 21st century [16]. Therefore, the study of future fire weather under different climate scenarios for the region would be the basis for adapting to future climate change.

There are many wildfires each year in Central Asia, especially in Kazakhstan [17]. Wildfires frequently occurred and damaged forest resources in these countries due to less fire management. The annual average burned forest area was 4000 ha in Kazakhstan during the 1985–1990 period and increased to 20,000 ha during the 1996–2000 period. In 1997, the annual burned area was 200,000 ha [18]. However, there were just 486 forest fires and 3915 ha of burned areas in 2012 in Kazakhstan [19]. Grassland fires occur frequently in Central Asia [20]. The mean annual burned areas of grasslands in the broader steppe-dominated region was 15 million ha during the 2001–2009 period, which mainly occurred in August and September [21]. Potential wildfires in the future will increase due to climate warming and may cause more burned areas and environmental pollution [9,22]. There have been some studies on the fire risk of Kazakhstan [23,24], but little research has been conducted on the fire regime of Central Asia. Therefore, it would be interesting to study the fire weather and fire regime in Central Asia and further evaluate fire danger under future climate scenarios.

The Fire Weather Index can effectively describe the relationship between weather and fire danger [25]. Bedia et al. found that the FWI (Fire Weather Index) in the grasslands of Kazakhstan was higher than that in other regions during the 1981–2000 period [26]. However, they only analyzed fire danger on a global scale by using WFDEI data. The FWI calculation requires inputs of noon temperature, relative humidity, wind speed, and 24-h accumulated precipitation. It is difficult to obtain enough weather observation data covering the vegetation areas in Central Asia. Some fire weather indices developed in the former Soviet Union, such as the Nesterov index (NI) [27], Zhdanko index (ZhI) [28], and Modified Nesterov index (MNI) [29], are widely used in Russia and countries in Central Asia [30–33]. These indices are calculated with mid-day temperature, dew point deficit, and precipitation and need fewer inputs than the FWI.

The objective of this study was to analyze fire weather by using MNI and fire regimes for each vegetation zone in Central Asia. We will define the fire seasons for each vegetation zone based on the process of the daily fire weather index and burned area and evaluate the potential fire danger in the 2030s and 2080s under future climate scenarios.

2. Methods

2.1. Study Area

Central Asia is located in central Eurasia and is adjacent to China in the west and the Caspian Sea in the west. Its geographical range is 35°08′–55°25′ N, 46°28′–87°29′ E. The total area is approximately 3970 million ha. It has a temperate continental climate and uneven distribution of rainfall. The annual precipitation ranges from 100–400 mm, and it can be more than 500 mm in high mountain areas and less than 200 mm on plains [7].

The forest coverage rate in the study area was 1.6%, which was mainly distributed in the northeast. Shrubs were mainly distributed in the western and central parts of the study area, with a coverage rate of 22.4%. Grassland and farmland accounted for 23.8% and 20.3%, respectively (Figure 1).

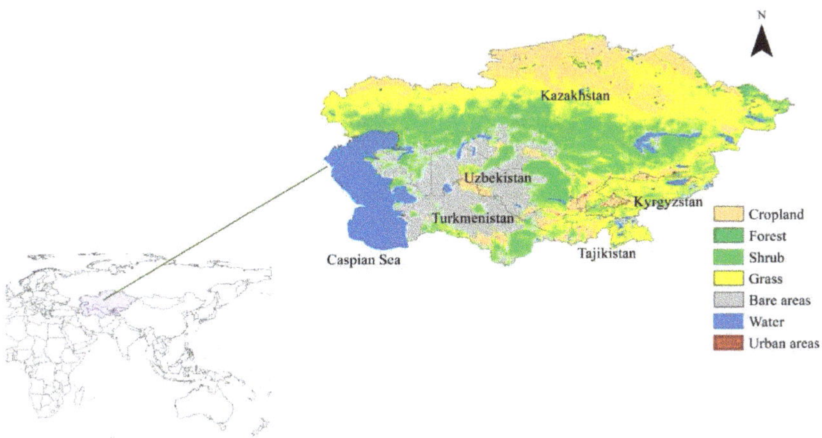

Figure 1. Location of the study area and vegetation types.

2.2. Data Sources

Land cover data (2001–2015, 300 m spatial resolution) were downloaded from the land cover state products of the European Space Agency (https://www.esa-landcover-cci.org, accessed on 17 January 2020), which include forest, grassland, shrub, and water. The historical daily meteorological data (2001–2015) came from the global historical meteorological network (https://data.nodc.noaa.gov), including eighteen weather stations located in the forest and grassland. The daily data included the maximum temperature, minimum temperature, 24 h precipitation, and dew point. MODIS-MCD64A1 products (2001–2015, 500 m spatial resolution and daily resolution in temporal) were obtained from NASA (https://earthdata.nasa.gov, accessed 17 January 2020).

The simulated climate data (1971–2099, 0.5° × 0.5° resolution) of the HadGEM2-ES global climate model with four climate scenarios (RCP2.6, RCP4.5, RCP6.0, and RCP8.5) (GCM) were downloaded from the Inter-Sectoral Impact Model Intercomparison Project (ISI-MIP) (https://www.isimip.org/, accessed 17 January 2020). The daily climate data included maximum temperature, minimum temperature, and precipitation.

2.3. Climate Data Processing

For the missing temperature and dew point temperature in historical observation data for the 2001–2015 period, we used the sliding average of the before and after five days to replace the missing data. Due to the lack of daily precipitation data, data from neighboring meteorological stations were used.

The precipitation frequency from the historical climate data was used to correct the simulated precipitation frequency of the simulated climate data. We assumed that the daily precipitation frequencies of the simulated data were identical to the historical data (2001–2015). We obtained the precipitation threshold based on its frequency and set the data less than the threshold to zero. Then, we used the coefficient between the simulated annual precipitation and reserved data to correct the daily precipitation.

The mid-day temperature was replaced with the maximum temperature minus the average difference (2 °C). The dew point temperature was calculated from the daily minimum temperature. The calculation formula was developed with the dew point temperature above 0 °C and the daily minimum temperature in each month of the historical observation data.

2.4. Fire Weather Indices Calculation

The modified Nesterov index was calculated with the following equation [29]:

$$MNI(n) = (MNI(n - 1) + T \times d) \times K(n) \tag{1}$$

where $MNI(n - 1)$ and $MNI(n)$ are the fire weather index on days $n - 1$ and n, respectively. T is the mid-day temperature, d is the dew point temperature, $K(n)$ is a scale coefficient that controls the index change when precipitation occurs on day n [33] (Table 1).

Table 1. The corresponding relationship between coefficient K and daily precipitation.

Rain/mm	0	0.1–0.9	1.0–2.9	3.0–5.9	6.0–14.9	15.0–19.9	≥20.0
K	1.0	0.8	0.6	0.4	0.2	0.1	0

2.5. Vegetation Zone

The study area can be divided into four zones based on the vegetation types, which include shrub (R1), grassland (R2), mountain forest (R3), and rare vegetation zone (R4) (Figure 2). This paper focuses on zones with vegetation, such as R1, R2, and R3.

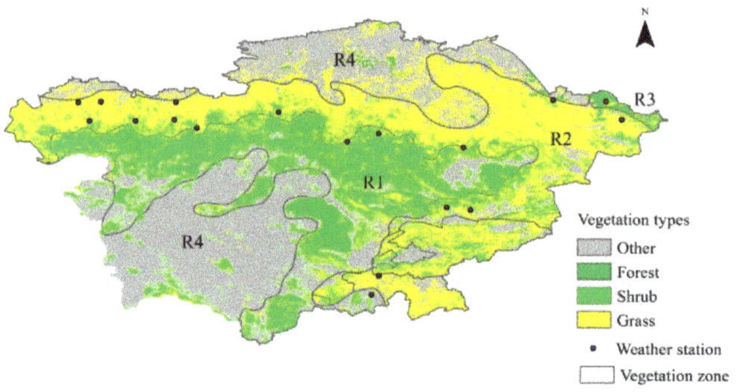

Figure 2. Vegetation zones and weather stations in the study area.

2.6. Burned Areas

The fire season and land cover data were used to filter the false burned areas of the MODIS-MCD64A1 products. Then, we obtained the daily burned areas of each vegetation zone.

2.7. Data Processing

SPSS software was used to analyze the correlation between the fire weather index and burned area. The MNI of each zone was interpolated with the kriging method.

3. Results

3.1. Fire seasons for Each Vegetation Zone

The daily MNI of each vegetation zone was roughly normally distributed. The MNI increased slowly for 90 days and decreased rapidly from the peak. The MNI of R1 increased from the 95th day (Julian day), reached a maximum on the 280th day, and remained at a very low level after the 325th day (Figure 3(a1)). In R2, the MNI increased from the 95th day (peak on the 283rd day) and was maintained at a very low value after the 342nd day. In R3, the MNI increased from the 125th day and decreased to a very low value after the 300th day (maximum on the 248th day).

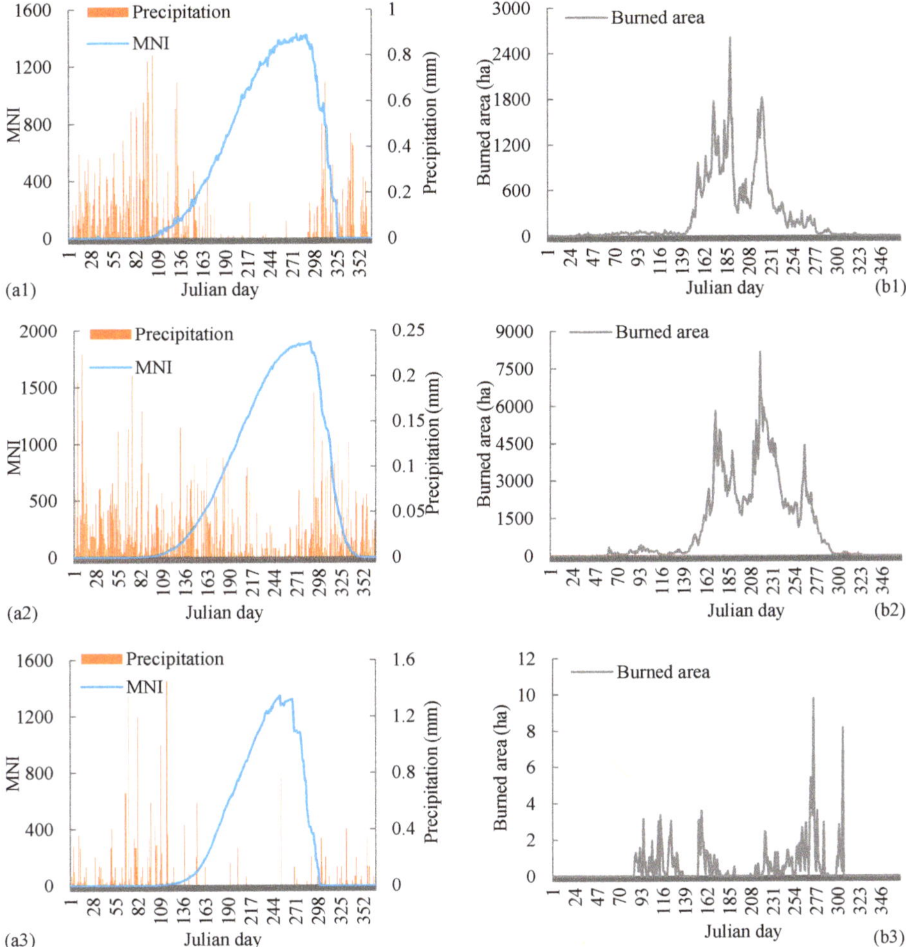

Figure 3. Mean daily modified Nesterov index (MNI), precipitation, and burned areas for the three vegetation zones from 2001–2015. (**a1**) the mean daily MNI and precipitation for vegetation zones R1; (**a2**) the mean daily MNI and precipitation for vegetation zones R2, (**a3**) the mean daily MNI and precipitation for vegetation zones R3; (**b1**) the mean daily burned areas in R1; (**b2**) the mean daily burned areas in R2; (**b3**) the mean daily burned areas in R3.

The fire season of each vegetation zone could be defined by the fire weather index and the daily burned areas. The daily burned areas for R1, R2, and R3 showed apparent changes during the periods of days 150–280, 140–300, and 90–305, respectively (Figure 3(b1–b3)). However, there were some grass fires during days 60–140. The dates of the increases in daily burned areas were earlier than that of the increases in MNI. In early spring, the fuels are dry and cured grass, which can burn at low MNI conditions. The date when the daily burned areas decreased to zero was also earlier than the date when the MNI reached a very low value. The MNI only indicates fire weather, which does not reflect the seasonal status of live fuel. Although daily burned areas showed a relationship with the MNI ($r > 0.23$), their changes were not completely consistent.

The fire season of each zone was defined based on the process of daily MNI and burned areas, which were 1 April–30 November for R1, 1 March–30 November for R2, and 1 April–31 October for R3.

During the period from 2001–2015, the mean MNI in the R1 fire season was 10,213. The high MNI values were distributed in the central areas, especially in the middle part near R2 (MNI > 12,000) (Figure 4). The mean MNI of R2 was 12,104, and the high MNI values were distributed in the northwestern and south-central parts. The mean MNI of R3 was 10,769 (Figure 4).

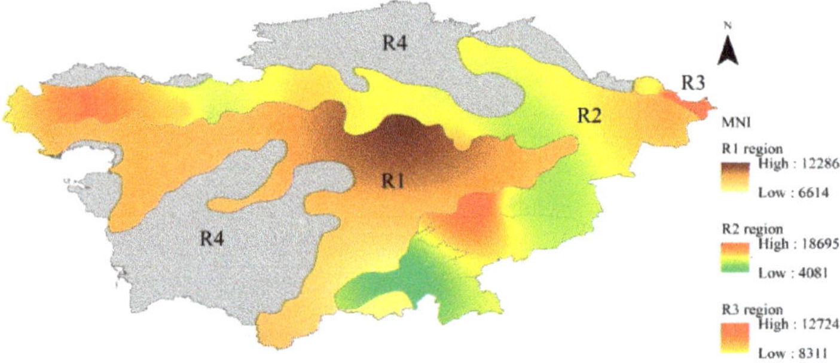

Figure 4. MNI distribution in three vegetation zones for the 2001–2015 period.

3.2. Distribution of Burned Areas in the 2001–2015 Period

The mean annual burned area was 87,812 ha in R1 during the 2001–2015 period. The burned areas in 2002, 2004, and 2005 were significantly higher than the average of the period ($\alpha = 0.05$), and the burned areas were significantly lower in 2001, 2003, 2011, 2012, and 2013 (Figure 5(a1)). The monthly maximum burned areas occurred in May, with 26,603 ha, and the minimum occurred in November, with 250 ha (Figure 5(b1)). The mean annual burned area in R2 was 389,020 ha, which was much higher than that in the other zones. The maximum monthly burned area was 133,452 ha in August, and the minimum was 1211 ha in November. The mean annual burned area in R3 was 150 ha. The months in the fire season with the maximum and minimum burned areas were September and June, respectively (Figure 5(b3)).

The monthly burned areas and MNI in the fire season for R1 and R3 did not show a strong correlation ($r = 0.36$), but they showed a strong correlation for R2 ($r = 0.6$) as follows:

$$B(n) \;=\; 2.4683MNI(n) \;+\; 13,101 \;(R^2 \;= 0.32) \tag{2}$$

where, $B(n)$ and $MNI(n)$ are the burned areas and mean MNI in month n.

Grass fires and shrub fires mainly occurred in the period from June to September, and they were usually distributed in the plains and hilly areas with elevations less than 500 m. There were no wildfires in the areas with elevations greater than 2000 m. The fires in mountain forests mainly occurred in the April–June and September–October periods and were usually distributed the areas with low elevation

(<500 m). A few wildfires also occurred in high-altitude areas (>1500 m ASL) during July and August (Figure 6).

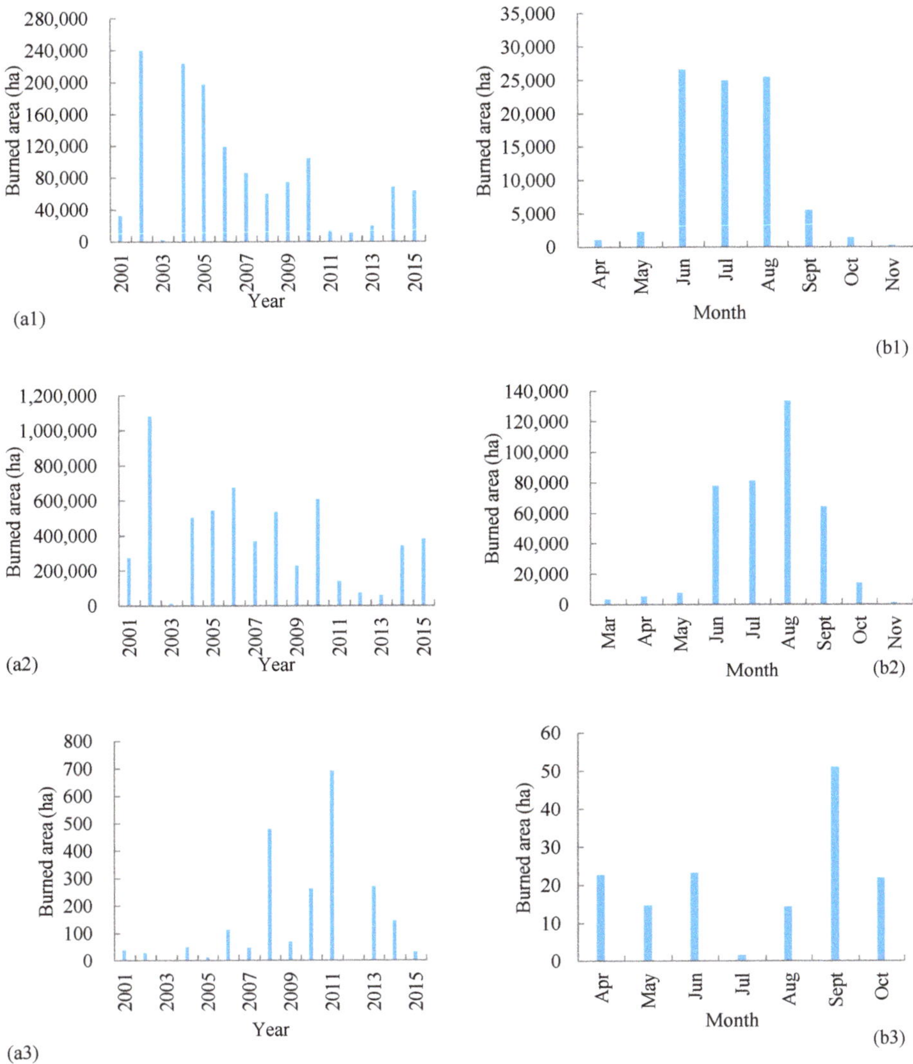

Figure 5. Annual and monthly burned areas for three vegetation zones during the 2001–2015 period. (**a1**) the mean annual burned areas in R1; (**a2**) the mean annual burned areas in R2; (**a3**) the mean annual burned areas in R3; (**b1**) the mean monthly burned areas in R1; (**b2**) the mean monthly burned areas in R2; (**b3**) the mean monthly burned areas in R3.

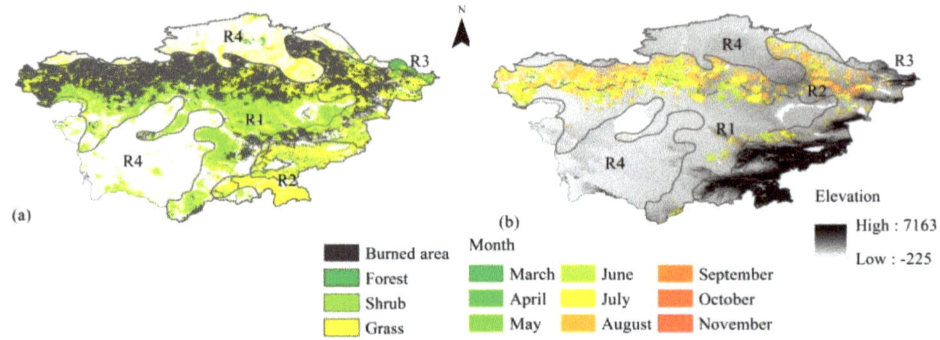

Figure 6. Burned area distribution in the study area during the 2001–2015 period.

3.3. Climate Change in the 2030s and 2080s

The mean daily maximum temperature in the fire season of R1 was 23.2 °C during the baseline period. It will be 26.4 °C band 28.2 °C in the 2030s and 2080s, respectively, which will be an increase of 14% and 21% compared with the baseline (p = 0.00). The precipitation in the fire season of R1 was 180 mm at the baseline and will be 219, 174, 198, and 180 mm under the RCP2.6, RCP4.5, RCP6.0, and RCP8.5 scenarios in the 2030s, respectively. The precipitation will increase by 7% to 193 mm in the 2030s. However, the increase was not significant (*F*-test, p = 0.12). In the 2080s, the precipitation will be 215, 195, 178, and 214 mm under the RCP2.6, RCP4.5, RCP6.0, and RCP8.5 scenarios, respectively. The mean precipitation will significantly increase by 11% to 200 mm in the 2080s (*F*-test, p = 0.04) (Figure 7).

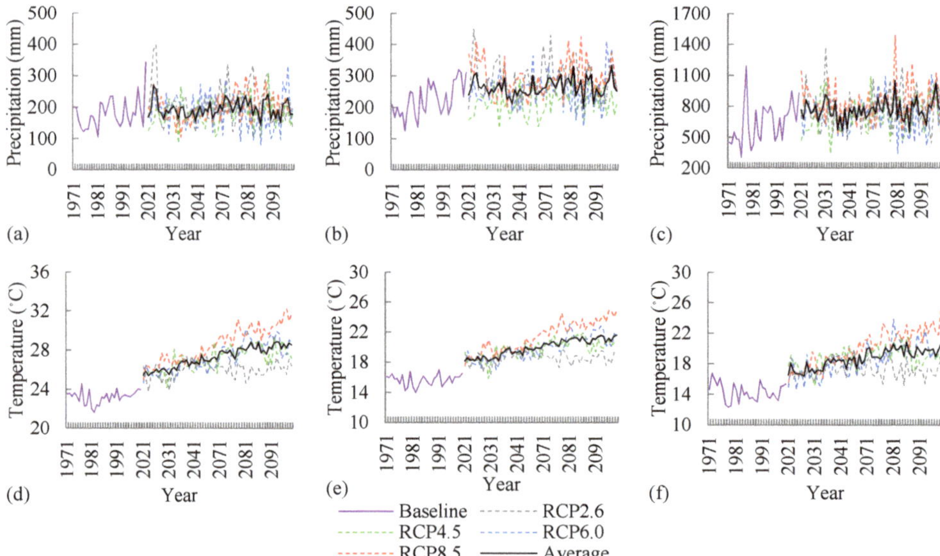

Figure 7. Precipitation and temperature in the fire season for three vegetation zones in the baseline, 2030s, and 2080s. (**a**) the precipitation during the fire season in the R1; (**b**) the precipitation during the fire season in the R2; (**c**) the precipitation during the fire season in the R3; (**d**) the mean daily maximum temperatures of the fire season in the R1; (**e**) the mean daily maximum temperatures of the fire season in the R2; (**f**) the mean daily maximum temperatures of the fire season in the R3.

The mean daily maximum temperature in fire season of R2 was 15.7 °C in the baseline period. It will significantly increase by 21% and 34% in 2030s and 2080s, respectively ($p = 0.00$). The precipitation in fire season of R2 was 226 mm at the baseline, and it will be 293, 206, 256, and 284 mm in the 2030s under the RCP2.6, RCP4.5, RCP6.0, and RCP8.5 scenarios, respectively. The mean precipitation will increase 15% in 2030s ($p = 0.00$). In addition, the precipitation in the fire season will significantly increase by 21% in 2080s ($p = 0.00$), and the precipitation will be 294, 231, 260, and 303 mm under the RCP2.6, RCP4.5, RCP6.0, and RCP8.5 scenarios, respectively.

In R3, the mean daily maximum temperature in fire season was 14.3 °C in baseline. It will increase by 23% to 17.7 °C in 2030s, when the temperature will be 17.3, 18.2, 17.2, and 18.2 °C under the RCP2.6, RCP4.5, RCP6.0, and RCP8.5 scenarios, respectively. In 2080s, the mean daily maximum temperature will significantly increase to 19.6 °C ($p = 0.00$), which will be 17.2, 19.4, 20.0, and 21.9 °C under the RCP2.6, RCP4.5, RCP6.0, and RCP8.5 scenarios, respectively. The precipitation in fire season was 646 mm at the baseline and will be 743, 696, 718, and 817 mm under RCP2.6, RCP4.5, RCP6.0, and RCP8.5 scenarios in the 2030s, respectively. In 2080s, the mean precipitation will increase significantly to 784 mm ($p = 0.00$), representing an increase of 21% from the baseline.

3.4. MNI Changes in the 2030s and 2080s

The mean MNI in the fire season of R1 was 3812 at the baseline, and it will increase by 33% and 63% in the 2030s and 2080s, respectively ($p = 0.00$) (Figure 8). In R2, the mean MNI of the fire season was 1759 at baseline, and it will increase by 42% in the 2030s and 73% in the 2080s ($p = 0.00$). The mean MNI of the fire season for R3 was 713 at baseline, and it will increase to 1195 in the 2030s and 1752 in the 2080s ($p = 0.00$).

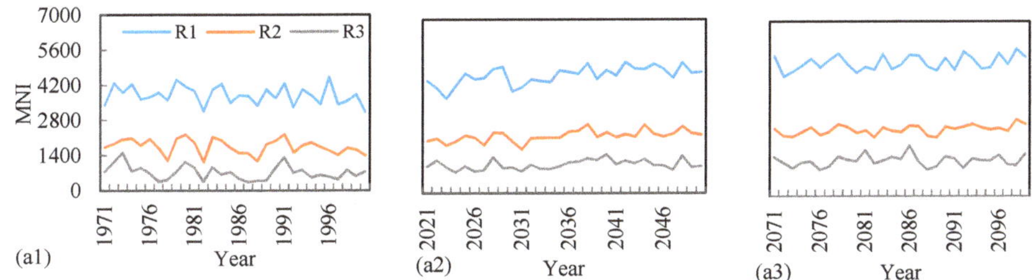

Figure 8. Mean MNI of fire season for three vegetation zones in different period. (**a1**) mean MNI of fire season in baseline; (**a2**) mean MNI of fire season in 2030s; (**a3**) mean MNI of fire season in 2080s.

Most areas of R1 showed low MNI values (<4691) at the baseline, and only southern areas had high MNI values (8735–17,156) (Figure 9(a1)). In the 2030s, the MNI values in western and northern R1 showed a slight increase but will increase from 23%–30% in central areas and 37%–59% in the south. The mean MNI will increase by 19%, 28%, 20%, and 33% in the 2030s under the RCP2.6, RCP4.5, RCP6.0, and RCP8.5 scenarios, respectively. The MNI will increase obviously under scenarios RCP4.5 and RCP8.5, and the maximum increase will reach 68% and 82%, respectively (Figure 9(b3,b5)). In the 2080s, MNIs will increase by more than 37% in central and southern R1, while MNIs in western R1 will increase by only 15% (Figure 9(c1)). The mean MNI will increase by 15%, 39%, 46%, and 64% over the baseline under the RCP2.6, RCP4.5, RCP6.0, and RCP8.5 scenarios, respectively. Under the RCP2.6 scenario, the MNI will increase significantly in the south, while the MNI will increase in the central and southern regions significantly under the other scenarios.

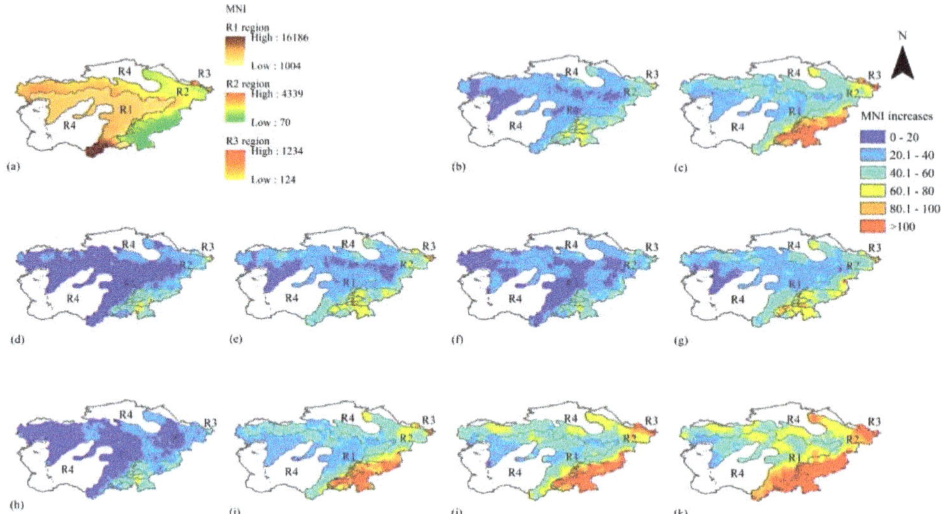

Figure 9. Distribution of MNI in the fire season for each vegetation zone. (**a**) Baseline; (**b**) the mean MNI increase in the 2030s; (**d**), (**e**), (**f**), and (**g**) are MNI increases under the RCP2.6, RCP4.5, RCP6.0, and RCP8.5 scenarios in the 2030s, respectively; (**c1**) the mean MNI increase in the 2080s; and (**h**), (**i**), (**j**), and (**k**) are MNI increases under the RCP2.6, RCP4.5, RCP6.0, and RCP8.5 scenarios in the 2080s, respectively.

The MNI values in western R2 were high at baseline. The mean MNI will increase by 23%, 35%, 26%, and 42% in the 2030s under the RCP2.6, RCP4.5, RCP6.0, and RCP8.5 scenarios, respectively. The increase is significant for the RCP4.5 and RCP8.5 scenarios. The MNI in the southern area will increase clearly (+50%), but the values in the western areas will increase by only 10%–38% in the 2030s. The MNI will increase by 20%, 51%, 61%, and 73% in the 2080s under the RCP2.6, RCP4.5, RCP6.0, and RCP8.5 scenarios, respectively. The MNI values in the western and southern R2 will increase significantly under RCP2.6 scenario, and the values will increase significantly in the western, central, and southern areas under the other scenarios.

The MNI values were high in the northwest of R3 (1163–1353) and low in the southeast (322–599) at the baseline. The mean MNI will increase by 49%, 77%, 49%, and 68% in the 2030s over the baseline under the RCP2.6, RCP4.5, RCP6.0, and RCP8.5 scenarios, respectively. In the 2080s, the mean MNI will increase by 32%, 84%, 127%, and 146% under the RCP2.6, RCP4.5, RCP6.0, and RCP8.5 scenarios, respectively. A greater increase in MNI will be distributed in the northwestern and central areas under the RCP2.6 scenario, while MNI will increase clearly in the northwest of R3 under the other scenarios.

4. Discussion

The MNI can effectively reflect drought weather and fuel moisture in Central Asia [33]. In the study area, there was less precipitation and higher temperatures from June–September, and the MNI reached its peak in August or September. Burned area is the most important indicator of fire regime. We used the daily burned areas from 2001–2015 from MODIS data to describe their distributions spatially and temporally [34]. There have been some studies on wildfires in Central Asia based on remote sensing data in recent years [21,35,36]. We determined the fire season for each vegetation zone according to the MNI and daily burned areas. We considered the characteristics of fire regime and fire weather. The defined fire season would be an important indicator of fire dynamics and fire management in the region.

Grassland fire and shrub fire mainly occurred in summer (June–September), and the largest burned areas in grassland usually occurred in August or September [21]. The fires in mountain forests in the low altitude regions mainly occurred in spring (April–June) and autumn (September and October). Nevertheless, the fires mainly occurred in summer (July and August) for the forests at high-altitude areas (>1500 m ALS). Fires in mountain forests are usually distributed in areas near farmland or towns, which indicates that human activities impact these occurrences. The references indicated that most forest fires were ignited by humans in low-altitude areas and fires in high-altitude areas in summer were mainly caused by lightening [4,18].

Although both the temperature and temperature in the fire season of the vegetation zones will increase in the 2080s under future climate scenarios, their MNIs will still increase clearly over the baseline. This indicates that fire danger will increase in the future, which is consistent with the results of Liu et al. [9]. They also believed that the mean daily maximum temperature and fire danger rating of Central Asia (HadCM3 model with A2a scenario) would increase from 2071–2100, but they projected that the annual precipitation would decrease.

Climate change will also affect the vegetation of the study area. In the study we did not simulate the vegetation change resulting from the future climate change. Vegetation distribution was affected by many factors, such as climate, anthropic activities, and natural disturbances. In fact, forest and grassland decreased during the period 1992–2003 and increased slightly during 2003–2015. However, vegetation types and their spatial distribution was no changed clearly in the period [37]. We assumed that the vegetation will not signification change in the coming decades. This point will not have influences on the judgment of fire weather changes under the future climate scenarios.

The fewer meteorological stations (18) available from 2001–2015, and their uneven spatial distribution, may affect the interpolated results of fire weather. However, each meteorological factor and fire weather index showed very similar processes in those years. The influence would not affect the reliability of the results.

Fire weather affects ignitions and fire spread. The MNI showed a positive correlation with the monthly burned areas ($r < 0.3$). Based on the correlation, we projected that the potential burned areas would increase in the future. The potential burned areas of R1, R2, and R3 in the 2030s will increase by 4%, 8%, and 2% over the baseline and will increase by 6%, 13%, and 3%, respectively, in the 2080s.

The wildfires were mainly distributed in shrub and grassland areas. The annual burned area has generally declined since 2010. This trend reflects the role of fire management activities. The governments in Central Asia established fire agencies and promulgated laws and regulations on wildfire management in this century [38]. The wildfires were still serious in some years (such as 2011). The vegetation in the study area plays an important role in the regional ecological environment and biodiversity protection [39]. It is necessary to strengthen wildfire management in the region to adapt to future climate change.

5. Conclusions

Fire seasons are different for each vegetation zone. Grassland has the longest fire season, and mountain forests have the shortest fire season. The fire seasons of grassland, shrub, and mountain forest are 1 March–30 November, 1 April–30 November, and 1 April–31 October, respectively. Most wildfires in the study area mainly occurred in shrub and grasslands. Most shrub and grass fires occurred in the period from June–September, and fires in mountain forests occurred mainly in the April–June and September–October periods.

The MNI index is a good indicator of fire danger for Central Asia. In the 2030s, the mean daily maximum temperature in the fire season for vegetation areas will increase significantly over the baseline, and the precipitation will increase by 7%–15%. The MNI will increase by 33%–68% for vegetation areas in the 2030s. The mean daily maximum temperature, precipitation and MNI of vegetation areas will increase significantly in the 2080s. The MNI will increase by 63%–146%, and the potential areas will increase by 3%–13% for each vegetation zone.

Author Contributions: X.Z. and X.T. conceived and designed the experiments; X.Z. wrote the original draft and edited the manuscript. X.T. wrote, reviewed and edited the manuscript; Y.Y. contributed on data collection and reviewed the manuscript. All authors have read and agreed to the published version of the manuscript.

Funding: This research was funded by (Strategic Priority Research Program of the Chinese Academy of Sciences) grant number (XDA20020202) and (Project of National Natural Science Foundation of China) grant number (31770695).

Conflicts of Interest: The authors declare no conflicts of interest.

References

1. Marlon, J.R.; Bartlein, P.J.; Carcaillet, C.; Gavin, D.G.; Harrison, S.P.; Higuera, P.E.; Prentice, I.C. Climate and human influences on global biomass burning over the past two millennia. *Nat. Geosci.* **2008**, *1*, 697–702. [CrossRef]

2. Girardin, M.P.; Portier, J.; Remy, C.C.; Ali, A.A.; Paillard, J.; Blarquez, O.; Bergeron, Y. Coherent signature of warming-induced extreme sub-continental boreal wildfire activity 4800 and 1100 years BP. *Environ. Res. Lett.* **2019**, *14*, 124042. [CrossRef]

3. Viatte, C.; Strong, K.; Patonwalsh, C.; Mendonca, J.; Oneill, N.T.; Drummond, J.R. Measurements of CO, HCN, and C2H6 Total Columns in Smoke Plumes Transported from the 2010 Russian Boreal Forest Fires to the Canadian High Arctic. *Atmos. Ocean* **2013**, *51*, 522–531. [CrossRef]

4. Arkhipov, V.; Moukanov, B.M.; Khaidarov, K.; Goldammer, J.G. Overview on forest fires in Kazakhstan. *Int. For. Fire News* **2000**, *22*, 40–48. Available online: http://hdl.handle.net/11858/00-001M-0000-0014-9566-B (accessed on 10 May 2020).

5. Keyser, A.R.; Westerling, A.L. Predicting increasing high severity area burned for three forested regions in the western United States using extreme value theory. *For. Ecol. Manag.* **2019**, *432*, 694–706. [CrossRef]

6. Tansey, K.; Grégoire, J.M.; Stroppiana, D.; Sousa, A.; Silva, J.M.; Pereira, J.M.; Peduzzi, P. Vegetation burning in the year 2000: Global burned area estimates from SPOT VEGETATION data. *J. Geophys. Res. Atmos.* **2004**, *109*, D14S03. [CrossRef]

7. Atkin, M. Inside Central Asia: A Political and Cultural History of Uzbekistan, Turkmenistan, Kazakhstan, Kyrgyzstan, Tajikistan, Turkey, and Iran by Dilip Hiro. *Int. J. Middle East Stud.* **2011**, *43*, 190–192. [CrossRef]

8. Kleine, M.; Colak, A.H.; Kirca, S.; Saghebtalebi, K.; Orozumbekov, A.; Lee, D.K. Rehabilitating degraded forest landscapes in West and Central Asia. *IUFRO World Ser.* **2009**, *20*, 5–26.

9. Liu, Y.; Stanturf, J.; Goodrick, S. Trends in global wildfire potential in a changing climate. *For. Ecol. Manag.* **2010**, *259*, 685–697. [CrossRef]

10. Flannigan, M.; Cantin, A.S.; De Groot, W.J.; Wotton, M.; Newbery, A.; Gowman, L.M. Global wildland fire season severity in the 21st century. *For. Ecol. Manag.* **2013**, *294*, 54–61. [CrossRef]

11. AghaKouchak, A.; Huning, L.S.; Chiang, F.; Sadegh, M.; Vahedifard, F.; Mazdiyasni, O.; Mallakpour, I. How do natural hazards cascade to cause disasters? *Nature* **2018**, *561*, 458–460. [CrossRef] [PubMed]

12. Liang, S.; Hurteau, M.D.; Westerling, A.L. Large-scale restoration increases carbon stability under projected climate and wildfire regimes. *Front. Ecol. Environ.* **2018**, *16*, 207–212. [CrossRef]

13. San-Miguel-Ayanz, J.; Durrant, T.; Boca, R.; Libertà, G.; Branco, A.; de Rigo, D.; Ferrari, D.; Maianti, P.; Artés Vivancos, T.; Costa, H.; et al. *Forest Fires in Europe, Middle East and North Africa 2017*; Publications Office of the European Union: Luxemburg, 2018; EUR 29318; ISBN 978-92-79-92831-4. [CrossRef]

14. Kovats, R.S.; Valentini, R.; Bouwer, L.M.; Georgopoulou, E.; Jacob, D.; Martin, E.; Rounsevell, M.; Soussana, J.F. Europe. In *Climate Change 2014: Impacts, Adaptation, and Vulnerability. Part B: Regional Aspects. Contribution of Working Group II to the Fifth Assessment Report of the Intergovernmental Panel on Climate Change*; Barros, V.R., Field, C.B., Dokken, D.J., Mastrandrea, M.D., Mach, K.J., Bilir, T.E., Chatterjee, M., Ebi, K.L., Estrada, Y.O., Genova, R.C., et al., Eds.; Cambridge University Press: Cambridge, UK, 2015.

15. Wotton, B.M.; Nock, C.A.; Flannigan, M.D. Forest fire occurrence and climate change in Canada. *Int. J. Wildland Fire* **2010**, *19*, 253–271. [CrossRef]

16. Torzhkov, I.O.; Kushnir, E.A.; Konstantinov, A.V.; Koroleva, T.; Efimov, S.V.; Shkolnik, I.M. Assessment of Future Climate Change Impacts on Forestry in Russia. *Russ. Meteorol. Hydrol.* **2019**, *44*, 180–186. [CrossRef]

17. Klein, I.; Gessner, U.; Kuenzer, C. Regional land cover mapping and change detection in Central Asia using MODIS time-series. *Appl. Geogr.* **2012**, *35*, 219–234. [CrossRef]

18. Goldammer, J.G.; Davidenko, E.P.; Kondrashov, L.G.; Ezhov, N.I. Recent trends of forest fires in Central Asia and opportunities for regional cooperation in forest fire management. In Proceedings of the Regional Forest Congress Forest Policy: Problems and Solutions, Bishkek, Kyrgyzstan, 25–27 November 2004.

19. Kazakhstan International Security Exhibition. 2012 Forest Fires in Kazakhstan up by 41%. Available online: https://www.aips.kz/en/home/9-press-center/news/137-forest-fires-in-kazakhstan-up-by-41 (accessed on 18 June 2020).

20. Cao, X.; Meng, Y.; Chen, J. Mapping grassland wildfire risk of the world. In *World Atlas of Natural Disaster Risk*; Springer: Berlin/Heidelberg, Germany, 2015; pp. 277–283.

21. Loboda, T.V.; Giglio, L.; Boschetti, L.; Justice, C.O. Regional fire monitoring and characterization using global NASA MODIS fire products in dry lands of Central Asia. *Front. Earth Sci.* **2012**, *6*, 196–205. [CrossRef]

22. Warneke, C.; Bahreini, R.; Brioude, J.; Brock, C.A.; De Gouw, J.A.; Fahey, D.W.; Veres, P.R. Biomass burning in Siberia and Kazakhstan as an important source for haze over the Alaskan Arctic in April 2008. *Geophys. Res. Lett.* **2009**, *36*, L02813. [CrossRef]

23. Babu, K.V.S.; Kabdulova, G.; Kabzhanova, G. Developing the Forest Fire Danger Index for the Country Kazakhstan by Using Geospatial Techniques. *J. Environ. Inf. Lett* **2019**, *1*, 48–59. [CrossRef]

24. Spivak, L.; Arkhipkin, O.; Sagatdinova, G. Development and prospects of the fire space monitoring system in Kazakhstan. *Front. Earth Sci.* **2012**, *6*, 276–282. [CrossRef]

25. Van Wagner, C.E.; Forest, P. *Development and Structure of the Canadian Forest Fire Weather Index System*; Forestry Technical Report; Canadian Forestry Service: Ottawa, ON, Canada, 1987; p. 35.

26. Bedia, J.; Herrera, S.; Gutiérrez, J.M.; Benali, A.; Brands, S.; Mota, B.; Moreno, J.M. Global patterns in the sensitivity of burned area to fire-weather: Implications for climate change. *Agric. For. Meteorol.* **2015**, *214*, 369–379. [CrossRef]

27. Nesterov, V.G. *Forest fire Potential and Methods of Its Determination*; Goslesbumizdat Publishing House: Moscow, Russia, 1949.

28. Zhdanko, V.A. Scientific basis of development of regional scales and their importance for forest fire management. In *Contemporary Problems of Forest Protection from Fire and Firefighting*; Melekhov, I.S., Ed.; Lesnaya Promyshlennost Publishing: Moscow, Russia, 1965.

29. Sherstyukov, B.G. *Index of Forest Fire. Yearbook of Weather, Climate and Ecology of Moscow*; Moscow State University Publishing: Moscow, Russia, 2002.

30. Ganatsas, P.; Antonis, M.; Marianthi, T. Development of an adapted empirical drought index to the Mediterranean conditions for use in forestry. *Agric. For. Meteorol.* **2011**, *151*, 241–250. [CrossRef]

31. Niu, R.; Zhai, P. Study on forest fire danger over Northern China during the recent 50 years. *Clim. Chang.* **2012**, *111*, 723–736. [CrossRef]

32. Karouni, A.; Daya, B.; Bahlak, S. A comparative study to find the most applicable fire weather index for Lebanon allowing to predict a forest fire. *J. Commun. Comput.* **2013**, *11*, 1403–1409. Available online: http://www.wanfangdata.com.cn/details/detail.do?_type=perio&id=David_20171208_2213 (accessed on 10 June 2020).

33. Groisman, P.Y.; Sherstyukov, B.G.; Razuvaev, V.N.; Knight, R.W.; Enloe, J.G.; Stroumentova, N.S.; Karl, T.R. Potential forest fire danger over Northern Eurasia: Changes during the 20th century. *Glob. Planet. Chang.* **2007**, *56*, 371–386. [CrossRef]

34. Mouillot, F.; Schultz, M.G.; Yue, C.; Cadule, P.; Tansey, k.; Ciais, P.; Chuvieco, E. Ten years of global burned area products from space borne remote sensing—A review: Analysis of user needs and recommendations for future developments. *Int. J. Appl. Earth Obs. Geoinf.* **2014**, *26*, 64–79. [CrossRef]

35. Hall, J.V.; Loboda, T.V.; Giglio, L.; Hall, J.V.; Loboda, T.V.; Giglio, L.; Mccarty, G.W. A MODIS-based burned area assessment for Russian croplands: Mapping requirements and challenges. *Remote Sens. Environ.* **2016**, *184*, 506–521. [CrossRef]

36. Zhu, C.; Kobayashi, H.; Kanaya, Y.; Zhu, C.; Kobayashi, H.; Kanaya, Y.; Saito, M. Size-dependent validation of MODIS MCD64A1 burned area over six vegetation types in boreal Eurasia. Large underestimation in croplands. *Sci. Rep.* **2017**, *7*, 1–9. [CrossRef]

37. Ruan, H.W.; Yu, J.J. Changes in land cover and evapotranspiration in the five CentralAsian countries from 1992 to 2015. *Acta Geogr. Sin.* **2019**, *74*, 1292–1304. [CrossRef]

38. Yang, G.; Teng, Y.; Shu, L.F.; Cai, H.Y.; Di, X.Y. Review of Forest and Grassland Fire Prevention Along "the Belt and Road". *World For. Res.* **2018**, *31*, 82–88. [CrossRef]
39. Global Forest Resources Assessment 2015. Available online: http://www.fao.org/3/a-au190e.pdf (accessed on 10 June 2020).

 *forests*

Perspective

Linking Forest Flammability and Plant Vulnerability to Drought

Rachael H. Nolan [1,2,*], **Chris J. Blackman** [1,3], **Víctor Resco de Dios** [4,5], **Brendan Choat** [1], **Belinda E. Medlyn** [1], **Ximeng Li** [1,6], **Ross A. Bradstock** [2,7] **and Matthias M. Boer** [1,2]

1   Hawkesbury Institute for the Environment, Western Sydney University, Locked Bag 1797, Penrith, NSW 2751, Australia; cj.blackman27@gmail.com (C.J.B.); b.choat@westernsydney.edu.au (B.C.); B.Medlyn@westernsydney.edu.au (B.E.M.); liximeng2009@hotmail.com (X.L.); m.boer@westernsydney.edu.au (M.M.B.)
2   NSW Bushfire Risk Management Research Hub, Wollongong, NSW 2522, Australia; rossb@uow.edu.au
3   INRA, UMR547, PIAF, 63100 Clermont-Ferrand, France
4   School of Life Science and Engineering, Southwest University of Science and Technology, Mianyang 621010, China; v.rescodedios@gmail.com
5   Department of Crop and Forest Sciences and Agrotecnio Centre, Universitat de Lleida, 25198 Lleida, Spain
6   College of Life and Environmental Science, Minzu University of China, Beijing 100081, China
7   Centre for Environmental Risk Management of Bushfires, University of Wollongong, Wollongong, NSW 2522, Australia
*   Correspondence: rachael.nolan@westernsydney.edu.au; Tel.: +61-2-4570-1267

Received: 10 June 2020; Accepted: 16 July 2020; Published: 20 July 2020

Abstract: Globally, fire regimes are being altered by changing climatic conditions. New fire regimes have the potential to drive species extinctions and cause ecosystem state changes, with a range of consequences for ecosystem services. Despite the co-occurrence of forest fires with drought, current approaches to modelling flammability largely overlook the large body of research into plant vulnerability to drought. Here, we outline the mechanisms through which plant responses to drought may affect forest flammability, specifically fuel moisture and the ratio of dead to live fuels. We present a framework for modelling live fuel moisture content (moisture content of foliage and twigs) from soil water content and plant traits, including rooting patterns and leaf traits such as the turgor loss point, osmotic potential, elasticity and leaf mass per area. We also present evidence that physiological drought stress may contribute to previously observed fuel moisture thresholds in south-eastern Australia. Of particular relevance is leaf cavitation and subsequent shedding, which transforms live fuels into dead fuels, which are drier, and thus easier to ignite. We suggest that capitalising on drought research to inform wildfire research presents a major opportunity to develop new insights into wildfires, and new predictive models of seasonal fuel dynamics.

Keywords: drought; flammability; fuel moisture; leaf water potential; plant traits; wildfire

---

1. Introduction

Fire has played an important role in determining the composition and distribution of ecosystems almost since the emergence of the first land plants [1]. In many regions, the frequency of wildfires is projected to increase under climate change due to changes in fuel (i.e., biomass) production, accelerated aboveground biomass turnover rates and fuel drying [2]. This increase in wildfire frequency has the potential to drive species extinctions and cause ecosystem state changes [2]. Indeed, conversion of forests to shrublands or grasslands due to increased fire frequency is already occurring in the Mediterranean Basin [3], the western United States [4] and south-eastern Australia [5]. The role of wildfires in the terrestrial carbon cycle [6], and subsequent feedbacks into the climate system, as well as potential implications for precipitation [7], highlight an urgent need to increase our understanding of the climate–fire–vegetation interactions underlying global fire regimes. At the same time, predicting the likelihood of wildfire at shorter time scales (weeks–months) is required for land managers to target suppression resources in order to protect people, property and infrastructure, as well as fire-sensitive ecosystems.

Large fires generally coincide with periods of high soil water deficit and atmospheric water demand in most forests and woodlands [8]. In these ecosystems, spatially continuous arrays of fuel (e.g., litter, foliage, twigs) are usually present, except for immediately after fire. However, these fuels are usually too wet to propagate fire. During drought or seasonal dryness, the moisture content of these fuels declines [8,9]. Low fuel moisture content increases the probability of ignition, rate of fire spread and fire intensity [10–12]. While low fuel moisture content is likely to be important for the probability of ignition and initial rate of spread, other factors such as fuel load, wind and terrain can be of greater importance for subsequent fire behaviour [13]. Thus, dry fuels are a prerequisite for large forest fires, along with weather and ignition sources ([14]; Figure 1).

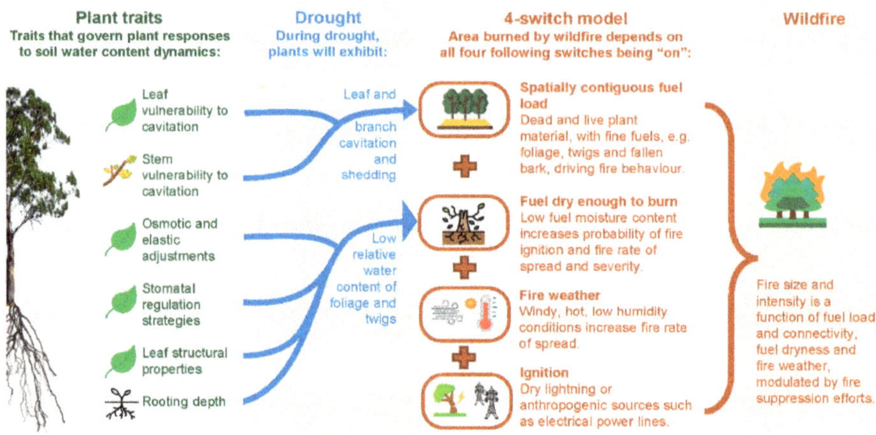

Figure 1. Conceptual model illustrating linkages between drought-related plant traits and the likelihood of wildfire. The 4-switch model is outlined in Bradstock [14].

In addition to causing declines in fuel moisture content, drought or seasonal dryness can also cause changes in the ratio of dead to live fuels [15,16]. Drought stress and subsequent mortality is potentially an important mechanism driving large wildfires, since the moisture content of dead fuels can decline far below that of live fuels (e.g., ~7–30% for dead fuels and ~50–200% for live fuels; [9]). For example, following drought-induced dieback in the Jarrah forests of south-western Australia, Ruthrof et al. [15] observed a large increase in surface fine fuel loads (i.e., litter). Additionally, tree mortality was associated with a more open canopy, which affected the microclimate of the forest floor, increasing temperature and vapour pressure deficit, and hence the rate of drying of the understory and litter fuels [15]. These changes in forest structure resulted in a 30% increase in predicted fire spread rates [15]. Thus, drought events can increase the probability of wildfire through multiple mechanisms, including changes to understory microclimate, fuel moisture, forest structure, the ratio of dead to live fuels in the canopy, and the amount of litter on the forest floor.

As drought events become more severe, there is increasing attention being paid to drought-induced tree mortality. This concern is driving a wave of research into plant vulnerability to drought and plant water relations under stress [17,18]. However, despite the co-occurrence of forest fires and drought, and recognition of the role of plant physiology and phenology in governing the moisture content of live fuels [8], current approaches to modelling fuel attributes, such as live fuel moisture content or the ratio of dead to live fuels (i.e., foliage), generally do not explicitly incorporate plant physiological responses to water stress (but see [19–22]).

There is a tremendous opportunity to capitalise on recent advances in drought research to inform our understanding and prediction of fuel attributes such as live fuel moisture content and dead:live fuel ratios. Currently, live fuel moisture content (LFMC, moisture content of foliage and twigs) is monitored through satellite remote sensing (e.g., [9,23]) or inferred from drought indices that require species-specific calibrations (e.g., [24,25]). However, remote sensing can at best only estimate LFMC in near real-time, and cannot be used to predict flammability under future (novel) climatic conditions. Drought indices lack the physical basis required to reliably quantify flammability outside of the ecosystems for which they were calibrated. Similarly, while there is recognition of the importance of dead to live fuel ratios on fire behaviour [26,27], these ratios are currently only inferred from drought indices (e.g., in the United States National Fire Danger Rating System, [28]), if at all.

Here, we demonstrate how drought-related research can be used to advance our understanding of forest flammability. We use the term "flammability" to refer to the general ability of vegetation to burn, following Gill and Zylstra [29]. We particularly focus on the potential applications of drought-related research to inform the prediction of live fuel moisture content and changes to the ratio of dead to live fuels in the forest canopy. These fuel attributes are fundamentally important constraints of wildfire [14,30,31] and underpin many fire behaviour models for forests. For example, many of the fire behaviour models used in Australia and North America require inputs of fuel load, particularly of surface fuels (i.e., litter) and fuel moisture content, which is often approximated by drought indices [32,33]. Here, we (i) present a conceptual model illustrating the links between plant responses to drought and critical fuel properties limiting the probability of landscape-scale fire (specifically live fuel moisture content and dead: live fuel ratios); (ii) demonstrate that established relationships between leaf water content and leaf water potential (pressure-volume curves) can be adapted to model live fuel moisture content; (iii) present a framework for modelling live fuel moisture content from soil water content and drought-related plant traits; and (iv) examine potential links between physiological drought stress, including leaf cavitation and shedding, and fuel attributes. We do this via a combination of literature review and analyses from a common garden experiment presented here as a case study. Our goal is to stimulate joint research on plant responses to drought and forest flammability.

2. Linking Fire with Drought: A Conceptual Model

For landscape-scale fires to occur, four conditions need to be met: (i) the presence of spatially contiguous fuel; (ii) that fuel being dry enough to burn; (iii) weather conditions favourable to the spread of fire; and (iv) an ignition source (e.g., lightning; [14]). These conditions have been characterised as switches, with all four needing to be activated for wildfires to occur ([14]; Figure 1). We posit that the second switch (fuel dryness) is influenced by plant responses to drought, which in turn are governed by plant traits (Figure 1). The first switch (fuel load) is also likely affected, to some extent, by plant responses to drought (Figure 1). For live fuels, moisture content is a function of soil water availability across the root zone, and the osmotic and elastic adjustments that determine the relationship between leaf water content and leaf water potential [34]. Moisture content is also a function of leaf structural properties, which set the limit on maximum water content [35].

For fuel load, we suggest that both the quantity and spatial arrangement are modified by drought-related plant traits. For example, during extreme drought which results in canopy dieback, there may be a large, temporary transformation of live fuels into dead fuels. When this senescing foliage is finally shed from plants, the density of live fuels in the canopy will decrease. At the same time, the influx of litter into the surface fuel layer will be relatively uncompacted, and thus well-aerated, and therefore more readily available to burn [36]. These relationships between plant responses to drought and fuel properties are conceptualised in Figure 1. We now explore these linkages between plant responses to drought and forest flammability in detail.

3. Common Garden Case Study

We use a common garden experiment as a case study to examine the hypothesised linkages between drought-related plant traits and wildfire risk outlined in Figure 1. Eight *Eucalyptus* tree species originating from across a strong gradient in rainfall (250–1125 mm), temperature (10–21 °C) and moisture index (0.2–1.1; ratio of mean annual precipitation to potential evapotranspiration) across New South Wales, Australia, were grown in a common garden. The eucalypts were sourced from a range of vegetation communities (wet sclerophyll forest, dry sclerophyll forest, grassy woodland and semiarid woodland). Further details of the study design and data were provided in Li et al. [37], Li et al. [38] and Blackman et al. [39]. Images of the leaves of each species are provided in Figure S1. Saplings of the eight eucalypt species were progressively dried and coupled measurements of leaf water potential (Ψ_{leaf}) and live fuel moisture content (LFMC) of foliage were taken periodically at pre-dawn and midday during the imposed drought. Measurements were undertaken on >56 leaves per species and >7 individuals per species. During the drought, the canopy leaf area that each plant lost progressively to leaf shedding was calculated from measurements of the dry weight of shed leaves and the mean leaf mass per area (LMA) of foliage sampled prior to the drought treatment. For each species, the Ψ_{leaf} value associated with initiation of leaf shedding was calculated by averaging Ψ_{leaf} values when some leaf shedding had occurred but >90% of the plant leaf area was still present. Leaf hydraulic vulnerability to drought-induced embolism was also measured (see [38]).

4. From Relative Water Content to Live Fuel Moisture Content

The foundation of the linkage between drought and fire is the moisture content of fine fuels (e.g., foliage, twigs). In the drought literature, the water content of foliage is characterised as relative water content (RWC), whereas in the fire literature it is characterised as live fuel moisture content (LFMC). However, these metrics are two sides of the same coin. Both RWC and LFMC quantify the mass of water in foliage, with RWC expressing this mass relative to saturated water content, whereas LFMC expresses this mass relative to foliar dry weight (Equations (1) and (2), respectively). The similarity of these two metrics means that the response of LFMC to drying soils can be modelled in the exact same way that RWC is modelled. In the drought literature, RWC is commonly modelled as a function of leaf water potential (Ψ_{leaf}). This relationship is characterised by the pressure–volume curve [34], which is a fundamental method of assessing drought tolerance [40].

$$\text{RWC} = \left(\frac{F_w - D_w}{T_w - D_w}\right)\cdot 100 \tag{1}$$

$$\text{LFMC} = \left(\frac{F_w - D_w}{D_w}\right)\cdot 100 \tag{2}$$

where F_w is the fresh weight (i.e., weight prior to rehydration), D_w is the dry weight and T_w is the turgid (or saturated) weight of the fuel (that is, leaf or shoot).

Pressure–volume curves are typically derived from repeated measurements of Ψ_{leaf} and RWC on a cut leaf or shoot dehydrating on a bench. As leaves dehydrate, cell volume shrinks, turgor pressure decreases and osmotic potential (Ψ_π), and thus Ψ_{leaf}, decline [34]. The curve is obtained by plotting $-1/\Psi_{\text{leaf}}$ as a function of RWC (Figure 2a). Above the turgor loss point (Ψ_{TLP}), the curve is non-linear, but it approaches a linear relationship as $-1/\Psi_{\text{leaf}}$ falls below the Ψ_{TLP} [34]. This relationship can be reformulated to express LFMC as a function of Ψ_{leaf} by simply replacing RWC with LFMC as follows (Figure 2b):

$$\text{LFMC} = m_a\,\Psi_{\text{leaf}} + c_a \text{ for } \Psi_{\text{leaf}} > \Psi_{\text{TLP}} \tag{3}$$

$$\text{LFMC} = \left(\left(\frac{-1}{\Psi_{\text{leaf}}}\right) - c_b\right)/m_b \text{ for } \Psi_{\text{leaf}} < \Psi_{\text{TLP}} \tag{4}$$

where m_a and c_a are the slope and intercept for the linear model of LFMC and Ψ_{leaf} above the Ψ_{TLP}, and m_b and c_b are regression coefficients for the non-linear model of LFMC and $-1/\Psi_{\text{leaf}}$ below the Ψ_{TLP}, respectively. Note, m_b and c_b can be calculated from a linear regression if Equation (4) is rearranged to model $-1/\Psi_{\text{leaf}}$ as a function of LFMC. Equations (3) and (4) will have the exact same form whether RWC or LFMC are used as the dependent variable, assuming there are no changes in leaf dry matter content (see Section 6).

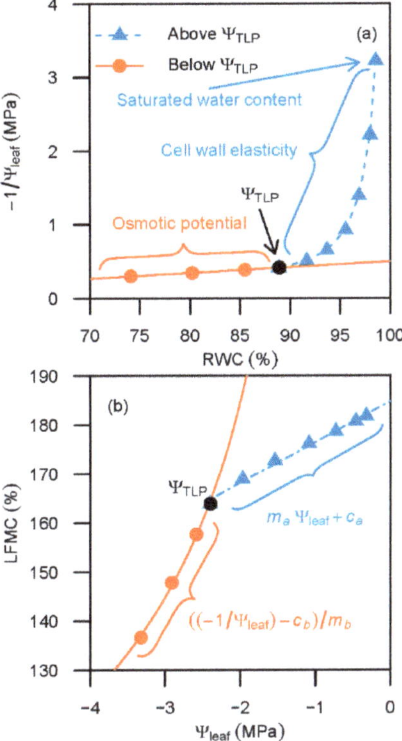

Figure 2. (**a**) Pressure–volume curve illustrating the non-linear relationship between $-1/\Psi_{\text{leaf}}$ and declining water status (RWC) (when $\Psi_{\text{leaf}} > \Psi_{\text{TLP}}$), and the linear relationship (when $\Psi_{\text{leaf}} < \Psi_{\text{TLP}}$). Additionally shown is the turgor loss point (Ψ_{TLP}), saturated water content, and region of the graph affected by cell wall elasticity and osmotic potential at full hydration. (**b**) Relationship between live fuel moisture content (LFMC) and Ψ_{leaf} above and below the Ψ_{TLP}, derived from pressure-volume curve relationships. Note, this is a theoretical relationship and not based on observations.

To date, there have been few studies modelling LFMC as a function of Ψ_{leaf} (but see [19,22]). Here, we use data from our case study (see Section 3) to demonstrate that declining LFMC during drought can be modelled from Ψ_{leaf} using Equations (3) and (4). Note, our data represent progressive measurements on multiple leaves during drought, rather than on a single leaf dehydrating on a bench. We modelled the decline in LFMC and Ψ_{leaf} using Equation (3) (for data $> \Psi_{\text{TLP}}$) and Equation (4) (for data $< \Psi_{\text{TLP}}$: $-1/\Psi_{\text{leaf}}$ versus LFMC). The transition between the two models (the Ψ_{TLP}) was estimated following Sack et al. [41], whereby the r^2 of the linear regression below the Ψ_{TLP} was maximised. The Ψ_{TLP} calculated in this way was similar to that calculated from traditional pressure-volume curves using excised leaves dehydrating on a bench (the mean absolute error was 0.19 MPa, Figure S2 in Supplementary Material).

For each of our eight species of eucalypt, we found that the model below the Ψ_{TLP} (i.e., Equation (4)) fit the data well: $r^2 = 0.77$–0.94, $p < 0.001$ (Figure 3). Above the Ψ_{TLP}, the regression slope was close to zero for many species, and so the fit of the linear models (i.e., Equation (3)) was relatively poor, as expected when regression slopes are at or near zero: $p > 0.05$ for five spp. and $p < 0.05$ for three spp.

($r^2 = 0.33$–0.44 for these spp.). However, the intercepts were always statistically significantly different from zero ($p < 0.001$). Despite the poor fit of the linear regression above the Ψ_{TLP} for some species, we think a linear model is still the best model to fit, since (a) there is a good theoretical basis for doing so [34]; and (b) the data do not exhibit a non-linear relationship. As discussed later in Section 7.2, modelling LFMC above ~150–200%, which is above the value of LFMC at the Ψ_{TLP} for these species, is of relatively minor importance for predicting critical periods of live fuel moisture content, as most wildfires occur well below this value.

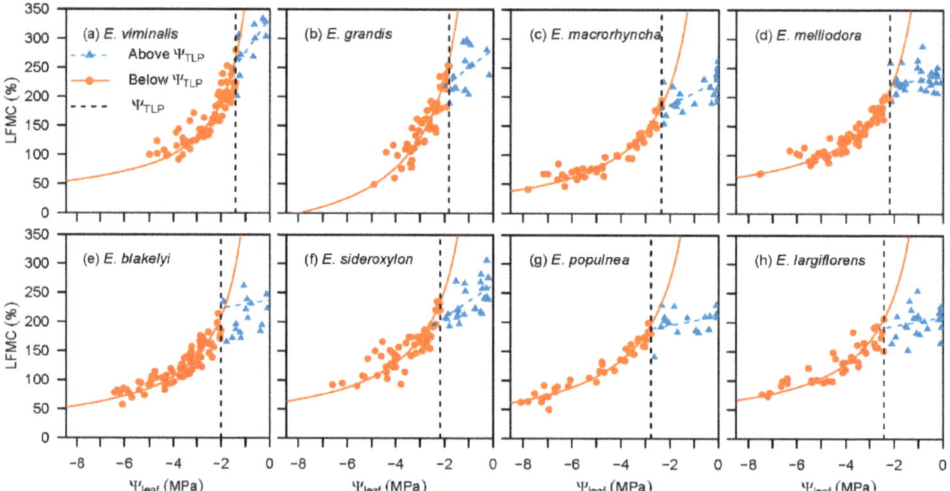

Figure 3. Decline in LFMC and Ψ_{leaf} modelled using Equation (3) (for data $> \Psi_{\mathrm{TLP}}$) and Equation (4) (for data $< \Psi_{\mathrm{TLP}}$: $-1/\Psi_{\mathrm{leaf}}$ versus LFMC), for saplings of eight *Eucalyptus* species. Species are ordered by increasing moisture index (ratio of precipitation to potential evapotranspiration) from climate of origin, i.e., *E. largiflorens* is from the most arid climate. Regressions for all species were statistically significant ($p < 0.05$), except the slope of the linear regression above the Ψ_{TLP} for the following species: *E. grandis*, *E. melliodora*, *E. blakelyi*, *E. populnea* and *E. largiflorens*. (**a**) *E. viminalis*; (**b**) *E. grandis*; (**c**) *E. macrorhyncha*; (**d**) *E. melliodora*; (**e**) *E. blakelyi*; (**f**) *E. sideroxylon*; (**g**) *E. populnea*; (**h**) *E. largiflorens*.

This case study demonstrates that LFMC can be modelled as a function of Ψ_{leaf} following the pressure–volume curve approach. There is an extensive literature that quantifies pressure–volume curve parameters, globally [40]. Utilising this literature to model LFMC offers the potential to rapidly develop models for the prediction of spatiotemporal change in LFMC across a range of ecosystems. While Ψ_{leaf} does not directly affect LFMC (rather, it is foliar water content that affects Ψ_{leaf}), developing a model of LFMC as a function of Ψ_{leaf} provides a framework for modelling LFMC from soil water content, which is discussed in the next section.

5. How Drought Models Can Inform Fire Models: Predicting Live Fuel Moisture Content

Leaf water potential is a key parameter for modelling carbon and water fluxes [42] and is now being implemented into land surface models [43]. These models have largely been developed to predict changes in carbon and water cycling due to drought but could be harnessed to predict live fuel moisture content. Figure 4 outlines the general framework for how Ψ_{leaf}, and subsequently LFMC, can be modelled from soil water content.

Plant water potential generally equilibrates with root-zone soil water potential (Ψ_{soil}) overnight [44]. For this reason, pre-dawn Ψ_{leaf} is frequently used as a proxy for Ψ_{soil}. Soil water potential can in turn be modelled from soil water content and basic soil hydraulic properties that govern the soil water retention curve, such as soil texture (see [42]). We note that in some circumstances, night-time transpiration, solute accumulation and other processes may affect the relationship between pre-dawn soil and leaf water potential [45]. Diurnal values of Ψ_{leaf} will fluctuate relative to Ψ_{soil} due to transpiration [42]. These Ψ_{leaf} fluctuations can be modelled from a soil-plant-atmosphere-continuum type of model. For example, Tuzet et al. [42] developed a coupled model of stomatal conductance, photosynthesis and transpiration that predicts diurnal values of Ψ_{leaf}. Required inputs for this model are Ψ_{soil}, vegetation attributes that control hydraulic conductance (leaf area index, canopy height, plant hydraulic resistance, canopy mixing length), stomatal conductance and atmospheric demand for water (Figure 4). A similar approach has been proposed for modelling LFMC of *Calluna*-dominated heathlands in the United Kingdom [46].

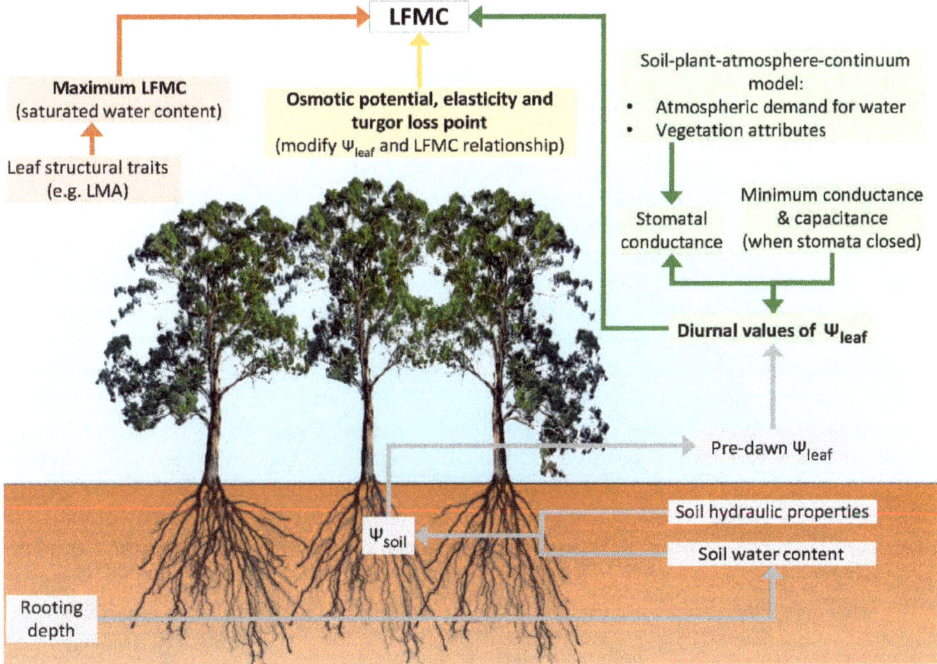

Figure 4. Overview of plant structural and physiological traits which modify the relationship between live fuel moisture content (LFMC) and soil water content.

While diurnal variation in LFMC is largely a function of the stomatal regulation of water loss, water loss still occurs following stomatal closure through the cuticle and incompletely closed stomata [47]. Thus, diurnal values of Ψ_{leaf} will be dependent on rates of minimum conductance following stomatal closure, which is particularly relevant during drought. However, the processes controlling leaf desiccation in very dry soil are poorly understood compared to the stomatal regulation of Ψ_{leaf} [18], and thus are a critical knowledge gap for modelling both drought vulnerability and diurnal LFMC under extreme drought conditions.

6. Drought-Related Plant Traits Determine the Response of Live Fuel Moisture Content to Drying Soil

Drought-related plant traits, such as rooting depth and the leaf traits which modify the relationship between RWC and Ψ_{leaf}, affect the development of critically low values of LFMC (Figure 2a).

Thus, existing models for the prediction of live fuel moisture content models may be improved by incorporating these traits (Figure 4). There is already recognition that plant traits play a role in forest flammability. For example, leaf size and shape affect the packing of litter beds, which in turn affects ignitability (i.e., ease of ignition; [36,48]). Similarly, leaf mass per area of live foliage affects ignitability [49,50]. We now discuss the role of plant traits in determining LFMC dynamics.

Rooting depth determines access to water resources, and will therefore influence seasonal and inter-annual LFMC dynamics. For example, in Mediterranean environments, tree species typically have access to deeper soil water or ground water reserves than co-occurring shrubs, and consequently exhibit less seasonal variation in LFMC [19,51]. In addition to lifeform, rooting depth is often related to post-fire regeneration strategy. For example, species that can resprout following high intensity fire typically have greater allocation to roots and deeper root systems than species lacking this capacity, and consequently exhibit less seasonal variation in LFMC than non-resprouting species [19,24].

One of the central leaf traits characterising physiological responses to soil dryness is the turgor loss point (Ψ_{TLP}), which defines the operating range of water potentials that plants use to control moisture content [52]. Above the Ψ_{TLP} the rate of decline of Ψ_{leaf} with RWC is largely dependent on cell wall elasticity [34]. Below the Ψ_{TLP}, cell walls are relaxed and the rate of decline in Ψ_{leaf} with RWC is dependent upon the concentration of solutes in cells, which is characterised by osmotic potential. While the Ψ_{TLP}, cell-wall elasticity and osmotic potential at full turgor control the rate of decline in LFMC with Ψ_{leaf}, saturated water content affects the absolute value of LFMC. Saturated water content is analogous to maximum LFMC and is negatively correlated with leaf structural properties, including leaf mass per area (LMA), leaf thickness and leaf density [35]. Here, we found that maximum LFMC from each of the eight *Eucalyptus* species in the common garden study declined with increasing LMA (Figure 5a). This relationship between LFMC and LMA is expected, since both traits incorporate leaf dry mass, and will therefore be auto-correlated. Thus, the key leaf traits that determine variation in the relationship between soil water content and LFMC are the Ψ_{TLP}, LMA, leaf elasticity and osmotic potential at full hydration.

These plant traits are known to vary along environmental gradients. The Ψ_{TLP}, leaf cell wall elasticity and osmotic potential at full hydration all generally decline with site water availability, enabling plants to continue gas exchange during periods of soil water deficit [40]. Our case study results are largely consistent with this observation. We examined the relationship between leaf traits and the climatic moisture index of the location of origin of each species (obtained from the Atlas of Living Australia website at http://www.ala.org.au). We found that: (i) the Ψ_{TLP} increased with the climatic moisture index (Figure 5b); (ii) above the Ψ_{TLP}, the slope of Ψ_{leaf} versus LFMC (indicative of cell wall elasticity) increased with the moisture index (Figure 5c); and (iii) below the Ψ_{TLP}, the slope of $-1/\Psi_{leaf}$ versus LFMC (indicative of osmotic potential at full hydration) largely increased with the moisture index, although this correlation was not significant (Figure 5d). Plants can exhibit some plasticity in these traits through solute accumulation during drought or from wet to dry seasons [53]. Thus, there may be some variability in the relationship between LFMC and Ψ_{leaf} through time due to osmotic adjustment.

LMA also varies along environmental gradients, particularly light, temperature and nutrient and water availability [54]. Here, we found that LMA from our case study *Eucalyptus* species increased with declining moisture availability from their climate of origin (Figure 5e). LMA also increases during leaf maturation [55]. This effect of leaf age has been associated with seasonal declines in conifer LFMC [56]. Therefore, we suggest that LFMC models may be improved by taking seasonal variation in LMA into account.

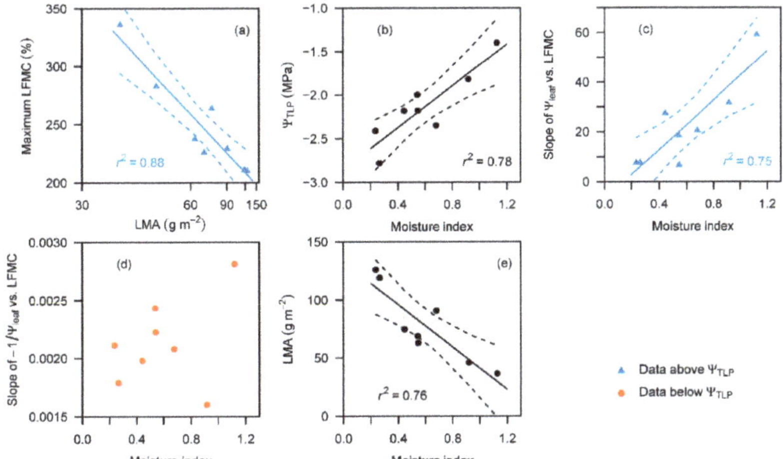

Figure 5. Relationship between (**a**) maximum live fuel moisture content (LFMC) and average leaf mass per area (LMA) in eight *Eucalyptus* species in a common garden experiment; and (**b–e**) plant ecophysiological traits and the moisture index from the climate of origin (calculated as the ratio of precipitation to potential evapotranspiration), including (**b**) the turgor loss point (Ψ_{TLP}), (**c**) the slope of Ψ_{leaf} versus LFMC above the Ψ_{TLP} (analogous to cell wall elasticity), (**d**) the slope of $-1/\Psi_{leaf}$ versus LFMC below the Ψ_{TLP} (analogous to osmotic potential at full hydration), and (**e**) average LMA. Note, the same LMA values are used in (**a,e**), but the data are transformed by a negative reciprocal transformation in (**a**). Dashed lines represent the 95% confidence interval of each regression.

7. How Does Physiological Drought Stress Affect Fuel Availability

During the early stages of drought, declining soil water content will affect live fuel moisture content, as well as the moisture content of surface dead fuels. As the drought progresses, the onset of leaf shedding and, eventually, tree death will have major implications for fuel properties. Currently, there is a concerted research effort to quantify and predict thresholds in leaf- and branch-level die-back, and whole forest mortality [17,18]. However, linkages between thresholds in drought mortality and thresholds in wildfire risk have not been explicitly examined to date. Thus, joint research on drought and forest flammability presents a major opportunity to inform predictive models of wildfire risk. Thresholds in plant vulnerability to drought are typically calculated from leaf or xylem water potential [18] or drought indices [57], while thresholds in wildfire risk are typically calculated from observed relationships between the area burnt by wildfire and fuel moisture content across large spatial areas (e.g., [9,58]).

7.1. Influence of Physiological Drought Stress on the Distribution of Dead Fuels

Plant vulnerability to cavitation is a major predictor of drought-induced mortality [18], and is therefore likely to affect the amount and spatial distribution of dead fuels within a forest. When cavitation is severe enough to trigger leaf death, it results in the transformation of live fuels into dead fuels. While they are retained within the canopy, these dead fuels decline to moisture contents well below those of live fuels. These senescent canopy fuels may therefore increase the probability of crown fire [9,10,36]. When these dead fuels are shed from the canopy, there is an influx of litter to the surface fuel bed. Initially, these litter fuels are likely to be relatively uncompacted, and thus well-aerated and available to burn [36]. Thus, leaf shedding may potentially increase the likelihood of surface fires, although the likelihood of crown fires may decrease due to the lowered overall fuel load within the tree canopy. Thus, we hypothesise that physiological drought stress can lead to an increase in the probability of large forest fires.

There is some evidence for this hypothesis. For example, leaf shedding is known to occur during infrequent drought events in eucalypt forests [59,60], and this coincides with periods of peak wildfire activity in these forests ([14]; Figure 6a). Similarly, in Amazonian forests, leaf shedding during severe drought events coincides with increased fire activity [61]. There is also some evidence that leaf death and subsequent shedding due to causes other than drought-induced cavitation can affect flammability. For example, in Mediterranean *Pinus halepensis* stands, needle senescence occurs during late June and mid-July, which coincides with peak fire activity in the region ([3]; Figure 6b). In contrast, in western North America, defoliation due to bark beetles does not result in an increase in the area burnt by wildfire [62], and may even reduce fire severity (i.e., fuel consumed) due to decreased canopy fuel loads [63]. Further, while seasonal leaf senescence may coincide with peak fire activity in many regions, this is not necessarily evidence that physiological drought stress is a causal factor of wildfires. Rather, physiological drought stress may simply be correlated with the causal factors of fire intensity, such as high atmospheric evaporative demand.

Figure 6. (**a**) Eucalypt woodland in north-eastern New South Wales, Australia, during severe drought in October 2019, illustrating the conversion of live fuels (i.e., foliage) into dead fuels. In this region, over five million hectares was subject to wildfire in 2019/2020. (**b**) Seasonal needle senescence in *Pinus halepensis* in Catalonia, Spain. Several days following this photo (22 June 2019), large fires occurred nearby. Image credit: Carles Arteaga.

7.2. Co-Occurrence of Thresholds in Physiological Drought Stress and Wildfire

Thresholds in drought vulnerability vary substantially among species, which may manifest in different thresholds in wildfire risk across biomes. The potential linkages between thresholds in drought vulnerability and wildfire risk have so far not been explicitly examined. A key barrier to investigating this linkage is the differing metrics and terminology used in the drought and fire research disciplines, e.g., RWC and LFMC. By developing a model of LFMC based on Ψ_{leaf}, we can overcome this barrier, and begin examining whether drought and fire thresholds co-occur. Examining the relationship between physiological drought stress and wildfire may potentially lead to new insights and hypotheses about the mechanisms underlying fire occurrence, and development of more reliable tools for predicting the risk of large wildfires.

Several studies have demonstrated that when fuel moisture content declines below a threshold value, there is a significant increase in the area burnt by wildfire. For live fuels, thresholds of 70–95% have been identified for Mediterranean shrublands [58,64,65] and 100–120% for forests [9,66]. We note that a recent discussion has emerged in the literature on the methods used to estimate these critical LFMC thresholds, with alternate methods likely to result in small changes in these threshold values (e.g., a ~10% difference in thresholds was observed in Mediterranean shrublands, depending on the method applied; [65]).

Using our common garden case study, we were able to calculate the value of LFMC when the eucalypt saplings were experiencing drought stress, and compare these values of LFMC with critical thresholds of LFMC identified for landscape-scale fires in eucalypt forests and woodlands across south-eastern Australia [9]. We calculated the value of LFMC at three metrics of physiological drought stress: (i) leaf P_{50}, which is a measure of the Ψ_{leaf} corresponding to a 50% decline in maximum leaf hydraulic conductance, and is commonly used to compare leaf hydraulic vulnerability among species [67]; (ii) leaf P_{88}, which is a measure of the Ψ_{leaf} corresponding to an 88% decline in maximum leaf hydraulic conductance and may better represent critical hydraulic failure [68]; and (iii) the value of LFMC associated with the initiation of leaf shedding (defined here as 10% of total canopy leaf shedding, see Section 3). We found that the value of LFMC at leaf P_{50} was largely above the critical threshold of ~102% identified for landscape-scale fire in eucalypt forests and woodlands ([9]; 91–120%, Figure 7a). However, the value of LFMC at leaf P_{88} was at or below this critical threshold (77–102%; Figure 7b). We also found that for all except the most mesic species, the LFMC value corresponding to the initiation of leaf shedding was at or below this same critical threshold of 102% (Figure 5c).

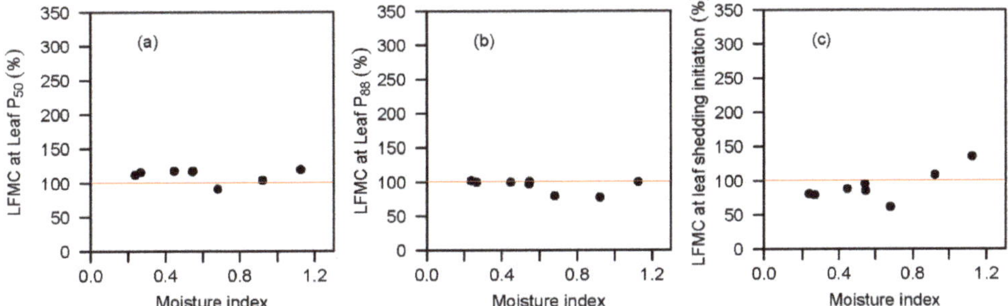

Figure 7. Value of LFMC at (**a**) leaf P_{50}, (**b**) leaf P_{88} and (**c**) initiation of leaf shedding (when up to 10% leaf shedding had occurred) for eight species of *Eucalyptus*. The *y*-axis range is the same as for Figure 3, to illustrate the range of LFMC values observed in the case study. The horizontal lines represent the LFMC threshold (~102%) associated with a step-change in fire activity for south-eastern Australian eucalypt forests and woodlands (see [9]).

It is remarkable that the value of LFMC at critical periods of drought stress in eucalypt saplings aligns so closely with the critical LFMC threshold, leading to a step-change in the area burnt by wildfire across south-eastern Australian forests and woodlands. These results support our hypothesis that physiological drought stress contributes to an increased probability of large forest fires. We note that other factors are important in contributing to these LFMC thresholds, in particular the increasing connectivity between patches of dry fuel that occurs across the landscape as LFMC declines, e.g., the drying of gullies which would otherwise act as a fire break [69].

8. Bridging the Gap between the Drought and Fire Literature

We have demonstrated that plant responses to drought affect fuel attributes, and thus may exert an important influence on the probability of wildfire. The establishment of better links between ecophysiological and fire behaviour research communities has the potential to transform knowledge of fire dynamics. We propose three major research directions:

1. Build connections between the drought and fire literature, in particular, identify ways to translate between the different measures used in each, e.g., relative water content and live fuel moisture content.
2. Utilise ecophysiological principles and metrics of drought vulnerability to develop new, predictive models of fuel dynamics.

3. Investigate the application of physiological knowledge of critical properties of plants for fire behaviour modelling.

The potential applications of this proposed research include (i) the development of new types of live fuel moisture content models which do not require species-specific calibrations; (ii) the capacity to model live fuel moisture content under future climatic conditions; (iii) the potential to derive new insights into the mechanisms underlying major wildfire events; and (iv) the development of physiologically based thresholds of forest flammability.

There is a large body of literature quantifying pressure–volume curve parameters (e.g., turgor loss point, osmotic potential and elasticity) globally [40]. Applying this research to model LFMC as a function of Ψ_{leaf} requires conversion from RWC to LFMC. One method to convert between RWC and LFMC is to develop a model of maximum LFMC, i.e., when RWC is 100%. The value of LFMC at critical values of RWC can then be readily calculated. For example, the value of RWC at the turgor loss point is commonly calculated in pressure–volume curves, and can be easily converted to LFMC if the maximum LFMC is known for a given species. Given that maximum LFMC is auto-correlated with LMA (see Section 6), and LMA is a common and easily measurable trait [54], quantifying the relationship between maximum LFMC and LMA would provide a pathway to rapidly convert between RWC and LFMC across many species globally. Thus, research on the leaf structural and environmental drivers of maximum LFMC is required to bridge the gap between the drought and fire literature.

While we have established that there is a connection between plant responses to drought and forest flammability, applying this research to fire behaviour modelling requires spatially explicit models of relevant plant traits, to model plant responses to variation in soil water content, which is not trivial. An important challenge to developing a physiologically based model of LFMC as a function of soil water content will be characterising rooting depth. In particular, characterising whether species have access to groundwater resources, which can buffer LFMC against seasonal variation in soil moisture content [19]. Vegetation access to groundwater can be inferred from remotely sensed observations of canopy greenness and surface temperature [70]. Characterising access to water resources among co-occurring species may be inferred from plant function types, e.g., trees versus shrubs, or post-fire resprouting versus non-resprouting species. We suggest further studies to test the generality of these relationships between plant functional types, access to water resources and seasonal variability in LFMC dynamics across biomes.

A further challenge will be modelling the leaf-level traits which govern responses to declining soil water content. We suggest modelling the key leaf-level traits as a function of environmental gradients, in particular aridity and soil nutrient content. However, many of these traits can vary seasonally and inter-annually, as a function of phenology or climate. For the purposes of modelling wildfire risk, these seasonal dynamics in leaf traits may be somewhat unimportant, given that wildfires typically only occur during particular seasons. Thus, research efforts to quantify these plant traits should prioritise measurements during the fire season.

We hypothesise that leaf death and subsequent shedding as a result of drought-induced cavitation may affect fire behaviour in multiple, opposing directions. Clearly, this hypothesis requires further research, and would benefit greatly from quantification of the extent and timing of leaf death and subsequent shedding. It has been hypothesised that leaf death due to drought occurs as a protective mechanism to delay dangerous cavitation within stems; however, this is not consistent among species [71]. Furthermore, the environmental cues that trigger seasonal leaf death and shedding have often not been well characterised. Thus, further studies on the mechanisms underlying leaf death, in addition to observations on the extent of leaf death and shedding during drought are suggested.

Further studies are also required to assess the linkages between physiological drought stress and wildfire occurrence. We suggest additional research focuses on quantifying the distribution of LFMC at critical periods of drought stress (i.e., critical thresholds of leaf/stem cavitation), and comparing these values with observed LFMC thresholds that lead to a step change in area burnt by wildfire. Establishing a physiological basis for thresholds in wildfire occurrence would enable the quantification and prediction of wildfire risk across biomes, without the need for local empirical modelling. Investigation of the relationship between physiological drought stress and wildfire occurrence may also be facilitated by remotely sensed metrics of vegetation water stress. An indicator of vegetation water stress is when evapotranspiration declines below seasonal averages [72]. Recent advances in remote sensing have enabled estimation of evapotranspiration over large spatial scales at high spatial resolution [73]. Linking remotely sensed evapotranspiration with drought metrics derived from precipitation and potential evapotranspiration would enable large-scale investigation of the relationships between vegetation drought stress and wildfire activity.

Supplementary Materials: The following are available online at http://www.mdpi.com/1999-4907/11/7/779/s1, Figure S1: Scanned images of the leaves of each species used in the common garden study. Not to scale.; Figure S2: Turgor loss point ($\Psi_{TLP} \pm 1SE$) of eight *Eucalyptus* species, calculated from standard pressure-volume curves (bench dehydration method) and Ψ_{TLP} calculated from progressive observations of leaf water potential (Ψ_{leaf}) and live fuel moisture content (LFMC) of plants subject to imposed drought. Mean absolute error is 0.19 MPa.

Author Contributions: R.H.N., C.J.B., V.R.d.D., B.C., B.E.M., R.A.B. and M.M.B. developed the ideas, C.J.M. and X.L. contributed the case study data. R.H.N. wrote the manuscript. All authors have read and agree to the published version of the manuscript.

Funding: We thank the New South Wales Government's Department of Planning, Industry and Environment for providing funds to support this research via the NSW Bushfire Risk Management Research Hub; the Spanish Government (RYC-2012-10970, AGL2015-69151-R); an Australian Research Council Linkage grant with the New South Wales Department of Planning, Industry and Environment (LP140100232); and an Australian Research Council Future Fellowship (FT130101115).

Conflicts of Interest: The authors declare no conflict of interest.

References

1. Pausas, J.G.; Keeley, J.E. A Burning Story: The Role of Fire in the History of Life. *Bioscience* **2009**, *59*, 593–601. [CrossRef]

2. Enright, N.J.; Fontaine, J.B.; Bowman, D.; Bradstock, R.A.; Williams, R.J. Interval squeeze: Altered fire regimes and demographic responses interact to threaten woody species persistence as climate changes. *Front. Ecol. Environ.* **2015**, *13*, 265–272. [CrossRef]

3. Karavani, A.; Boer, M.M.; Baudena, M.; Colinas, C.; Diaz-Sierra, R.; Peman, J.; de Luis, M.; Enriquez-de-Salamanca, A.; de Dios, V.R. Fire-induced deforestation in drought-prone Mediterranean forests: Drivers and unknowns from leaves to communities. *Ecol. Monogr.* **2018**, *88*, 141–169. [CrossRef]

4. Tepley, A.J.; Thompson, J.R.; Epstein, H.E.; Anderson-Teixeira, K.J. Vulnerability to forest loss through altered postfire recovery dynamics in a warming climate in the Klamath Mountains. *Glob. Chang. Biol.* **2017**, *23*, 4117–4132. [CrossRef] [PubMed]

5. Fairman, T.A.; Nitschke, C.R.; Bennett, L.T. Too much, too soon? A review of the effects of increasing wildfire frequency on tree mortality and regeneration in temperate eucalypt forests. *Int. J. Wildland Fire* **2016**, *25*, 831–848. [CrossRef]

6. Balshi, M.S.; McGuire, A.D.; Duffy, P.; Flannigan, M.; Kicklighter, D.W.; Melillo, J. Vulnerability of carbon storage in North American boreal forests to wildfires during the 21st century. *Glob. Chang. Biol.* **2009**, *15*, 1491–1510. [CrossRef]

7. De Sales, F.; Okin, G.S.; Xue, Y.K.; Dintwe, K. On the effects of wildfires on precipitation in Southern Africa. *Clim. Dyn.* **2019**, *52*, 951–967. [CrossRef]

8. Nelson, R.M.J. Water relations of forest fuels. In *Forest Fires: Behavior and Ecological Effects*; Johnson, E.A., Miyanishi, K., Eds.; Academic Press: London, UK, 2001; pp. 79–149.

9. Nolan, R.H.; Boer, M.M.; Resco de Dios, V.; Caccamo, G.; Bradstock, R.A. Large-scale, dynamic transformations in fuel moisture drive wildfire activity across southeastern Australia. *Geophys. Res. Lett.* **2016**, *43*, 4229–4238. [CrossRef]

10. Van Wagner, C.E. Conditions for the start and spread of crown fire. *Can. J. For. Res.* **1977**, *71*, 23–34. [CrossRef]

11. Rothermel, R.C. *A Mathematical Model for Predicting Fire Spread in Wildland Fuels*; Research Paper INT-115; USDA: Ogden, UT, USA, 1972.

12. Rossa, C.G.; Fernandes, P.M. Live fuel moisture content: The 'pea under the mattress' of fire spread rate modeling? *Fire* **2018**, *1*, 43. [CrossRef]

13. Coen, J.L.; Stavros, E.N.; Fites-Kaufman, J.A. Deconstructing the King megafire. *Ecol. Appl.* **2018**, *28*, 1565–1580. [CrossRef] [PubMed]

14. Bradstock, R.A. A biogeographic model of fire regimes in Australia: Current and future implications. *Glob. Ecol. Biogeogr.* **2010**, *19*, 145–158. [CrossRef]

15. Ruthrof, K.X.; Fontaine, J.B.; Matusick, G.; Breshears, D.D.; Law, D.J.; Powell, S.; Hardy, G. How drought-induced forest die-off alters microclimate and increases fuel loadings and fire potentials. *Int. J. Wildland Fire* **2016**, *25*, 819–830. [CrossRef]

16. Stephens, S.L. Fuel loads, snag abundance, and snag recruitment in an unmanaged Jeffrey pine-mixed conifer forest in Northwestern Mexico. *For. Ecol. Manag.* **2004**, *199*, 103–113. [CrossRef]

17. Allen, C.D.; Breshears, D.D.; McDowell, N.G. On underestimation of global vulnerability to tree mortality and forest die-off from hotter drought in the Anthropocene. *Ecosphere* **2015**, *6*, 129. [CrossRef]

18. Choat, B.; Brodribb, T.J.; Brodersen, C.R.; Duursma, R.A.; López, R.; Medlyn, B.E. Triggers of tree mortality under drought. *Nature* **2018**, *558*, 531–539. [CrossRef] [PubMed]

19. Nolan, R.H.; Hedo, J.; Arteaga, C.; Sugai, T.; Resco de Dios, V. Physiological drought responses improve predictions of live fuel moisture dynamics in a Mediterranean forest. *Agric. For. Meteorol.* **2018**, *263*, 417–427. [CrossRef]

20. Jolly, W.M.; Hadlow, A.M.; Huguet, K. De-coupling seasonal changes in water content and dry matter to predict live conifer foliar moisture content. *Int. J. Wildland Fire* **2014**, *23*, 480–489. [CrossRef]

21. Jolly, W.; Johnson, D. Pyro-ecophysiology: Shifting the paradigm of live wildland fuel research. *Fire* **2018**, *1*, 8. [CrossRef]

22. Pivovaroff, A.L.; Emery, N.; Sharifi, M.R.; Witter, M.; Keeley, J.E.; Rundel, P.W. The effect of ecophysiological traits on live fuel moisture content. *Fire* **2019**, *2*, 28. [CrossRef]

23. Yebra, M.; Dennison, P.E.; Chuvieco, E.; Riaño, D.; Zylstra, P.; Hunt, E.R.; Danson, F.M.; Qi, Y.; Jurdao, S. A global review of remote sensing of live fuel moisture content for fire danger assessment: Moving towards operational products. *Remote Sens. Environ.* **2013**, *136*, 455–468. [CrossRef]

24. Ruffault, J.; Martin-StPaul, N.; Pimont, F.; Dupuy, J.-L. How well do meteorological drought indices predict live fuel moisture content (LFMC)? An assessment for wildfire research and operations in Mediterranean ecosystems. *Agric. For. Meteorol.* **2018**, *262*, 391–401. [CrossRef]

25. Caccamo, G.; Chisholm, L.A.; Bradstock, R.A.; Puotinen, M.L.; Pippen, B.G. Monitoring live fuel moisture content of heathland, shrubland and sclerophyll forest in south-eastern Australia using MODIS data. *Int. J. Wildland Fire* **2012**, *21*, 257–269. [CrossRef]

26. Fernandes, P.M.; Cruz, M.G. Plant flammability experiments offer limited insight into vegetation-fire dynamics interactions. *New Phytol.* **2012**, *194*, 606–609. [CrossRef] [PubMed]

27. Plucinski, M.P.; Anderson, W.R.; Bradstock, R.A.; Gill, A.M. The initiation of fire spread in shrubland fuels recreated in the laboratory. *Int. J. Wildland Fire* **2010**, *19*, 512–520. [CrossRef]

28. Schlobohm, P.; Brain, J. *Gaining an Understanding of the National Fire Danger Rating System*; National Wildfire Coordinating Group, United States Department of Agriculture, United States Department of the Interior, National Association of State Foresters: Boise, ID, USA, 2002.

29. Gill, A.M.; Zylstra, P. Flammability of Australian forests. *Aust. For.* **2005**, *68*, 87–93. [CrossRef]

30. Boer, M.M.; Bowman, D.; Murphy, B.P.; Cary, G.J.; Cochrane, M.A.; Fensham, R.J.; Krawchuk, M.A.; Price, O.F.; De Dios, V.R.; Williams, R.J.; et al. Future changes in climatic water balance determine potential for transformational shifts in Australian fire regimes. *Environ. Res. Lett.* **2016**, *11*, 13. [CrossRef]

31. Moritz, M.A.; Parisien, M.A.; Batllori, E.; Krawchuk, M.A.; Van Dorn, J.; Ganz, D.J.; Hayhoe, K. Climate change and disruptions to global fire activity. *Ecosphere* **2012**, *3*, 22. [CrossRef]

32. Cruz, M.G.; Gould, J.S.; Alexander, M.E.; Sullivan, A.L.; McCaw, W.L.; Matthews, S. CSIRO Land and Water Flagship, Canberra, ACT, and AFAC, Melbourne, Australia. In *A Guide to Rate of Fire Spread Models for Australian Vegetation*; CSIRO Land and Water Flagship, Canberra, ACT, and AFAC: Melbourne, VIC, Australia, 2015; p. 123.

33. Stocks, B.J.; Lawson, B.D.; Alexander, M.E.; Van Wagner, C.E.; McAlpine, R.S.; Lynham, T.J.; Dube, D.E. Canadian Forest Fire Danger Rating System: An overview. *For. Chron.* **1989**, *65*, 258–265. [CrossRef]

34. Tyree, M.T.; Hammel, H.T. The Measurement of turgor pressure and water relations of plants by pressure bomb technique. *J. Exp. Bot.* **1972**, *23*, 267–281. [CrossRef]

35. Prior, L.D.; Eamus, D.; Bowman, D. Leaf attributes in the seasonally dry tropics: A comparison of four habitats in northern Australia. *Funct. Ecol.* **2003**, *17*, 504–515. [CrossRef]

36. Cornelissen, J.H.C.; Grootemaat, S.; Verheijen, L.M.; Cornwell, W.K.; van Bodegom, P.M.; van der Wal, R.; Aerts, R. Are litter decomposition and fire linked through plant species traits? *New Phytol.* **2017**, *216*, 653–669. [CrossRef] [PubMed]

37. Li, X.; Blackman, C.J.; Choat, B.; Duursma, R.A.; Rymer, P.D.; Medlyn, B.E.; Tissue, D.T. Tree hydraulic traits are coordinated and strongly linked to climate-of-origin across a rainfall gradient. *Plant Cell Environ.* **2018**, *41*, 646–660. [CrossRef] [PubMed]

38. Li, X.; Chris, B.J.; Peters, J.M.R.; Choat, B.; Rymer, P.D.; Medlyn, B.E.; Tissue, D.T. More than iso/anisohydry: Hydroscapes integrate plant water-use and drought tolerance traits in ten eucalypt species from contrasting climates. *Funct. Ecol.* **2019**, *33*, 1035–1049. [CrossRef]

39. Blackman, C.J.; Li, X.; Choat, B.; Rymer, P.D.; De Kauwe, M.G.; Duursma, R.A.; Tissue, D.T.; Medlyn, B.E. Desiccation time during drought is highly predictable across tree species from contrasting climates. *New Phytol.* **2019**. [CrossRef] [PubMed]

40. Bartlett, M.K.; Scoffoni, C.; Sack, L. The determinants of leaf turgor loss point and prediction of drought tolerance of species and biomes: A global meta-analysis. *Ecol. Lett.* **2012**, *15*, 393–405. [CrossRef] [PubMed]

41. Sack, L.; Pasquet-Kok, J.; Nicotra, A. Leaf Pressure-Volume Curve Parameters. Available online: http://prometheuswiki.org/tiki-index.php?page=Leaf+pressure-volume+curve+parameters&highlight=pressure%20volume%20curves (accessed on 30 October 2018).

42. Tuzet, A.; Perrier, A.; Leuning, R. A coupled model of stomatal conductance, photosynthesis and transpiration. *Plant Cell Environ.* **2003**, *26*, 1097–1116. [CrossRef]

43. Kennedy, D.; Swenson, S.; Oleson, K.W.; Lawrence, D.M.; Fisher, R.; Lola da Costa, A.C.; Gentine, P. Implementing Plant Hydraulics in the Community Land Model, Version 5. *J. Adv. Model. Earth Syst.* **2019**, *11*, 485–513. [CrossRef]

44. Hinckley, T.M.; Lassoie, J.P.; Running, S.W. Temporal and spatial variations in water status of forest trees. *For. Sci.* **1978**, *24*, 1–72.

45. Donovan, L.A.; Linton, M.J.; Richards, J.H. Predawn plant water potential does not necessarily equilibrate with soil water potential under well-watered conditions. *Oecologia* **2001**, *129*, 328–335. [CrossRef]

46. Davies, G.M.; Legg, C.J. Developing a live fuel moisture model for moorland fire danger rating. In *Modelling, Monitoring and Management of Forest Fires*; DeLasHeras, J., Brebbia, C.A., Viegas, D., Leone, V., Eds.; Wit Press: Southampton, UK, 2008; Volume 119, pp. 225–236.

47. Duursma, R.A.; Blackman, C.J.; López, R.; Martin-StPaul, N.K.; Cochard, H.; Medlyn, B.E. On the minimum leaf conductance: Its role in models of plant water use, and ecological and environmental controls. *New Phytol.* **2019**, *221*, 693–705. [CrossRef] [PubMed]

48. Anderson, H.E. Forest fuel ignitability. *Fire Technol.* **1970**, *6*, 312–319. [CrossRef]

49. Murray, B.R.; Hardstaff, L.K.; Phillips, M.L. Differences in Leaf Flammability, Leaf Traits and Flammability-Trait Relationships between Native and Exotic Plant Species of Dry Sclerophyll Forest. *PLoS ONE* **2013**, *8*, e079205. [CrossRef] [PubMed]

50. Grootemaat, S.; Wright, I.J.; van Bodegom, P.M.; Cornelissen, J.H.C.; Cornwell, W.K. Burn or rot: Leaf traits explain why flammability and decomposability are decoupled across species. *Funct. Ecol.* **2015**, *29*, 1486–1497. [CrossRef]

51. Viegas, D.X.; Pinol, J.; Viegas, M.T.; Ogaya, R. Estimating live fine fuels moisture content using meteorologically-based indices. *Int. J. Wildland Fire* **2001**, *10*, 223–240. [CrossRef]

52. Blackman, C.J. Leaf turgor loss as a predictor of plant drought response strategies. *Tree Physiol.* **2018**, *38*, 655–657. [CrossRef]

53. Bartlett, M.K.; Zhang, Y.; Kreidler, N.; Sun, S.W.; Ardy, R.; Cao, K.F.; Sack, L. Global analysis of plasticity in turgor loss point, a key drought tolerance trait. *Ecol. Lett.* **2014**, *17*, 1580–1590. [CrossRef]

54. Poorter, H.; Niinemets, U.; Poorter, L.; Wright, I.J.; Villar, R. Causes and consequences of variation in leaf mass per area (LMA): A meta-analysis. *New Phytol.* **2009**, *182*, 565–588. [CrossRef]

55. Wujeska-Klause, A.; Crous, K.Y.; Ghannoum, O.; Ellsworth, D.S. Leaf age and eCO2 both influence photosynthesis by increasing light harvesting in mature *Eucalyptus tereticornis* at EucFACE. *Environ. Exp. Bot.* **2019**, *167*, 103857. [CrossRef]

56. Jolly, W.M.; Hintz, J.; Linn, R.L.; Kropp, R.C.; Conrad, E.T.; Parsons, R.A.; Winterkamp, J. Seasonal variations in red pine (*Pinus resinosa*) and jack pine (*Pinus banksiana*) foliar physio-chemistry and their potential influence on stand-scale wildland fire behavior. *For. Ecol. Manag.* **2016**, *373*, 167–178. [CrossRef]

57. Mitchell, P.J.; O'Grady, A.P.; Hayes, K.R.; Pinkard, E.A. Exposure of trees to drought- induced die- off is defined by a common climatic threshold across different vegetation types. *Ecol. Evol.* **2014**, *4*, 1088–1101. [CrossRef]

58. Dennison, P.E.; Moritz, M.A. Critical live fuel moisture in chaparral ecosystems: A threshold for fire activity and its relationship to antecedent precipitation. *Int. J. Wildland Fire* **2009**, *18*, 1021–1027. [CrossRef]

59. Pook, E.W. Canopy dynamics of *Eucalyptus maculata* Hook. IV contrasting responses to two severe droughts. *Aust. J. Bot.* **1986**, *34*, 1–14. [CrossRef]

60. Duursma, R.A.; Gimeno, T.E.; Boer, M.M.; Crous, K.Y.; Tjoelker, M.G.; Ellsworth, D.S. Canopy leaf area of a mature evergreen *Eucalyptus* woodland does not respond to elevated atmospheric CO2 but tracks water availability. *Glob. Chang. Biol.* **2016**, *22*, 1666–1676. [CrossRef] [PubMed]

61. Nepstad, D.C.; Verissimo, A.; Alencar, A.; Nobre, C.; Lima, E.; Lefebvre, P.; Schlesinger, P.; Potter, C.; Moutinho, P.; Mendoza, E.; et al. Large-scale impoverishment of Amazonian forests by logging and fire. *Nature* **1999**, *398*, 505–508. [CrossRef]

62. Hart, S.J.; Schoennagel, T.; Veblen, T.T.; Chapman, T.B. Area burned in the western United States is unaffected by recent mountain pine beetle outbreaks. *Proc. Natl. Acad. Sci. USA* **2015**, *112*, 4375–4380. [CrossRef]

63. Meigs, G.W.; Zald, H.S.J.; Campbell, J.L.; Keeton, W.S.; Kennedy, R.E. Do insect outbreaks reduce the severity of subsequent forest fires? *Environ. Res. Lett.* **2016**, *11*, 045008. [CrossRef]

64. Jurdao, S.; Yebra, M.; Guerschman, J.P.; Chuvieco, E. Regional estimation of woodland moisture content by inverting Radiative Transfer Models. *Remote Sens. Environ.* **2013**, *132*, 59–70. [CrossRef]

65. Pimont, F.; Ruffault, J.; Martin-StPaul, N.K.; Dupuy, J.-L. A Cautionary Note Regarding the Use of Cumulative Burnt Areas for the Determination of Fire Danger Index Breakpoints. *Int. J. Wildland Fire* **2019**. [CrossRef]

66. Agee, J.K.; Wright, C.S.; Williamson, N.; Huff, M.H. Foliar moisture content of Pacific Northwest vegetation and its relation to wildland fire behavior. *For. Ecol. Manag.* **2002**, *167*, 57–66. [CrossRef]

67. Blackman, C.J.; Gleason, S.M.; Chang, Y.; Cook, A.M.; Laws, C.; Westoby, M. Leaf hydraulic vulnerability to drought is linked to site water availability across a broad range of species and climates. *Ann. Bot.* **2014**, *114*, 435–440. [CrossRef]

68. Urli, M.; Porté, A.; Cochard, H.; Guengant, Y.; Burlett, R.; Delzon, S. Xylem embolism threshold for catastrophic hydraulic failure in angiosperm trees. *Tree Physiol.* **2013**, *33*. [CrossRef] [PubMed]

69. Caccamo, G.; Chisholm, L.A.; Bradstock, R.A.; Puotinen, M.L. Using remotely-sensed fuel connectivity patterns as a tool for fire danger monitoring. *Geophys. Res. Lett.* **2012**, *39*. [CrossRef]

70. Eamus, D.; Fu, B.; Springer, A.E.; Stevens, L.E. Groundwater Dependent Ecosystems: Classification, Identification Techniques and Threats. In *Integrated Groundwater Management*; Jakeman, A.J., Barreteau, O., Hunt, A.J., Barreteau, O., Hunt, R.J., Rinaudo, J.D., Ross, A., Eds.; Springer International Publishing: Cham, Switzerland, 2016.

71. Wolfe, B.T.; Sperry, J.S.; Kursar, T.A. Does leaf shedding protect stems from cavitation during seasonal droughts? A test of the hydraulic fuse hypothesis. *New Phytol.* **2016**, *212*, 1007–1018. [CrossRef]

72. Chang, K.Y.; Xu, L.Y.; Starr, G.; Paw, U.K.T. A drought indicator reflecting ecosystem responses to water availability: The Normalized Ecosystem Drought Index. *Agric. For. Meteorol.* **2018**, *250*, 102–117. [CrossRef]

73. Fisher, J.B.; Lee, B.; Purdy, A.J.; Halverson, G.H.; Dohlen, M.B.; Cawse-Nicholson, K.; Wang, A.; Anderson, R.G.; Aragon, B.; Arain, M.A.; et al. ECOSTRESS: NASA's Next Generation Mission to Measure Evapotranspiration From the International Space Station. *Water Resour. Res.* **2020**, *56*, 20. [CrossRef]

Article

Needle Senescence Affects Fire Behavior in Aleppo Pine (*Pinus halepensis* Mill.) Stands: A Simulation Study

Rodrigo Balaguer-Romano [1,*], Rubén Díaz-Sierra [1], Javier Madrigal [2,3], Jordi Voltas [4,5] and Víctor Resco de Dios [4,5,6,*]

[1] Mathematical and Fluid Physics Department, Faculty of Sciences, Universidad Nacional de Educación a Distancia (UNED), 28040 Madrid, Spain; sierra@ccia.uned.es
[2] Department of Forest Dynamics and Management, INIA–CIFOR, Ctra. A Coruña Km 7.5, 28040 Madrid, Spain; incendio@inia.es
[3] ETSI Montes, Forestal y del Medio Natural, Universidad Politécnica de Madrid (UPM), Ramiro de Maeztu, 28040 Madrid, Spain
[4] Department of Crop and Forest Sciences, Universitat Lleida, 25198 Lleida, Spain; jordi.voltas@udl.cat
[5] Oint Research Unit CTFC-AGROTECNIO, Universitat de Lleida, 25198 Lleida, Spain
[6] School of Life Science and Engineering, Southwest University of Science and Technology, Mianyang 621010, China
* Correspondence: rodrigo.balaguer.romano@gmail.com (R.B.-R.); v.rescodedios@gmail.com (V.R.d.D.)

Received: 21 August 2020; Accepted: 28 September 2020; Published: 29 September 2020

Abstract: *Research Highlights:* Pre-programmed cell death in old Aleppo pine needles leads to low moisture contents in the forest canopy in July, the time when fire activity nears its peak in the Western Mediterranean Basin. Here, we show, for the first time, that such needle senescence may increase fire behavior and thus is a potential mechanism explaining why the bulk of the annual burned area in the region occurs in early summer. *Background and Objectives:* The brunt of the fire season in the Western Mediterranean Basin occurs at the beginning of July, when live fuel moisture content is near its maximum. Here, we test whether a potential explanation to this conundrum lies in Aleppo pine needle senescence, a result of pre-programmed cell death in 3-years-old needles, which typically occurs in the weeks preceding the peak in the burned area. Our objective was to simulate the effects of needle senescence on fire behavior. *Materials and Methods:* We simulated the effects of needle senescence on canopy moisture and structure. Fire behavior was simulated across different phenological scenarios and for two highly contrasting Aleppo pine stand structures, a forest, and a shrubland. Wildfire behavior simulations were done with BehavePlus6 across a wide range of wind speeds and of dead fine surface fuel moistures. *Results:* The transition from surface to passive crown fire occurred at lower wind speeds under simulated needle senescence in the forest and in the shrubland. Transitions to active crown fire only occurred in the shrubland under needle senescence. Maximum fire intensity and severity were always recorded in the needle senescence scenario. *Conclusions:* Aleppo pine needle senescence may enhance the probability of crown fire development at the onset of the fire season, and it could partly explain the concentration of fire activity in early July in the Western Mediterranean Basin.

Keywords: fire behavior; crown fire; fire modeling; senescence; foliar moisture content; canopy bulk density

---

1. Introduction

Pine-dominated ecosystems are one of the major landscape types in the Mediterranean Basin, where they cover 25% of the forest surface [1]. One of the most abundant and widespread pine species in the Mediterranean Basin is *Pinus halepensis* Mill. (Aleppo pine), which covers 6.8 Mha, at low altitudes (<500 m) and near the coastline [2]. Aleppo pine is a fire-embracer species meaning that it depends, at least partly, on fires for seed release from serotinous cones and consequent regeneration [3,4]. Post-fire regeneration often results in dense thickets that show a high accumulation of ladder fuels leading to vertical fuel continuity [5]. *P. halepensis* shows a low degree of self-pruning, and these forests are thus particularly prone to crown fires. Approximately one-third of the total annual burned area in the Mediterranean Basin occurs in *P. halepensis* stands [6].

There are different types of crown fires, ranging from individual tree torching, active crown fires and, under exceptional circumstance, independent crown fires that become decoupled from surface fuels may also occur [7]. Wildfire in *P. halepensis* stands often show potential for developing active crown fires beyond extinction capacity [8]. The high rate of spread and intensity of crown fires in *P. halepensis* stands, combined with long range spotting are characteristics that pose a serious threat to life and property [9].

In order to understand potential wildfire behavior, mathematical models have been developed to account for the various interacting processes that drive fire behavior [10]. In North America and Europe, different models that link [11,12] surface and crown fire rate of spread predictions with [7,13] crown fire transition and propagation criteria have often been used [14], including BehavePlus (USDA, Missoula, MT, USA) [15], FlamMap (USDA, Missoula, MT, USA) [16] or NEXUS (USDA, Missoula, MT, USA) [17].

In these semi-empirical approaches, the onset of a crown fire is defined by the transition of a wildfire from surface to canopy fuels. This transition occurs when the surface fire intensity attains or exceeds a certain critical surface intensity (I_0), which, in turn, is determined by the interaction between foliar moisture content (FMC) and the canopy base height (CBH) [7]:

$$I_0 = (0.01\, \text{CBH}\, (460 + 25.9\, \text{FMC}))^{1.5} \tag{1}$$

After the transition from the surface to the canopy layer, a certain canopy bulk density (CBD) is needed to develop and maintain a solid flame front. If this CBD is not reached, the crown fire will passively torch isolated trees (or groups of trees), but it will not spread across the canopy [17]. Consequently, for active crown fire development, a critical minimum spread rate (R_0), which depends on CBD, is needed to maintain continuous crowning [12]:

$$R_0 = \frac{3}{\text{CBD}} \tag{2}$$

Characterization of the fuel structure and its relevance for fire behavior has been the topic of much research [18]. Variations in live fuel moisture are often taken into account, although some discussions are still active on its role in fire propagation [19]. However, an aspect that has seldom been considered is the role of pre-programmed needle senescence, despite its potential to increase crown fire intensity and severity [19,20].

Needle lifespan in *P. halepensis* is approximately three years, and three-years-old needles typically become dry and senesce towards the end of June or start of July (Figure 1A). This is immediately before the peak of the fire season in the Western Mediterranean basin, which often occurs in the first half of July [21] (Figure 1B). Consequently, pre-programmed needle senescence (a developmental process that allows nutrient recycling in old leaves before shedding) potentially leads to one-third of the canopy (that is, all 3 years-old leaves) being dry right before the peak fire season [22].

Some studies have addressed the role of FMC on fire behavior [23]. Others have addressed how canopy drying, following bark beetle attacks, for instance, impacts fire behavior [24–26]. However, to the best of our knowledge, the effects of partial canopy drying after needle senescence on crown flammability have not been quantified so far [19,20,22].

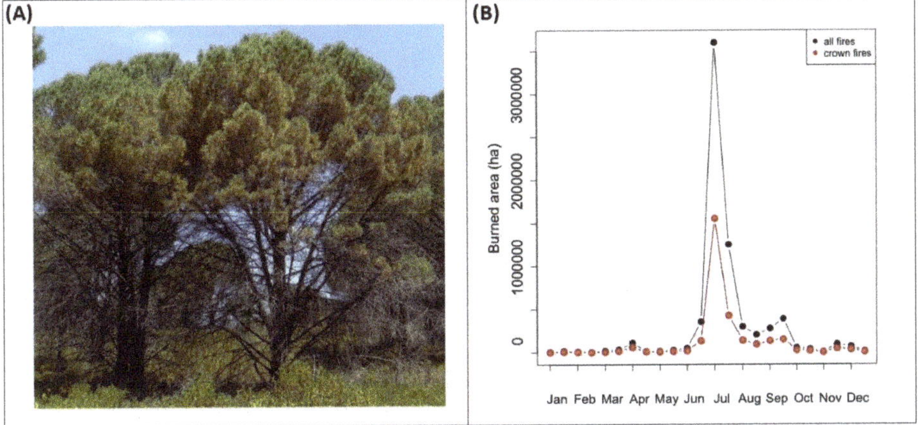

Figure 1. (**A**) Needle senescence in *P. halepensis* affecting the old leaves cohort (3-years-old) typically occurs between the end of June or July, and it drastically modifies the moisture of the canopy. Photo by Carles Arteaga. (**B**); Temporal pattern of long-term average (1968–2015) burned area (black, all fires; red, crown fires) in the *Pinus halepensis* forests of the Mediterranean regions of Spain. (Data from the Estadística General de Incendios Forestal provided by the Ministry of Agriculture, Fishing, and Food).

Temporal and spatial coincidence of low foliar moisture content and high canopy bulk density creates optimal conditions to increase the probability of crown fire occurrence as well as their intensity and severity. High-intensity crown fires burn canopies by convection, leading to widespread defoliation and, consequently, plant death. Preprogrammed old needle senescence may thus enhance Aleppo pine mortality rates after wildfires, if it does affect intensity fire behavior [19]. However, this effect only lasts for a few weeks, until leaf dropping [20]. After shedding of senesced needles, the probability of crown fire activity declines as the weighted foliar moisture content increases and the canopy bulk density decreases. Consequently, surface fires may become more intense after needle shedding due to an increase in surface fuel loads, but surface fires seldom reach intensities beyond extinction capacity.

It is currently unknown why the brunt of the fire season occurs in early July in the Western Mediterranean Basin [19]. During this time, FMC in Mediterranean trees, shrubs, and grasses is near its seasonal maximum [27] and fires occurring in late August, under much lower FMC, often burn at lower intensity [19]. Aleppo pine needle senescence could thus offer at least a partial explanation to such conundrum.

Here we seek to quantify the potential effects of needle senescence on fire behavior in *P. halepensis* stands. We simulated four scenarios that recreated the major annual physiological and structural changes in relation to needle senescence (that is, before, during and after leaf senescence and later in the year after the onset of litter decomposition in the autumn). Each of the four simulations was ran for two highly contrasting *P. halepensis* fuel structures (representatives of very high and very low crown fire likelihood) that are dominant in Valencia (E Spain), one of the most fire-prone regions in Mediterranean Spain. We wanted to test the potential effects of needle senescence on crown and surface fire behavior in contrasting stand types, and also to establish its dependence and interactions with wind speed and dead fuel moisture, two well-known drivers of fire behavior. More specifically, we wanted to test: (i) whether needle senescence increases the likelihood of transition from surface to crown fire; (ii) whether once the transition to crown fire has occurred, the likelihood to develop an active crown fire increases with needle senescence in widely contrasting stand structures; (iii) whether needle senescence increases mortality rates after wildfire activity; and (iv) what is the importance of the effect of needle senescence on crown fire likelihood relative to wind speed and dead fuel moisture.

2. Materials and Methods

2.1. Senescence Scenarios

Aleppo pine presents a tetracyclic annual shoot elongation process. Once senescence is active (end of June-beginning of July) needles have developed two thirds of the total annual elongation in current year shoots [28]. Thus, considering a three-year needle life span, pre-programmed senescence leads to 1/3.6th, or 28%, of the dried canopy (or dead mass fraction, f_d) if all 3 years old needles senesce at once. To simulate annual canopy physiological and structural changes caused by needle senescence, four phenological scenarios were created. The first one, scenario-A (Table 1), represents spring leaf sprout. At this time there is an increase in canopy bulk density, canopy cover and foliar moisture content. Scenario-B (Table 1) represents the time of needle senescence, when about 28% of the canopy is composed of dead matter at the beginning of July. To introduce these changes in FMC, canopy live matter moisture (M_l) and canopy dead matter moisture (M_d) were weighed (M_w) considering f_d as in [29]:

$$\text{FMC} = M_w = f_d M_d + (1 - f_d)M_l \tag{3}$$

Table 1. Parameters values in shrub and forest fuel types for each scenario: A, before senescence; B, during senescence; C, after shedding; D, in autumn.

Forest (TU-3)	A	B	C	D
Canopy Cover (%)	35	35	35	35
Canopy Height (m)	8	8	8	8
Canopy Base Height (m)	1.5	1.5	1.5	1.5
Canopy Bulk Density (kg/m^3)	0.15	0.15	0.1	0.1
Fine Fuel Load (t/ha)	2.5	2.5	3	2.5
1-h Dead Surface Fuel Moisture (%)	6	5	5	9
10-h Dead Surface Fuel Moisture (%)	7	6	6	10
100-h Dead Surface Fuel Moisture (%)	8	7	7	11
Foliar Moisture Content (%)	105	74	100	100
Shrub (SH-9)	**A**	**B**	**C**	**D**
Canopy Cover (%)	100	100	100	100
Canopy Height (m)	5	5	5	5
Canopy Base Height (m)	1	1	1	1
Canopy Bulk Density (kg/m^3)	0.22	0.22	0.15	0.15
Fine Fuel Load (t/ha)	10	10	10.7	10
1-h Dead Surface Fuel Moisture (%)	6	5	5	9
10-h Dead Surface Fuel Moisture (%)	7	6	6	10
100-h Dead Surface Fuel Moisture (%)	8	7	7	11
Foliar Moisture Content (%)	105	74	100	100

Scenario-C simulates the time when needles have been shed, which reduce canopy bulk density. The reduction of dry needles in the crown increases weighted foliar moisture content but needle shedding increases surface fine fuel loads. Finally, scenario-D (Table 1) corresponds to autumn and winter periods when surface fine fuel load decreases due to litter decomposition.

2.2. Stand Structures and Fuel Features

Forest structure and fuel loads play a critical role in fire behavior and crown fire susceptibility. We obtained fuel structure data from the fuel models developed by the Fire Service in Valencia, Spain [30]. The Valencian fuel model catalogue adapts the models from Scott and Burgan [31] to E Spain conditions. We used models SH-9 (shrubland from now on; Tables 1 and 2) and TU-3 (forest from now on; Tables 1 and 2). We will refer to SH-9 as a shrub fuel type, in the sense that it is short stature vegetation, but we note that it has two separate fuel layers (canopy fuels begin at 1 m above ground). It represents stands with a low proportion of large trees, extremely high tree density, and horizontal fuel continuity. In contrast, TU-3 is a forest fuel type representing stands with two separated layers, high proportion of large trees, moderate tree density and moderate to low vertical and horizontal fuel continuity. Both models are considered as dynamic fuels, thus live herbaceous fuels become dead depending on their moisture content [31]. For initial model simulations, dead fuel moisture for scenarios A, B, and C were established according to the low moisture values recorded after heat wave periods [32,33]. As scenario-D represents autumn, dead fuel moisture values are higher due to more benign conditions. We obtained M_l from [34] and M_d from [33]. Additionally, in order to understand the effect of leaf senescence relative to other drivers of fire behavior, we conducted a sensitivity analysis on how different values of 1-h dead surface fuel moisture affected fire behavior. Canopy bulk density, canopy height, and canopy base height were established according to [35]. Changes in canopy bulk density were established considering a reduction of 28% among scenarios before and after senescence, as previously argued. Canopy base height values were considered stable among scenarios because the differences in height between 3 and 2 years-old needles are negligible (<10 cm) for the purpose of these simulations.

Table 2. Fuel models SH-9 (shrub) and TU-3 (Forest) parameters values.

Fuel Parameters	Fuel Model TU-3	Fuel Model SH-9
1-h Dead Fuel Load	2.5 t/ha	10 t/ha
10-h Dead Fuel Load	0.34 t/ha	5.5 t/ha
100-h Dead Fuel Load	0.56 t/ha	0 t/ha
Live Herbaceous Fuel Load	1.5 t/ha	3.5 t/ha
Live Woody Fuel Load	2.5 t/ha	16 t/ha
1-h SAV Ratio	59.05 cm^2/cm^3	24.60 cm^2/cm^3
Live Herbaceous SAV Ratio	52.49 cm^2/cm^3	59.05 cm^2/cm^3
Live Woody SAV Ratio	45.93 cm^2/cm^3	49.21 cm^2/cm^3
Fuel Bed Depth	40 cm	134 cm
Dead Fuel Moisture of Extinction	30%	40%
Dead Fuel Heat Content	18,622.3 kJ/kg	18,622.3 kJ/kg
Live Fuel Heat Content	18,622.3 kJ/kg	18,622.3 kJ/kg

2.3. Fire Behavior Modelization

Wildland fire behavior simulation was done using BehavePlus6 [15] and crown fire was calculated using Scott and Reinhardt [17] as input option. The input values used in each stand type and each scenario are detailed in Tables 1 and 2. Slope steepness was set to 0% and 10 m open wind speed was established in a range from 0 to 30 km/h. Final figures were created using R.3.6.1. (Lucent Technologies, Murray Hill, NJ, USA) [36]. Assessment of fire severity were performed using the lethal thresholds (LD) developed by [19]. Thus, a crown fraction burned (CFB) between 0.4–0.8 eliminates 50% of the population (LD_{50}), and CFB higher than 0.8–0.9 completely eliminates the population (LD_{100}). When CFB remains below 0.4 CFB mortality is negligible (LD_0) [19].

2.4. Dead Mass Fraction Sensitivity Analysis

We also conducted a sensitivity analysis to assess how a varying proportion of f_d affected the transition ratio from a surface to crown layer. This is important because, assuming that the biomass of each cohort is constant, our previously estimated 28% of f_d would constitute a maximum potential value: needle senescence may start earlier in the year such that different values of f_d may occur when the fire season starts. Surface fire intensity was established from the mean surface intensity across scenarios with an intermediate wind speed of 15 km/h.

3. Results

We observed that maximum fire intensity and severity occurred in scenario-B under all wind speeds and fuel types (Table 3). Fire intensity and severity values were higher in the shrub than in the forest fuel model. The highest estimated value of Rate of Spread (ROS) in scenario-B for the forest fuel type was 14.6 m/min at a wind speed of 30 km/h. This value was between 2 and 3 times higher than the peak ROS in the other scenarios (Figure 2A). In the shrub fuel type, the highest ROS was 17.7 m/min, a value that was also reached in scenario-B with a wind speed of 30 km/h. ROS in scenario-B in the shrub fuel type was at least 1.4 times higher than in other scenarios (Figure 3A). The highest fire line intensity reached in scenario-B was 5924 kW/m in the forest stand and 17,179 kW/m in the shrub stand. Peak fire line intensity in scenario-B was 2–3 times higher in the forest fuel type and 1.5 times higher in the shrub fuel type compared to other scenarios (Table 3). The highest flame length occurred in scenario-B and took values of 8.7 m in the forest stand and 17.7 m in the shrub stand. Flame length remained between 2–3.3 m for the forest stand and between 10.1–14.4 m in the shrub stand in the other three scenarios (Table 3).

Table 3. Simulated Rate of Spread (m/min), Fire Line Intensity (kW/m), Flame Length (m) and Crown Fraction Burned for each scenario (A, B, C, D) under four 10 m open wind speeds (0, 10, 20, 30 km/h).

FOREST (TU-3)	**Wind Speed (km/h)**	**A**	**B**	**C**	**D**
Rate of Spread (m/min)	0	0.3	0.5	0.4	0.3
	10	0.9	1.2	1.1	0.9
	20	1.7	5.1	2.6	1.7
	30	5.8	14.6	6.9	3.9
Fire Line Intensity (kW/m)	0	48	74	69	45
	10	130	200	183	121
	20	259	1384	462	240
	30	1393	5924	1585	653
Flame Length (m)	0	0.5	0.6	0.5	0.4
	10	0.7	0.9	0.9	0.7
	20	1	3.3	1.6	1
	30	3.3	8.7	3.6	2
Crown Fraction Burned	0	0	0	0	0
	10	0	0	0	0
	20	0	0.35	0.06	0
	30	0.30	0.81	0.32	0.13
SHRUB (SH-9)	**Wind Speed (km/h)**	**A**	**B**	**C**	**D**
Rate of Spread (m/min)	0	0.7	1	0.8	0.7
	10	2.1	3.1	2.2	1.8
	20	5.7	8.6	5.5	4.4
	30	12.6	17.7	11.6	9.1
Fire Line Intensity (kW/m)	0	560	765	586	490
	10	1752	2615	1679	1330
	20	5208	8228	4510	3402
	30	12,562	17,179	10,074	7372
Flame Length (m)	0	1.8	2.2	1.9	1.7
	10	3.9	5.1	3.8	3.2
	20	8.0	10.9	7.3	6
	30	14.4	17.7	12.4	10.1
Crown Fraction Burned	0	0.13	0.19	0.1	0.08
	10	0.34	0.44	0.27	0.23
	20	0.63	0.79	0.49	0.43
	30	0.95	1	0.75	0.65

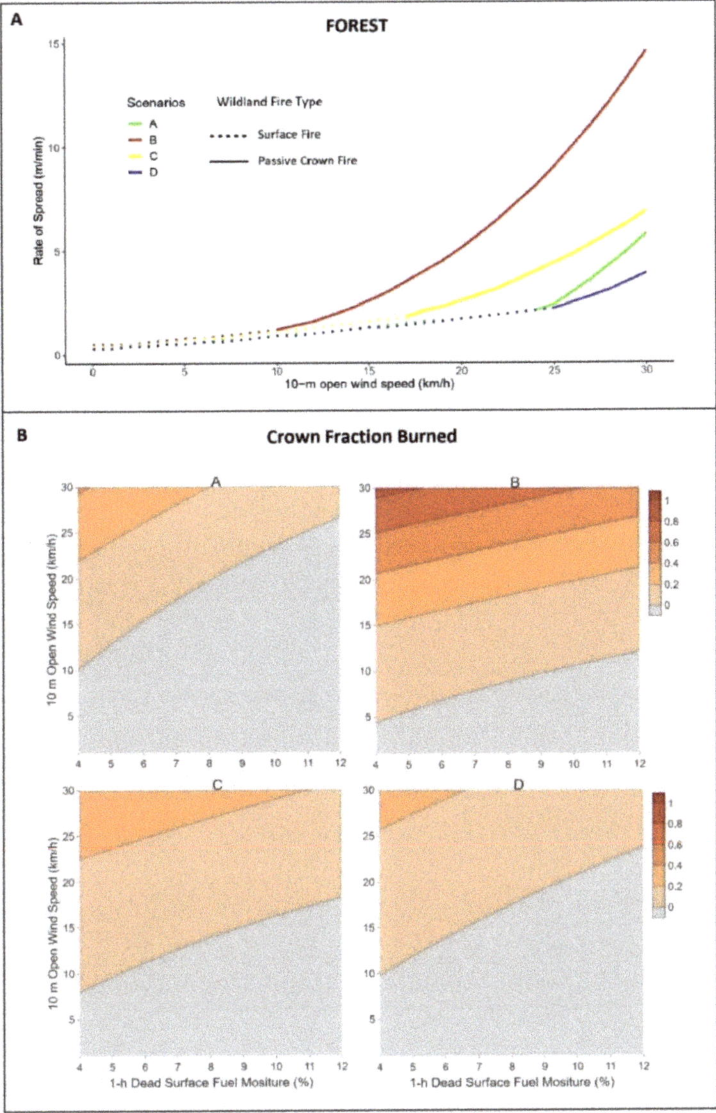

Figure 2. (**A**) Rate of Spread (m/min) in each scenario as a function of 10 m open wind speed in forest stands (TU-3 fuel model type). Dotted lines refer to surface fires, solid lines to passive crown fires. (**B**) Crown Fraction Burned values as a function of 10 m open wind speed (km/h) and 1-h dead surface fuel moisture (%) for each scenario: (**A**), before senescence; (**B**), during senescence; (**C**), after shedding; (**D**), in autumn.

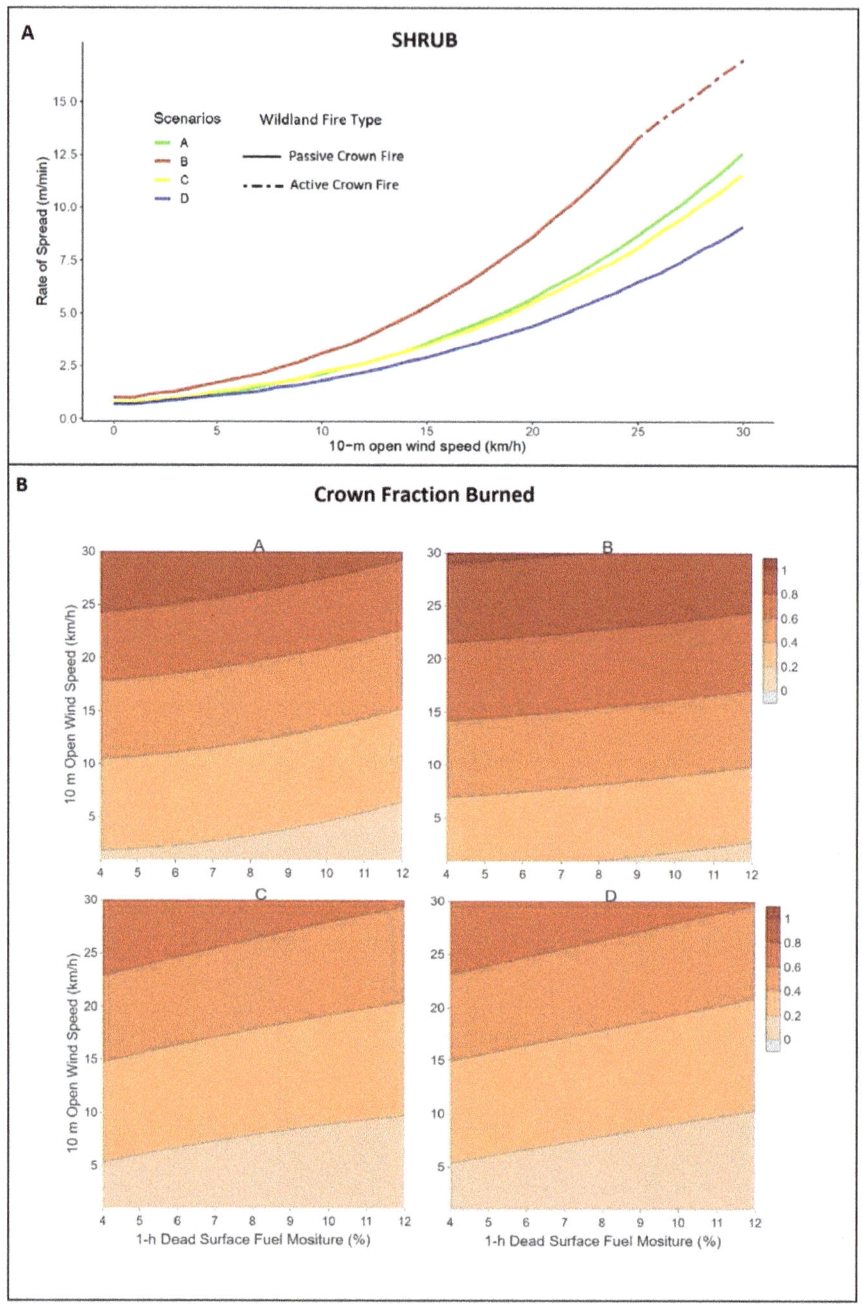

Figure 3. (**A**) Rate of spread (m/min) in each scenario as a function of 10 m open wind speed in the shrub stand (SH-9 model type). Solid lines to passive crown fires and dot-dash lines to active crown fires. (**B**) Crown Fraction Burned as a function of 10 m open wind speed (km/h) and 1-h dead surface fuel moisture (%) for each scenario: (**A**), before senescence; (**B**), during senescence; (**C**), after shedding; (**D**), in autumn.

The transition from surface to crown fire in the forest stand occurred with wind speeds higher than 11 km/h in scenario-B. For scenarios A, C and D, the wind speed thresholds necessary for crown fire development were 25, 18 and 26 km/h, respectively. However, it is important to note that we only observed a transition to passive crown fire development, not to active crown fires, in the forest fuel model TU-3. In the shrub fuel model SH-9, passive crown fires developed under all wind speed conditions. Active crown fire only developed in scenario-B, when wind speeds were larger than 25 km/h.

Regarding fire severity, crown fraction burned (CFB) values were always higher in scenario-B for both fuel types and under all wind speed conditions (Figures 2B and 3B). The relative effect of fuel type on CFB was higher in the forest stand than in the shrub stand since maximum CFB was six times larger in scenario-B (0.81) than in scenario-D (0.13). Importantly, the effect on CFB varied markedly with the moisture content of 1-h dead surface fuels. For instance, in the forest, a CBD leading to (LD_{100}) in the scenario-B occurred either under a wind speed of 25 km/h and a 1 h dead surface fuel moisture of 4% or with a wind speed of 30 km/h and 1-h dead surface fuel moisture of 10%. LD_{50} was similarly reached with wind speeds above 15 km/h under minimum 1 h dead surface fuel moisture (4%). In the remaining forest scenarios (scenarios A, C, and D), increasing wind speed and lowering 1-h dead surface fuel moisture led to increases in CFB, but they always remained below LD_{50}.

In shrublands (Figure 3B), at least some crown damage was recorded in all scenarios under any wind speed and 1-h dead surface fuel moisture conditions. CFB values ranged from 1 in scenario-B to 0.65 in scenario-D under the highest wind speed, indicating important differences depending on fuel phenology. Regarding lethal thresholds (LD), LD_{50} was reached in scenario-B, under a wind speed of 12 km/h when 1-h dead surface fuel moisture was at 12%, or under 8 km/h when 1-h dead surface fuel moisture was at 4%. Further increases in wind speed in this scenario would lead to LD_{100}. In the other scenarios, LD_{50} was recorded under an intermediate wind speed of 20 km/h and under critical wind speed conditions (30 km/h), LD_{100} also occurred in scenario A.

Finally, the sensitivity analysis on the effect of a varying f_d on the transition ratio was only performed in forest stands as critical transition to crown fires always occurred in the shrub fuel under any wind speed. Our simulations indicated that the critical surface intensity to crown fire transition under a wind speed of 15 km/h occurred with a minimum fraction of 0.17 of the canopy composed of dead foliar fuels (Figure 4).

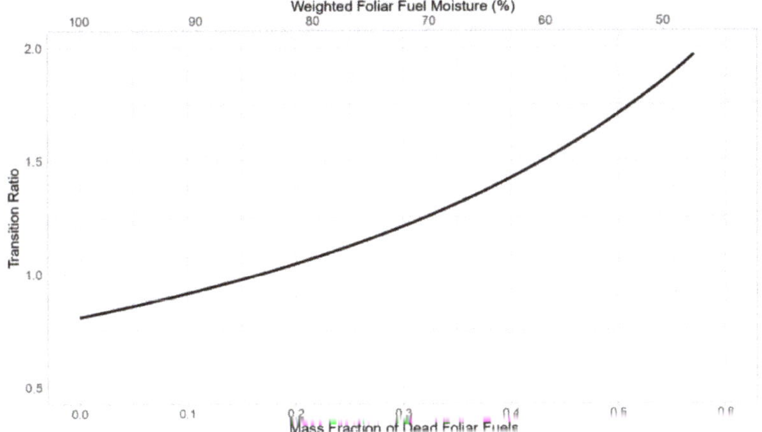

Figure 4. Sensitivity analysis on the effects of a varying mass fraction of dead foliar fuels (f_d) and associated weighted foliar moisture on the Transition Ratio from a surface fire to the canopy layer on forest stands. Fire transition occurs when the transition ratio between the surface fire intensity (250 kW/m) and critical surface intensity (I_0, Equation (1)) becomes equal or higher than 1.

4. Discussion

Our results suggest that Aleppo pine needle senescence significantly affects potential crown fire behavior. Simulations showed important differences in wildfire intensity and severity due to the physiological and structural changes caused by needle drying and shedding. However, the effect of needle senescence on fire behavior differed depending on fuel type and its interaction with wind speeds and 1-h dead surface fuel moisture. In other words, needle senescence by itself does not lead to active crown fire, but its presence lowers the critical wind speeds and 1-h dead fuels moisture values necessary to reach such transition point.

We observed stronger crown fire activity under scenario-B in both stand types (Figures 2 and 3). This scenario represents the process of needle senescence leading to a few-weeks period typically occurring towards the end of June or beginning of July [22], during which about one third of pine stand canopy is composed of dry needles (Figure 1A). Spatial and temporal coincidence of low foliar moisture content and high canopy bulk density favors the development of more intense and severe crown fires at lower wind speed conditions, particularly for the shrub fuel type, where active crown fires may develop only under needle senescence. These results indicate that needle senescence could be a contributing factor to increasing fire intensity in Aleppo pine stands. Consequently, this mechanism could partly explain why the peak in the burned area observed in the Western Mediterranean basin, where fires predominantly affect *P. halepensis*, occurs in early July (Figure 1B).

We also observed that the relative effect of needle senescence was more noticeable in the forest fuel model than in the shrub fuel model. This is likely due to the fact that baseline flammability in shrublands is already very high: this fuel type presents a lower canopy base height, which reduces, to some extent, the dependence of critical transitions to crown fire on foliar moisture (Equation (1)). Increasing needle flammability in the shrubland stand would thus have, comparatively speaking, a smaller relative effect for extreme fire behavior than on the forest stand. In fact, crown fires would develop under any wind speed and canopy moisture in shrublands (Figure 3A). However, needle senescence did increase the probability of active crown fires. That is, the development of active fires in the shrubland stand only occurred under canopy senescence. These differences observed between fire behavior in shrub and forest stands are consistent with other studies [5,37,38].

Needle senescence may influence crown fire behavior in at least two ways: affecting FMC and CBD. In our forest stand simulation, we recorded that the wind speed necessary for crown fire development decreased from 25 km/h to 10 km/h between scenarios A and B (Figure 1A) because of decreasing foliar moisture from 105% to 74% (Table 1). A lower FMC reduces the influx of energy required to start the ignition, because a smaller amount of water needs to be evaporated. Needle senescence may thus enhance crown fire development, by reducing foliar moisture content and hence the critical surface intensity threshold value at which surface fires become crown fires. Furthermore, as we observed in the sensitivity analysis (Figure 4), the critical surface intensity to cause the transition from a surface fire to the canopy layer occurred as the dead foliar fractions increased over 17%.

We need to acknowledge that the actual role of FMC in affecting the fire rate of spread is currently being discussed. Some authors argue that the role of FMC is exaggerated in fire behavior models because the high convective and radiative fluxes produced by the flame are several orders of magnitude higher than the energy required to dry the fuel, which would render FMC inconsequent [23]. However, other studies consider that the effect of FMC as a driver of fire spread has actually been underestimated [29,39]. Furthermore, empirical evidence across many biomes support that increases in burnt area occur under decreasing FMC [40–43]. A full discussion on this issue would be out of scope, and the reader is referred to a recent review of this issue for more details [19].

The effect of needle senescence on fire behavior was dependent on 1-h dead surface fuel moisture. As we observed in Figures 2B and 3B, senescence effects interact with variation in 1-h dead surface fuel moisture such that critical CFB values were reached in the senescence scenarios under low 1-h dead surface fuel moisture values. As previously stated, fire behavior is more affected, in relative terms, by the structural and physiological effects caused by needle senescence in forest stands compared to shrublands. Simulations showed that lethal thresholds varied from LD_0, which indicates negligible mortality in all forest scenarios, to LD_{100}, which represents the death of the entire population in scenario-B under a wind speed of 30 km/h (Table 3). These changes in tree mortality rates among scenarios were also noticeable in shrublands, where simulations showed that LD_{100} occurred in scenario-B after wind speeds as low as 21 km/h under low 1-h dead fuel moisture values. In the other shrubland scenarios, LD_{100} only occurred in scenario-A, under a critical wind speed condition of 30 km/h. Therefore, while needle senescence is, by itself, not enough to reach critical fire severity thresholds, it lowers the need for critical wind speeds and 1-h dead fuel moisture values necessary to reach LD_{50} or LD_{100}.

We recognize that a problem with our study is the way in which the effects of needle senescence on FMC were inputted into the model. We used a weighted average of FMC whereas, in reality, senesced leaves may form a layer of fuel that is effectively independent of live FMC. Future research should concentrate on building more realistic descriptions of needle arrangement such that fuel moisture within a whorl can change with time. We conducted additional simulations considering only the CBD of dead canopy fuels, but the resulting CBD (0.05 kg/m^3 for forests and 0.07 kg/m^3 for shrublands) was not high enough to produce canopy fires (data not shown).

Another problem with our study lies on the limitations of fire modeling. Considering the complex dynamics behind wildland fires processes, fire models are very simplified, and this could lead to misleading predictions. Furthermore, considering climate change, it is difficult to predict extreme fire conditions accurately. There is some anecdotal evidence that needle senescence enhances crown flammability (M. Castellnou pers. comm.), but further work should confirm experimentally that needle senescence does enhance canopy flammability.

An important yet unresolved aspect is whether needle senescence serves an evolutionary role. It has been reported that pre-programmed needle senescence in the oldest cohort, at least in some temperate and boreal conifers, increases as new leaves develop [44]. This would be a mechanism to recycle nutrients from old leaves into new, developing leaves. In our case, needle senescence co-occurs with the flush of current-year growth, and it could thus serve to support new needle growth. However, needle senescence also occurs as summer drought stress is starting to be important. Consequently, needle senescence could also serve as a water-saving mechanism that decreases transpirational area, at the expense of a transient increase in flammability [22]. However, as climate change intensifies summer drought and wildfire activity, needle senescence could turn maladaptive by enhancing crown fire likelihood. Further efforts towards quantifying the phenology of needle senescence and understanding its underlying drivers should be at the forefront of our research efforts.

We have shown that not considering needle senescence can lead to misleading predictions on fire risk, potentially misestimating wildfire behavior in Aleppo pine stands and this could potentially lead to the application of suboptimal forest and fire management activities. While simulations are routinely performed in order to decide forest management and fire prevention operations, these simulations could incorporate the role of needle senescence because it significantly lowers the threshold for catastrophic fire behavior. To date, needle senescence effects may be underrated in fire behavior simulations due to the relatively short period that it represents each year. However, they occur at a critical time of the year and, as such, its cascading effects on fire behavior may be rather important, as we have anticipated in this work.

An increased probability of extreme events has been forecasted for the next decades as a result of global change. According to predictions, fire seasons may be longer and drier, thereby producing more intense and severe wildfires [19]. Changes in fire regimes represent a challenge to fire-prone species and ecosystems. Aleppo pine post-fire regeneration strategy can be hard-pressed if wildfires return intervals become shorter than the time needed for trees to reach sexual maturity or to produce enough serotinous cones [45]. Also, extremely high wildfire intensity can damage serotinous seeds causing the decline of seedling recruitment and leading to populations collapse [22]. We can thus expect important changes in ecosystem structure in the coming decades, which would have important interactions with changes in the fire regime. Furthermore, it would be relevant to simulate Aleppo pine-woods responses to predicted future climate conditions for the different scenarios tested in this study. A better understanding of pyrophysiology should, therefore, be at the forefront of our research.

5. Conclusions

We have shown evidence, for the first time to our knowledge, of enhanced crown fire behavior in Aleppo pine driven by needle senescence through altered canopy structure and foliage in a period that is coincidental with the brunt of the fire season. Regarding our initial questions, changes in physiological and structural conditions following senescence enhance the probability of more intense and severe crown fires development and concentrate extreme tree mortality rates in senescence periods. Furthermore, in a fuel type with enough canopy bulk density, senescence effects may lead to development of active crown fires. Finally, it is important to consider that senescence, by itself, may not be enough to lead to extreme fire behavior. That is, needle senescence should be viewed as a contributing factor that may favor extreme fire behavior when environmental conditions (e.g., high wind speed) and 1-h dead surface fuel moisture are also at critical levels. We argue for further research to better understand and quantify the drivers of needle senescence and its effects on fire behavior in the field. A lack of consideration of this phenomenon in crown fire modeling systems may provide incomplete predictions leading to the application of unsatisfactory forest and fire management activities.

Author Contributions: Conceptualization, V.R.d.D., R.D.-S., J.M. and R.B.-R.; methodology, V.R.d.D., R.D.-S. and R.B.-R.; software, R.B.-R.; writing—original draft preparation, R.B.-R.; writing—review and editing, V.R.d.D., R.D.-S., J.M., J.V. and R.B.-R. All authors have read and agreed to the published version of the manuscript.

Funding: We acknowledge funding from the Natural Science Foundation in China (31850410483), the talent proposals from Sichuan Province (2020JDRC0065), the Southwest University of Science and Technology talents fund and the Spanish MICINN (AGL2015-69151-R). R.B.-R. acknowledges the Community of Madrid for the predoctoral contract PEJD-2019-PRE/AMB-15644 funded by the Youth Employment Initiative (YEI).

Acknowledgments: We appreciate useful discussion with JL Soriano, MA Botella and the Forest Fire Unit in VAERSA.

Conflicts of Interest: The authors declare no conflict of interest.

References

1. Barbero, M.; Loisel, R.; Quézel, P.; Romane, F.; Richardson, D.M. Pines of the Mediterranean Basin. In *Ecology and Biogeography of Pinus*; Cambridge University Press: Cambridge, UK, 1998; ISBN 9780521789103.
2. Mauri, A.; Di Leo, M.; de Rigo, D.; Caudullo, G. Pinus halepensis and Pinus brutia in Europe: Distribution, habitat, usage and threats. *Eur. Atlas For. Tree Species* **2016**, *1*, 122–123.
3. Keeley, J.E. Ecology and evolution of pine life histories. *Ann. For. Sci.* **2012**, *69*, 445–453. [CrossRef]
4. Resco de Dios, V.; Arteaga, C.; Hedo, J.; Gil-Pelegrín, E.; Voltas, J. A trade-off between embolism resistance and bark thickness in conifers: Are drought and fire adaptations antagonistic? *Plant Ecol. Divers.* **2018**, *11*, 253–258. [CrossRef]
5. Palmero-Iniesta, M.; Domènech, R.; Molina-Terrén, D.; Espelta, J.M. Fire behavior in Pinus halepensis thickets: Effects of thinning and woody debris decomposition in two rainfall scenarios. *For. Ecol. Manag.* **2017**, *404*, 230–240. [CrossRef]
6. Quézel, P. Taxonomy and biogeography of Mediterranean pines (Pinus halepensis and P. brutia). In *Ecology, Biogeography and Management of Pinus halepensis and Pinus brutia Forest Ecosystems in the Mediterranean Basin*; Néeman, G., Trabaud, L., Eds.; Backhuys Publishers: Leiden, The Netherlands, 2000; pp. 1–12.
7. Van Wagner, C.E. Conditions for the start and spread of crown fire. *Can. J. For. Res.* **1977**, *7*, 23–34. [CrossRef]
8. Cruz, M.G.; Alexander, M.E. Modelling the rate of fire spread and uncertainty associated with the onset and propagation of crown fires in conifer forest stands. *Int. J. Wildl. Fire* **2017**, *26*, 413–426. [CrossRef]
9. Dimitrakopoulos, A.P.; Mitsopoulos, I.D.; Raptis, D.I. Nomographs for predicting crown fire initiation in Aleppo pine (Pinus halepensis Mill.) forests. *Eur. J. For. Res.* **2007**, *126*, 555–561. [CrossRef]
10. Hoffman, C.M.; Canfield, J.; Linn, R.R.; Mell, W.; Sieg, C.H.; Pimont, F.; Ziegler, J. Evaluating Crown Fire Rate of Spread Predictions from Physics-Based Models. *Fire Technol.* **2016**, *52*, 221–237. [CrossRef]

11. Rothermell, R.C. *A Mathematical Model for Predicting Fire Spread*; US Department of Agriculture: Washington, DC, USA, 1972.

12. Rothermel, R.C. *Predicting Behavior and Size of Crown Fires in the Northern Rocky Mountains*; U.S. Department of Agriculture, Forest Service, Intermountain Forest and Range Experiment Station: Washington, DC, USA, 1991; Res. Pap.; Volume 438, p. 46.

13. Van Wagner, C.E. Prediction of crown fire behavior in two stands of jack pine. *Can. J. For. Res.* **1993**, *23*, 442–449. [CrossRef]

14. Alexander, M.E.; Cruz, M.G. *Crown Fire Dynamics in Conifer Forests*; USDA Forest Service: Washington, DC, USA, 2016; Gen. Tech. Rep. PNW-GTR; Volume 2016, pp. 163–258.

15. Andrews, P.L. Current status and future needs of the BehavePlus Fire Modeling System. *Int. J. Wildl. Fire* **2014**, *23*, 21–33. [CrossRef]

16. Finney, M.A. An overview of FlamMap fire modeling capabilities. In Proceedings of the Fuels Management-How to Measure Success: Conference Proceedings, Portland, OR, USA, 28–30 March 2006; pp. 213–220.

17. Scott, J.H.; Reinhardt, E.D. *Assessing Crown Fire Potential by Linking Models of Surface and Crown Fire Behavior*; USDA Forest Service: Washington, DC, USA, 2001; Res. Pap. RMRS-RP; pp. 1–62. [CrossRef]

18. Keane, R.E. *Wildland Fuel Fundamentals and Applications*; Springer: Cham, Switzerland, 2015; ISBN 9783319090153.

19. Resco de Dios, V. *Plant-Fire Interactions*; von Gadow, K., Pukkala, T., Tomé, M., Managing, F., Eds.; Springer: Cham, Switzerland, 2020; ISBN 9783030411916.

20. Nolan, R.H.; Blackman, C.J.; De Dios, R.; Choat, B.; Medlyn, B.E.; Li, X.; Bradstock, R.A.; Boer, M.M. Linking Forest Flammability and Plant Vulnerability to Drought. *Forests* **2020**, *11*, 779. [CrossRef]

21. López-Santalla, A.; López-Garcia, M. *Los Incendios Forestales en España. Decenio 2006–2015*; Ministerio de Agricultura Pesca y Alimentación, Secretaria General Técnica de Impresión: Madrid, Spain, 2019; p. 166.

22. Karavani, A.; Boer, M.M.; Baudena, M.; Colinas, C.; Díaz-Sierra, R.; Pemán, J.; de Luis, M.; Enríquez-de-Salamanca, Á.; Resco de Dios, V. Fire-induced deforestation in drought-prone Mediterranean forests: Drivers and unknowns from leaves to communities. *Ecol. Monogr.* **2018**, *88*, 141–169. [CrossRef]

23. Alexander, M.E.; Cruz, M.G. Assessing the effect of foliar moisture on the spread rate of crown fires. *Int. J. Wildl. Fire* **2013**, *22*, 415–427. [CrossRef]

24. Jenkins, M.J.; Page, W.G.; Hebertson, E.G.; Alexander, M.E. Fuels and fire behavior dynamics in bark beetle-attacked forests in Western North America and implications for fire management. *For. Ecol. Manag.* **2012**, *275*, 23–34. [CrossRef]

25. Talucci, A.C.; Krawchuk, M.A. Dead forests burning: The influence of beetle outbreaks on fire severity and legacy structure in sub-boreal forests. *Ecosphere* **2019**, *10*, e02744. [CrossRef]

26. Reiner, A.L. *Fire Behavior in Beetle-killed Stands: A Brief Review of Literature Focusing on Early Stages after Beetle Attack*; US Forest Service Pacific Southwest Region: Vallejo, CA, USA, 2017; Volume 5, pp. 1–5.

27. Nolan, R.H.; Hedo, J.; Arteaga, C.; Sugai, T.; Resco de Dios, V. Physiological drought responses improve predictions of live fuel moisture dynamics in a Mediterranean forest. *Agric. For. Meteorol.* **2018**, *263*, 417–427. [CrossRef]

28. Hover, A.; Buissart, F.; Caraglio, Y.; Heinz, C.; Pailler, F.; Ramel, M.; Vennetier, M.; Prévosto, B.; Sabatier, S. Growth phenology in Pinus halepensis Mill: Apical shoot bud content and shoot elongation. *Ann. For. Sci.* **2017**, *74*. [CrossRef]

29. Rossa, C.G.; Fernandes, P.M. On the effect of live fuel moisture content on fire rate of spread. *For. Syst.* **2017**, *26*. [CrossRef]

30. Generalitat Valenciana; Servei de Prevenció D'Incendis Forestals Clave para la identificación de los modelos de combustible de la comunitat valenciana. *Conselleria de Agricultura, Desarrollo Rural, Emergencia Climática y Transición Ecológica.* **2020**, *1*, 1–38.

31. Scott, J.H.; Burgan, R.F. *Standard Fire Behavior Fuel Models: A Comprehensive Set for Use with Rothermel's Surface Fire Spread Model*; USDA Forest Service: Washington, DC, USA, 2005; Gen. Tech. Rep. RMRS-GTR; pp. 1–76. [CrossRef]

32. Boer, M.M.; Nolan, R.H.; De Dios, V.R.; Clarke, H.; Owen, F.; Bradstock, R.A. Changing Weather Extremes Call for Early Warning of Potential for Catastrophic Fire. *Earths Future.* **2017**. [CrossRef]

33. Jervis, F.X.; Rein, G. Experimental study on the burning behaviour of Pinus halepensis needles using small-scale fire calorimetry of live, aged and dead samples. *FIRE Mater.* **2016**, *40*, 385–395. [CrossRef]

34. Soriano Sanchez, J.; Quilez Moraga, R. Análisis de la humedad del combustible vivo en la Comunitat Valenciana 1/13. *Soc. Española Cienc. For.* **2017**, 1–14.

35. Mitsopoulos, I.D.; Dimitrakopoulos, A.P. Canopy fuel characteristics and potential crown fire behavior in Aleppo pine (Pinus halepensis Mill.) forests. *Ann. For. Sci.* **2007**, *64*, 287–299. [CrossRef]

36. R Development Core Team. *R Development Core Team, R: A Language and Environment for Statistical Computing*; R Foundation for Statistical Computing: Vienna, Austria, 2020; ISBN 3900051070.

37. Alvarez, A.; Gracia, M.; Retana, J. Fuel types and crown fire potential in Pinus halepensis forests. *Eur. J. For. Res.* **2012**, *131*, 463–474. [CrossRef]

38. Keeley, J.E.; Pfaff, A.H.; Safford, H.D. Fire suppression impacts on postfire recovery of Sierra Nevada chaparral shrublands. *Int. J. Wildl. Fire* **2005**. [CrossRef]

39. Pimont, F.; Ruffault, J.; Martin-StPaul, N.K.; Dupuy, J.L. Why is the effect of live fuel moisture content on fire rate of spread underestimated in field experiments in shrublands? *Int. J. Wildl. Fire* **2019**, *28*, 127–137. [CrossRef]

40. Dennison, P.E.; Moritz, M.A.; Taylor, R.S. Evaluating predictive models of critical live fuel moisture in the Santa Monica Mountains, California. *Int. J. Wildl. Fire* **2008**, *17*, 18–27. [CrossRef]

41. Nolan, R.H.; Boer, M.M.; Resco De Dios, V.; Caccamo, G.; Bradstock, R.A. Large-scale, dynamic transformations in fuel moisture drive wildfire activity across southeastern Australia. *Geophys. Res. Lett.* **2016**, *43*, 4229–4238. [CrossRef]

42. Luo, K.; Quan, X.; He, B.; Yebra, M. Effects of live fuel moisture content on wildfire occurrence in fire-prone regions over southwest China. *Forests* **2019**, *10*, 887. [CrossRef]

43. Pimont, F.; Ruffault, J.; Martin-Stpaul, N.K.; Dupuy, J.L. A cautionary note regarding the use of cumulative burnt areas for the determination of fire danger index breakpoints. *Int. J. Wildl. Fire* **2019**. [CrossRef]

44. Kimmins, J.P. *Forest Ecology—A Foundation for Sustainable Forest Management and Environmental Ethics in Forestry*; Pearson: London, UK, 2004.

45. Pausas, J. Fuego y evolución en el Mediterráneo. *Investig. Cienc.* **2010**, *407*, 56–63.

Article

Effects of Live Fuel Moisture Content on Wildfire Occurrence in Fire-Prone Regions over Southwest China

Kaiwei Luo [1], Xingwen Quan [1,]*, Binbin He [1,]* and Marta Yebra [2,3]

[1] School of Resources and Environment, University of Electronic Science and Technology of China, Chengdu 611731, China; luokaiwei@std.uestc.edu.cn

[2] Fenner School of Environment and Society, The Australian National University, Canberra ACT 2601, Australia; marta.yebra@anu.edu.au

[3] Bushfire & Natural Hazards Cooperative Research Centre, Melbourne VIC 3002, Australia

* Correspondence: xingwen.quan@uestc.edu.cn (X.Q.); binbinhe@uestc.edu.cn (B.H.)

Received: 9 September 2019; Accepted: 6 October 2019; Published: 8 October 2019

Abstract: Previous studies have shown that Live Fuel Moisture Content (LFMC) is a crucial driver affecting wildfire occurrence worldwide, but the effect of LFMC in driving wildfire occurrence still remains unexplored over the southwest China ecosystem, an area historically vulnerable to wildfires. To this end, we took 10-years of LFMC dynamics retrieved from Moderate Resolution Imaging Spectrometer (MODIS) reflectance product using the physical Radiative Transfer Model (RTM) and the wildfire events extracted from the MODIS Burned Area (BA) product to explore the relations between LFMC and forest/grassland fire occurrence across the subtropical highland zone (Cwa) and humid subtropical zone (Cwb) over southwest China. The statistical results of pre-fire LFMC and cumulative burned area show that distinct pre-fire LFMC critical thresholds were identified for Cwa (151.3%, 123.1%, and 51.4% for forest, and 138.1%, 72.8%, and 13.1% for grassland) and Cwb (115.0% and 54.4% for forest, and 137.5%, 69.0%, and 10.6% for grassland) zones. Below these thresholds, the fire occurrence and the burned area increased significantly. Additionally, a significant decreasing trend on LFMC dynamics was found during the days prior to two large fire events, Qiubei forest fire and Lantern Mountain grassland fire that broke during the 2009/2010 and 2015/2016 fire seasons, respectively. The minimum LFMC values reached prior to the fires (49.8% and 17.3%) were close to the lowest critical LFMC thresholds we reported for forest (51.4%) and grassland (13.1%). Further LFMC trend analysis revealed that the regional median LFMC dynamics for the 2009/2010 and 2015/2016 fire seasons were also significantly lower than the 10-year LFMC of the region. Hence, this study demonstrated that the LFMC dynamics explained wildfire occurrence in these fire-prone regions over southwest China, allowing the possibility to develop a new operational wildfire danger forecasting model over this area by considering the satellite-derived LFMC product.

Keywords: critical LFMC threshold; forest/grassland fire; radiative transfer model; remote sensing; southwest China

---

1. Introduction

Wildfire is a natural phenomenon for many ecosystems since fire affects nutrient cycling, vegetation succession patterns, and resistance to pests [1]. It also poses a great threat to the ecological environment, economic development, as well as human life and property [2–6]. There are three major factors that relate to the incidence of wildfires, spatially continuous and dry enough to burn fuel (biomass), meteorological conditions conducive to the spread of fire, and ignitions [7–9]. In this context, fuel moisture content (FMC), defined as the proportion of water content to dry mass within the fuel and traditionally divided into FMC of live (LFMC) and dead fuels (DFMC) [10], is an important driver affecting fuel ignition and fire spread rate [11–13]. Additionally, FMC has also been proven to explain fire occurrence at a large scale as the burned area tends to increase as FMC decreases [14–17]. This is due to the fact that fuels with higher moisture content require more energy for water evaporation, slowing down the fire spread rate and decreasing the flame length [18].

Three methods are normally used to estimate FMC: on-ground field measurements, meteorological data, and satellite imagery [19]. Ground field measurements commonly achieve a high accuracy level if a standard protocol is followed and have been used to investigate the relation of FMC with wildfire occurrence, particularly for the Mediterranean region. For example, Chuvieco et al. [17] demonstrated FMC measurements to have a predictive effect on fire occurrence in central Spain where grassland FMC change was a significant factor explaining the numbers of fires, and shrub FMC was highly associated with large fires. Schoenberg et al. [20] showed that the burned area tended to increase when the field measured FMC was lower than 90%. Dennison et al. revealed an FMC threshold of 70%–80% [21] and 79% [15] in chamise and southern California, that explained the largest fires. However, despite the locally high accuracy level for FMC taken at the field, the high time- and cost-consuming make the large-scale and spatial–temporal FMC mapping unfeasible.

Meteorological indices such as the Keetch–Byram Drought Index (KBDI) and Cumulative Water Balance Index (CWBI) have been commonly used as indicators for FMC variations [22–24]. Ruffault et al. [25] predicted LFMC quantitative variations and critical values by evaluating the capacity of six drought indices (Duff Moisture Code (DMC), Drought Code (DC), KBDI, the Nesterov Index (NI) and the Relative Water Content (RWC_L and RWC_H)). FMC estimated from meteorological data allowed for long-term and large-scale mapping, however, the coarse spatial resolution and interpolation of meteorological data also introduce additional errors. Moreover, the LFMC estimate from meteorological data is still challenged because living plants can utilize moisture stored in the soil and have multiple drought adaptation strategies [19,26].

Remote sensing techniques are the only way to date to spatially and temporally understand the FMC dynamics at regional to continental scale. Methods based on remote sensing for FMC mapping can be broadly classified into two categories: empirical methods and the radiative transfer model (RTM) based methods [19,27–29]. The former techniques are known to use statistical formulas based on FMC measurements and corresponding reflectance or vegetation indices derived from remote sensing images. For example, with optimally averaged Enhanced Vegetation Index, Myoung et al. [30] developed an empirical model function of LFMC from MODIS satellite data for wildfire danger assessment in southern California USA. These statistical approaches are simple and have a known local accuracy, and their effect on wildfire occurrence has also been explored and analyzed in previous studies [31]. Jurdao et al. [31] suggested that the critical FMC that supported 90% of grassland and shrubland fire occurrence was 127.12% and 105.51%, respectively, by extracting the burned pixels from the MODIS Thermal Anomalies product (MOD14) and retrieving pre-fire FMC from empirical models applied to satellite images. Nolan et al. [9] determined DFMC thresholds of forest and woodland (14.6% and 9.9%, respectively) across eastern Australia, based on an empirical formula of vapor pressure deficit estimated from interpolated weather station measurements [32], and determined the LFMC threshold values that explained fire occurrence in eastern Australia (156.1% and 101.5%) by estimating LFMC using an empirically exponential model based on MODIS derived vegetation index. They also demonstrated the "switch" hypothesis [7] that flammable and non-flammable states can change quickly from one to another when the temporal dynamics of FMC are close to the thresholds. However, these empirical methods are known to lack reproducibility due to the shortcomings of sensor-specificity and position-dependence [33,34]. Alternatively, RTM-based methods have proven to be more reproducible for LFMC retrievals given they are based on physical laws that provide explicit connections between surface parameters and leaf and/or canopy spectra [35,36]. Furthermore, RTM-based methods for LFMC retrieval have been demonstrated to be robust, not site-specific, and easy to generalize [37].

This study aims to explore the effect of LFMC dynamics on wildfire occurrence in the fire-prone regions over southwest China between 2007 and 2016 using LFMC estimates derived from satellite data and RTM. Specifically, senescence in grasses results in the conversion of live fuel to dead fuel over time. Grassland LFMC in the sections below, is considered as the average moisture content of both live and dead fuel components, and the degree of curing is not explicitly accounted for. This study is novel as it (i) provides the first analysis of the effects of LFMC on wildfire occurrence over the southwest China fire-prone regions which are historically vulnerable to wildfire, and (ii) was entirely based on optical remote sensing data while RTM-derived LFMC offers a unique way to monitor LFMC dynamics at large scale. The materials and methods mentioned here can be applied to other fire-prone areas due to the generalization potential and reproducibility of the RTM-based LFMC retrieving methods used. The overarching objective of this study is to contribute to the development of an operational system over this region by considering the satellite- and RTM- based LFMC product. This new system will allow wildfire danger early prediction, suppression, and response, as well as improved awareness of fire risk to life and property.

2. Materials and Methods

2.1. Study Area

The study area (101° E–107° E, 22° N–27° N) is located in southwest China, which is part of the Yunnan–Guizhou plateau (Figure 1a), with most areas are 1500–2000 meters above sea level. According to the IGBP (International Geosphere–Biosphere Programme) classification scheme of the MODIS Land Cover product MCD12Q1 [38], evergreen broadleaf forests, mixed forests, woody savannas, grasslands, croplands, and cropland/natural vegetation mosaics are dominant vegetation types in this area (Figure 1b) (Table 1). Under the Koppen climate classification [39], the study area lies within the subtropical highland zone (Cwb) and humid subtropical zone (Cwa), with mild to warm winters, and tempered summers (Figure 1b). The annual average temperature of the study area is 15–18 °C, with an annual temperature difference between 12 °C and 16 °C. The annual precipitation of the study area ranges from 1000 mm to 1200 mm. The precipitation in May to October accounts for 80%–90% of the whole year, whereas November to the next April is the dry season with little precipitation, leading to a high frequency of wildfire occurrence during this period. Figure 2 shows the burned area per month from 2007 to 2016 within the study area extracted from the MODIS Burned Area (BA) product MCD64A1 [40], which also illustrated that the wildfires commonly occurred during the dry season and peak in January to April. Here, we defined the fire season as spanning from September (month with highest LFMC value) to the next year August (e.g., the 2009/2010 fire season starts from September 2009 to August 2010) in this study in terms of the annual LFMC dynamics. Additionally, two large fire (burned area greater than 10 km^2 following Arganaraz et al. [41]), Qiubei forest fire (104.42° E, 24.41° N) on the 1th February 2010 and burned 18.2 km^2 for around two weeks (Figure 1c), and Lantern Mountain grassland fire (103.23° E, 23.89° N) on the 13th February 2016 and burned about 35.4 km^2 for around three days (Figure 1d), were selected as the case studies to explore the relation between LFMC and fire occurrences.

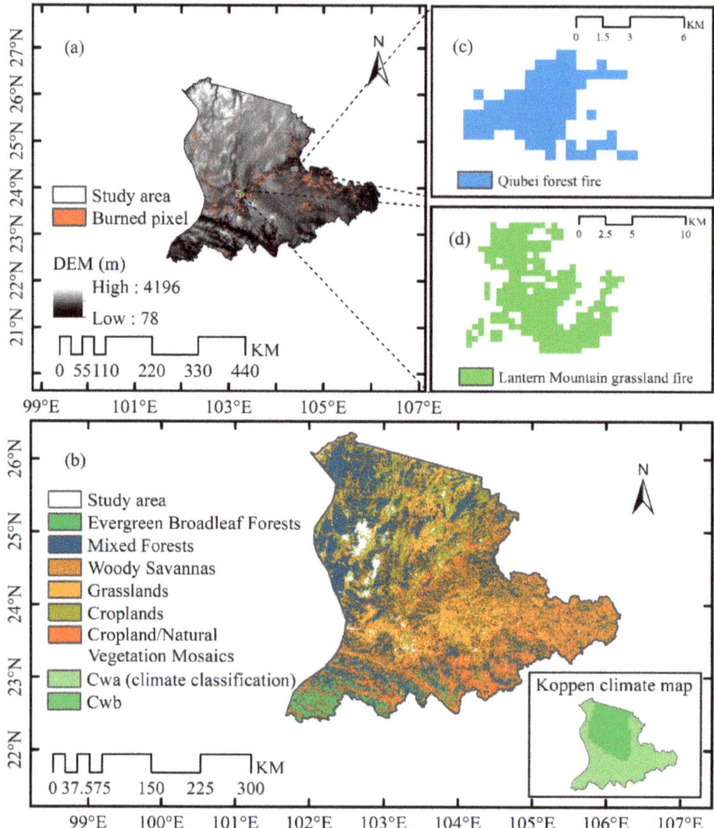

Figure 1. (**a**) Study area showing the historical (2007–2016) burned areas extracted from MCD64A1 Burned Area product and the DEM (Digital Elevation Model) as the background image. (**b**) Dominated vegetation types and Koppen climate classification. The location and burned pixels of (**c**) the Qiubei forest fire event on the 1st February 2010, and (**d**) Lantern Mountain grassland fire event on the 13th February 2016. Cwa: Subtropical Highland Zone; Cwb: Humid Subtropical Zone.

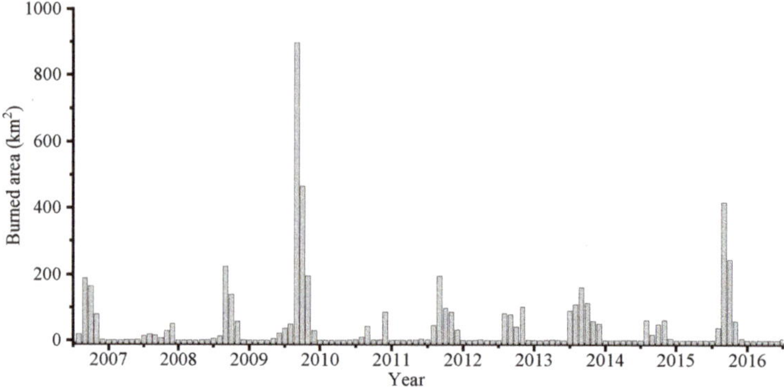

Figure 2. Burned area per month from 2007 to 2016 extracted from MCD64A1 product over southwest China.

2.2. LFMC Measurements

LFMC measurements taken at west China were used to validate the accuracy of the LFMC retrieved by RTM. The field campaigns were conducted at three areas during four periods, (i) Ruoergai Grassland (102.46° E–102.67° E, 33.38° N–33.98° N), 2013, (ii) Qinghai Lake (98.30° E– 101.09° E, 36.38° N–38.25° N), 2014, (iii) Qinghai Lake (98.48° E–101.07° E, 36.34° N–37.78° N), 2015, and (iv) Wenquan Town (102.44° E–102.46° E, 24.99° N–25.00° N), 2016. A total of 192 sampling plots (50 plots in 2013, 62 plots in 2014, 70 plots in 2015, and 10 plots in 2016) covering grass and forest areas were selected and sampled. The positions for each plot were measured through a global positioning system (GPS, Trimble Geo 3000). In each plot of grassland area (30 m × 30 m), the plants (0.5 m × 0.5 m) were randomly measured and destructively sampled. The forest samples were taken within 20 m from the center of the plots (i.e., the area of the plot is around 40 m × 40 m), and we used a telescopic scissor to sample the tree canopy leaves that were not easily reached. The fresh samples were first sealed in plastic bags to prevent the loss of water and then transported to the laboratory, weighed, then oven-dried, and finally weighed again to determine the LFMC. For details on the sampling protocol, please refer to refs. [36,42].

2.3. Satellite Data

Satellite products from two different sensors were used in this study, MODIS products and Landsat 8 OLI product. MODIS products were supplied by the Land Processes Distributed Active Archive Center (LPDAAC) at the U.S. Geological Survey (USGS) Earth Resources Observation and Science Center (EROS). Landsat 8 OLI product provided by USGS via Google Earth Engine (GEE).

2.3.1. Land Cover

The IGBP classification scheme integrated into the MODIS land cover product MCD12Q1 Collection 5 [38] was selected and used to re-classify the vegetation types into three fuel classes (Table 1): forest, grassland, and shrubland following Yebra et al. [43]. Since MCD12Q1 Collection 5 was only available from 2001 to 2013, we continued to use the year 2013 for the years 2014 to 2016. Notably, since little shrubland (around 0.46% of total vegetation coverage area) was identified in the study area, and the corresponding cumulative burned area was less than 10 km^2, we masked those shrubland pixels out and did not estimate their LFMC.

Table 1. Re-classified fuel classes based on the IGBP classification scheme and its corresponding coverage area and cumulative burned area.

Fuel Class	IGBP	Coverage Area (km^2)	Cumulative Burned Area (km^2)
Forest	Evergreen Needleleaf Forests	28.9	*
	Evergreen Broadleaf Forests	4263.6	117.7
	Deciduous Needleleaf Forests	1.7	*
	Deciduous Broadleaf Forests	10.7	*
	Mixed Forests	26,680	1880.2
Grassland	Woody Savannas,	22,384	2079.2
	Savannas	30	*
	Grasslands	5784.7	417.8
	Permanent Wetlands	108.9	*
	Croplands	9794.9	429
	Cropland/Natural Vegetation Mosaics	5687.9	281.5
Shrubland	Closed Shrublands	139.6	*
	Open Shrublands	206.4	*

The coverage area was calculated based on IGBP in 2013. The cumulative burned area was calculated from 2007 to 2016 based on IGBP in 2013. * denotes the cumulative burned area of this vegetation type from 2007 to 2016 is less than 5 km^2.

2.3.2. Reflectance

The MODIS MCD43A4 Collection 5 product [44] provides 8-day temporal resolution and 500 m surface reflectance adjusted by Nadir Bidirectional Reflectance Distribution Function (BRDF), and the MODIS MCD43A2 Collection 5 product records the quality information of the MCD43A4 pixel reflectance. MCD43A4 is based on a 16-day period, which makes LFMC retrieval less influenced by clouds or shadows [45]. Moreover, adjustment by BRDF makes the observed reflectance closer to RTM simulations that were based on zero zenith angles [46].

Directly validating satellite-derived LFMC with LFMC measurements is unreasonable since the LFMC was measured at 30/40 m scale which was mismatched with the spatial resolution of MCD43A4 (500 m). Therefore, Landsat 8 Operational Land Imager (OLI) products with spatial resolution of 30 m and acquired at the closest sampling periods were used to filter out the LFMC measurements sampled in heterogeneous areas (Section 2.4.1).

2.3.3. Burned Area

The MODIS Burned Area (BA) product MCD64A1 Collection 6 [40] was proven to be of high accuracy and large-scale BA product [47] was selected as the measure of fire occurrence in this study. The product is generated monthly at a spatial resolution of 500 m. All pixels with values greater than zero in the "Burn Date" layer were identified as burned pixels and were used to determine the burn locations and burn dates in the study area.

2.4. Data Analysis

2.4.1. LFMC Retrieval and Validation

LFMC dynamics were retrieved and mapped from MCD43A4 based on the Look-Up Table (LUT) algorithm following Quan et al. [28] (grassland) and Quan et al. [36] (forest). In these studies, the PROSAIL RTM (PROSPECT [48] + SAILH [49,50]) was used for the LFMC retrieval for grassland, and the PROSAIL RTM coupled with PROGeoSAIL RTM (PROSPECT + GeoSAIL [51]) was used for forest. The latter was coupled to better represent a two-layered forest characteristic with upper tree species and understory of grass. To validate the approach, LFMC field measurements (see Section 2.2) were used, however, directly validating estimations with field measurements may cause errors because of scale discrepancy [52] (Figure 3). To alleviate the scale discrepancy between sampling plots (30 m for grassland and 40 m for forest) and MCD43A4-derived LFMC pixels (500 m) only homogeneous plots within the MODIS footprint were selected. The homogeneity of the plots was assessed using the standard deviation of NDVI (Normalized Difference Vegetation Index, Equation (1), Figure 3) derived from the Landsat 8 OLI pixels within a 500 m × 500 m buffer (SD_{NDVI}) and the CV_{NDVI} (Equation (2)). We argue that plants within the 500 m × 500 m MODIS pixel size buffer should be more homogeneous in both species composition and moisture condition when the CV_{NDVI} (ranges from 0.05 to 0.15) and SD_{NDVI} (ranges from 0.15 to 0.30) are lower than a certain threshold [53]. Finally, the threshold values that resulted in high R^2 and low RMSE were selected for the final methodology. Noteworthily, 152 field data were finally obtained after calculating the average value of field LFMC measurements (originally 192 LFMC measurements) at MODIS scale (Figure 3b).

$$\text{NDVI} = \frac{\rho_{NIR} - \rho_{red}}{\rho_{NIR} + \rho_{red}} \tag{1}$$

$$CV_{NDVI} = \frac{SD_{NDVI}}{MEAN_{NDVI}} \tag{2}$$

where ρ_{NIR} and ρ_{red} are the near-infrared and red reflectance of Landsat 8 OLI product, respectively.

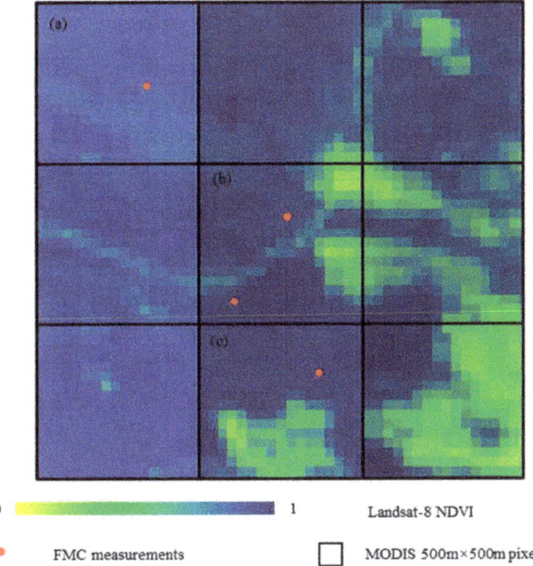

Figure 3. A case showing the spatial location of field plot, Landsat 8 OLI pixel and MODIS pixel. The black boxes indicate the 3×3 MODIS grid (1.5 km $\times$ 1.5 km). (**a**) homogeneous and (**c**) heterogeneous LFMC measurement plots of vegetation at MODIS spatial resolution and (**b**) an example of a MODIS pixel containing multiple LFMC measurement plots, and 152 field data were finally obtained after calculating the average value of the field LFMC measurements in such a case.

The processing of the Landsat data was conducted using the Google Earth Engine [54] and the LFMC retrieval algorithm was implemented in MATLAB (R2017a version, The MathWorks, Natick, Massachusetts, United States of America).

2.4.2. Critical LFMC Thresholds and Their Relation to Fire Occurrence over Southwest China

The MCD64A1 burned area product was used to extract the historical wildfire location and date which, however, were almost provided at the pixel level, rather than specific fire events (normally with the burned area more than one pixel). The KD-Tree based DBSCAN (Density-Based Spatial Clustering of Applications with Noise) algorithm [55] was used to extract the total burned area for each of the studied fire events. Figure 1c,d show the cluster results of the Qiubei forest fire and the Lantern Mountain grassland fire. To analyze the influence of pre-fire LFMC on the burned area and frequency, the critical LFMC thresholds should be determined beforehand. We identified the burn dates of all pixels within each fire event based on the information of "Burn Date" layer of MCD64A1 BA product. The pre-fire LFMC value of each pixel was equal to the LFMC prior to the burning date of that pixel and the median value of all pre-fire LFMC values was used to characterize the overall LFMC condition before the fire broke out. The cumulative burned area by fire event was therefore calculated as a function of decreasing the pre-fire median LFMC value following Dennison and Moritz [15] and Nolan et al. [9]. A segmented regression model [51] was then applied to fit the relation between pre-fire LFMC and cumulative burned area for all the fire events, thus identifying critical LFMC thresholds. The model with the lowest Akaike's Information Criterion (AIC) was selected as the optimal [9]. Breakpoints that indicate a significant increase in the cumulative burned area were finally identified as the critical threshold and other breakpoints were discarded because of the small significance. Here, we divided the study area into four areas (forest across Cwa and Cwb, grassland across Cwa and Cwb), and this methodology based on cumulative burned area was applied to each area. Moreover, the LFMC dynamics before and after the Qiubei forest fire and Lantern Mountain grassland fire were analyzed as two case studies to illustrate the effect of LFMC critical thresholds on fire occurrence.

Finally, the median LFMC values for forest and grassland over southwest China were calculated for each of the LFMC maps from 2007 to 2016, and then a boxplot was used to characterize the overall LFMC distribution on each DOY (Day of Year) at 8-day temporal resolution through the ten years. Because of the similarity of critical LFMC thresholds in Cwa and Cwb climate zones (see in Section 3.2), we did not distinguish here two climatic zones and analyzed the relation between LFMC climatology and fire occurrence based on different fuel classes (forest and grassland). We additionally analyzed the median LFMC and corresponding burned area dynamics for forest in 2015/2016 fire season and grassland in 2009/2010 fire season. Those periods were selected because of the occurrence of the large wildfires subject to investigation in this research. In this case, temporal resolution for the burned area was re-calculated to 8-days according to the burn date extracted from the MCD64A1 product.

3. Results

3.1. LFMC Validation and Mapping

The accuracy in the LFMC retrievals improved (R^2 increased and RMSE decreased) when decreasing the CV_{NDVI} and SD_{NDVI} threshold values used to filter out heterogeneous plots (Figure 4). Specifically, R^2 increased from 0.52 to 0.67, and RMSE slightly decreased from 41.8% to 40.5% as the CV_{NDVI} and SD_{NDVI} increased.

A result of this study is a multi-temporal LFMC dataset over the study area from 2007 to 2016. An example of the monthly LFMC distribution during the 2009/2010 fire season is shown in Figure 5. The LFMC in the study area was low from Nov 2009 to May 2010, particularly during the months of January to April. These months coincide with the dry season and peaks in the burned area in the study area (Figure 2). The rainy season generally begins in June and lasts until October. Consequently, high LFMC values are observed during this period but also more data gaps due to frequent cloudy and rainy weather (Figure 5).

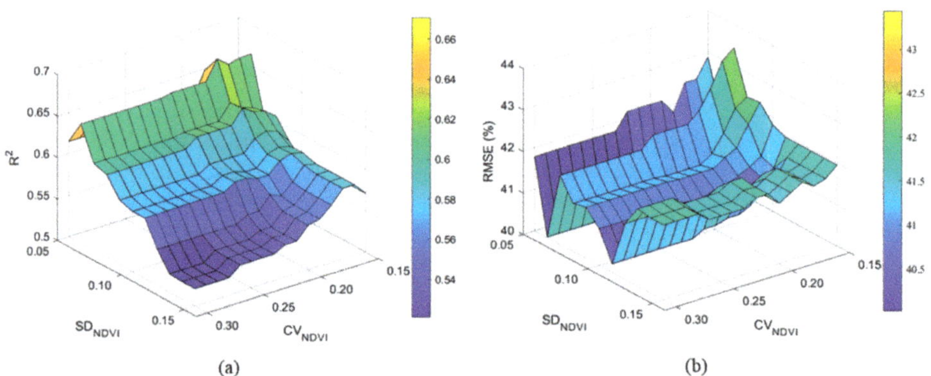

Figure 4. Three-dimensional maps showing R^2 (**a**) and RMSE (**b**) between LFMC observations and estimations with the variation of the threshold values of CV_{NDVI} and SD_{NDVI} used to filter out sites that are heterogeneous within the MODIS footprint.

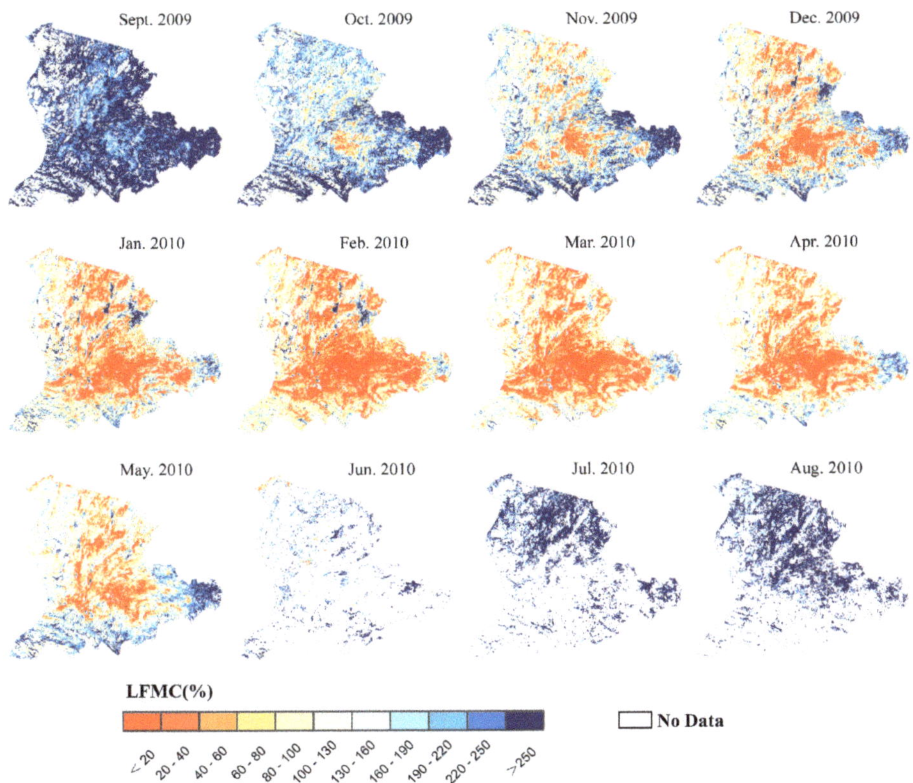

Figure 5. Multi-temporal mapping of monthly LFMC over southwest China throughout the 2009/2010 fire season.

3.2. Critical LFMC Thresholds and Their Relation to Fire Occurrence over Southwest China

The relations between the burned area and pre-fire LFMC were non-linear for both grassland and forest (Figure 6). However, the critical pre-fire LFMC thresholds that explain an increase in wildfire occurrence were different for these fuel classes. The burned area increased significantly when the LFMC was below the thresholds (Figure 6). Moreover, the thresholds of the same fuel class were similar under different climatic zones. More specifically, three LFMC thresholds were observed for the forest under the Cwa climate, 151.3% (95% Confidence Interval, CI: 146.8%–155.9%), 123.1% (95% CI: 121.8%–124.3%) and 51.4% (95% CI: 51.2%–51.7%). The ratio of the burned area below these thresholds to the total burned area accounted for 93.1%, 86.5%, and 34.2% of the total burned area, respectively (Figure 6a, Table 2). Ten large forest fires (burned area >10 km^2) were also detected in this region, and out of which 10, 9, and 5 occurred below the corresponding LFMC thresholds (Table 2). There were three additional breakpoints identified at 101.8% (95% CI: 100.2%–103.4%), 48.3% (95% CI: 47.9%–48.6%) and 39.8% (95% CI: 38.5%–41.1%), which were not identified as thresholds and were discarded since they did not indicate a significant increase in burned area (e.g., the slope between 51.4% and 101.8% was lower than the slope between 101.8% and 123.1%). Three LFMC thresholds were also observed for grassland across the Cwa climate (138.1% (95% CI: 134.1%–142.0%), 72.8% (95% CI: 70.8%–74.8%), and 13.1% (95% CI: 12.1%–14.1%)) (Figure 6b). Similarly, 17, 14, and 4 out of 21 large grassland fires occurred below the corresponding threshold (Table 2).

Two thresholds were found for forest in the Cwb zone (115.0% (95% CI: 113.6%–116.3%) and 54.4% (95% CI: 53.6%–55.2%)) (Figure 6c). These two threshold values were close to those found for the second (123.1%) and third (51.4%) thresholds for forest in the Cwa zone. The maximum LFMC supporting forest fire occurrence in this region was 124.3% and no wildfire was detected at the wetter range of pre-fire LFMC, thus, no threshold was identified. For grassland across Cwb, 137.5% (95% CI: 129.4%–145.6%), 69.0% (95% CI: 66.5%–71.4%), and 10.6% (95% CI: 10.2%–11.0%) were identified as critical LFMC thresholds (Figure 6d). The three thresholds were similar to the LFMC thresholds for grassland across the Cwa zone.

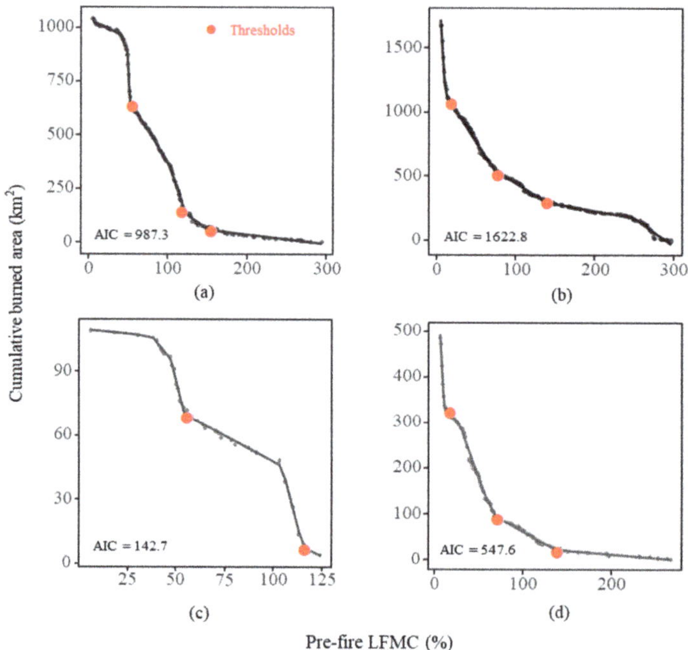

Figure 6. Relationship between pre-fire LFMC and cumulative burned area for forest (**a**) and grassland (**b**) across the Cwa climate zone, and forest (**c**) and grassland (**d**) across Cwb.

Table 2. Value and range of critical LFMC thresholds, and proportion of burned area and the number of large fires below the corresponding threshold for forest and grassland across Cwa and Cwb zones.

Fuel Class	Climate Zone	Threshold (%)	95% CI (%)	Burned Area Proportion (%)	Large Fire Number
Forest	Cwa	151.3	146.8–155.9	93.1	10/10
		123.1	121.8–124.3	86.5	9/10
		51.4	51.2–51.7	34.2	5/10
	Cwb	115.0	113.6–116.3	92.2	2/2
		54.4	53.6–55.2	34.1	0/2
Grassland	Cwa	138.1	134.1–142.0	81.6	17/21
		72.8	70.8–74.8	67.5	14/21
		13.1	12.1–14.1	33.7	4/21
	Cwb	137.5	129.4–145.6	94.4	2/2
		69.0	66.5–71.4	81.1	2/2
		10.6	10.2–11.0	30.7	0/2

95% CI (%) represents the range of this threshold under the 95% confidence interval. Burned area proportion represents the ratio of the burned area below this threshold to the total burned area. Large fire number denotes the number of large fires (greater than 10 km^2) below this threshold and the total number of large fires for forest or grassland across Cwa (Subtropical Highland Zone) or Cwb (Humid Subtropical Zone).

The Qiubei forest fire, which occurred on the 1st February 2010, affected 75% of the forested area and 25% of the grassland area in the region. In six months, the median LFMC within the burned area gradually decreased from a maximum of 179.7% (YEAR-DOY 2009-241) to a minimum of 49.8% (2010-025), and a fire broke out when the LFMC was below the 51.4% threshold reported for forest across the Cwa zone (Figure 7a). After the fire, the median LFMC value across the burned area stabilized around 49% for more than one month. Similarly, the median LFMC for the vegetation within the final burned area of the Lantern Mountain fire (78% grassland and 22% forest) declined from 286.8% (2015-249) to 151.7% (2015-345) (Figure 7b). Two months before the fire occurred, the median LFMC (125.2%, 2015-353) was already below the 138.1% critical LFMC threshold found for grassland across Cwa zone and further decreased to 17.3% (20156-041), which was slightly higher than the 13.1% grassland LFMC threshold, just before the fire broke out. Different to the Qiubei forest fire, the LFMC recovered quickly after the fire probably due to a quicker regrowth of the grass in this region.

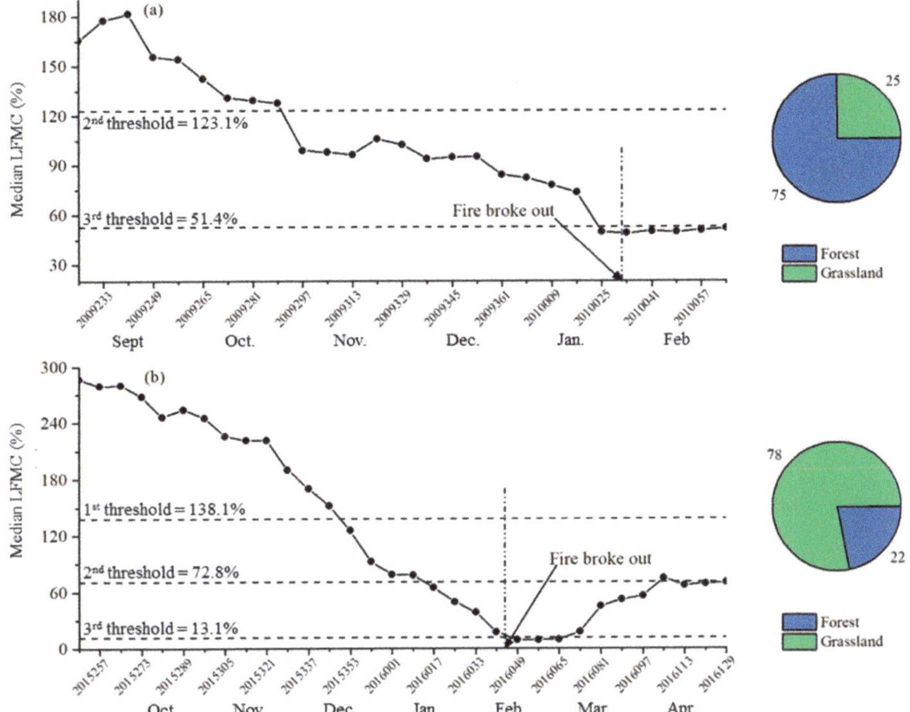

Figure 7. Temporal LFMC dynamics before and after the Qiubei (**a**) and the Lantern Mountain (**b**) fire events across the Cwa zone. The dotted lines represent the critical LFMC thresholds. The pie chart shows the percentage of the total burned area per land cover.

The dynamics variation of the median grassland (Figure 8b) and forest (Figure 8a) LFMC across the study area are similar in terms of alignment with the fire season (see Section 2.1). The lowest LFMC were observed from DOY 337 to DOY 113 which coincided with the dry season and the months with the highest fire occurrence (Figure 8). Additionally, the median LFMC for grassland and forest of the study area (red lines) reached the critical LFMC thresholds (dotted lines) earlier during the highest fire activity fire seasons (2009/2010 and 2015/2016) than for the whole time period (shown as boxplots). Moreover, the median LFMC values during these two fire seasons were almost lower than the first quartile value of the boxplot on the same DOY and were much more likely to be observed as the low outliers (dots traversed by the red line). For example, 14 and 21 median LFMC values were observed as the outliers in the 2015/2016 (Figure 8a) and 2009/2010 (Figure 8b) fire seasons, respectively. Furthermore, the period of lower LFMC coincided with the larger burned area (Figure 8). This suggests that the critical LFMC thresholds effectively explain the burned area. For example, almost all grassland fires in 2009/2010 occurred when the regional median LFMC was below the 70.9% threshold (Figure 8b).

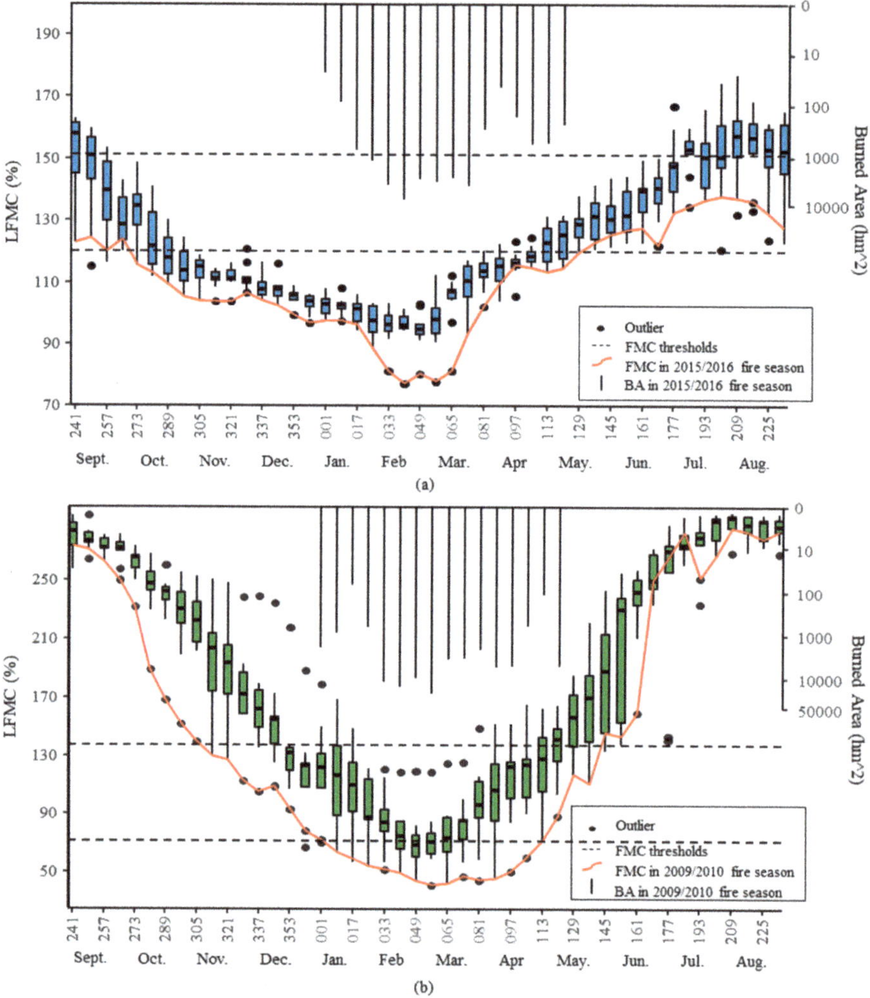

Figure 8. Dynamics of LFMC and burned area over forest (**a**) and grassland (**b**) in southwest China. Each boxplot indicates the distribution of median LFMC values of the forest or grassland on the same DOY over ten years (2007 to 2016). Red lines indicate the LFMC dynamics during the fire seasons when the studied fires broke down (**a**) 2015/2016 and (**b**) 2009/2010. Dotted lines denote the critical LFMC thresholds. The bar chart indicates the re-calculated 8-day burned area for (**a**) 2015/2016 and (**b**) 2009/2010.

4. Discussion

High-quality spatial information on LFMC is needed to explore the effect of LFMC on fire occurrence at a regional scale. In this study, we followed the methodologies by Quan et al. [28,36] to retrieve LFMC over southwest China using the MCD43A4 product. There was a statistically significant (p <0.01) agreement between retrieved and measured LFMC at field sites used to evaluate the inverse RTM method. The R^2 and RMSE of LFMC improved from 0.52 to 0.67 and from 41.8% to 40.5% when we removed pixels heterogeneous within the MODIS footprint. This indicated the heterogeneity in the sampling site highly influences the accuracy assessment of the LFMC retrievals and therefore it should be considered when it comes to model evaluation.

In this study, cumulative burned area methodology was applied to identify breakpoints, and thus to determine the thresholds in forest and grassland classes across Cwa and Cwb climatic regions. Pimont et al. [56] suggested that this kind of methodology should be considered with caution because it is biased by the frequency distribution of LFMC values. However, we argue that the LFMC threshold is not only affected by the distribution of pre-fire LFMC but also by the corresponding burned area. For example, we found that in low LFMC condition (under 15%, Figure 6b,d), cumulative burned area significantly increases, mainly because of the occurrence of several fire events with a large burned area. Therefore, it is reasonable to use this methodology to find LFMC thresholds values below which fires are prone to break out and burn large areas. On the other hand, Pimont et al. [56] also found that fire activity saturated at low LFMC values when cumulative burned area methodology was applied. For example, the cumulative burned area basically did not increase when LFMC was lower than 37.2% (Figure 6c). However, this saturation does not suggest that a low LFMC corresponds to a low fire occurrence. It suggests that it is enough to support fire when the LFMC condition is slightly higher than 37.2%. Therefore, a higher LFMC threshold of 54.4% (Figure 6c) was considered.

Similar to these previously reported studies [9,15,21,31], we found that most wildfires occurred under low LFMC conditions and when LFMC crossed critical LFMC thresholds. Conversely, when pre-fire LFMC exceeded 200% only 0.29% (forest across Cwa) and 0.19% (grassland across Cwb) of the total burned area occurred (Figure 6a,d). However, four large fires with pre-fire LFMC greater than 200% were unexpectedly detected for grassland across Cwa (Figure 6b). Yebra et al. [43] also found high pre-fire LFMC (232%–256%) in the Linksview Road Grassland Fire which occurred in New South Wales on 16th October 2013. This suggests that FMC is not the only driver of wildfire occurrence and therefore, other factors such as meteorological data (e.g., temperature, precipitation, air humidity, etc.) should be accounted for full characterization. For both forest and grassland over southwest China, the first threshold (137.5%–151.3%) is similar to the LFMC threshold identified by Nolan et al. [9] at 156.1% in eastern Australia forest and woodland. The second thresholds changed between different fuel classes with 115.0%–123.1% for forest and 69.0%–72.8% for grassland. Among them, 115.0%–123.1% for forest is similar to the threshold identified by Nolan et al. [9] at 113.6% in eastern Australia, and 69.0%–72.8% for grassland is similar to those reported in other researches [9,15,21]. The third threshold (51.4%–54.4%) for forest is close to the lowest value of LFMC that results in wildfire occurrence in other studies [9,15,17,20,21]. The third threshold (10.6%–13.2%) for grassland is consistent with the previously reported results that the occurrence of wildfires increases when the DFMC decreases to 12.4%–15.1% [9].

LFMC dynamics over a 10-year time series showed that median forest and grassland LFMC values were significantly lower during fire seasons with relatively higher fire activity than other fire seasons. Additionally, those LFMC values were commonly detected as the outliers of the 10-year time series. This suggested that the LFMC is an effective driver and an early indicator of wildfire occurrence over southwest China.

We used optical remote sensing data for LFMC mapping in near real-time. However, because of the weak penetration ability of optical spectral signals through clouds, its applications over cloudy areas are largely limited [57,58]. For example, the computed LFMC maps had many data gaps in June, July, and August over the study area (Figure 5) in spite of the monthly image composite of MCD43A4 data used. Fortunately, this has little impact on analyzing the effects of LFMC on fire occurrences because seldom have wildfires been recorded during this period over the past 10 years (Figure 2). Nevertheless, it poses the question that wildfires may be highly frequent in cloudy areas (e.g., the tropics are inevitably affected by fire [59] and are often covered by clouds [60]). Microwave remote sensing data has been demonstrated to have better potential application in LFMC inversion because of its strong penetration ability [61] and high sensitivity to surface moisture [62–65]. The combination of multi-source optical and microwave remote sensing may be an effective way to alleviate the missing data problem caused by weather conditions, allowing high quality and long-term LFMC products for wildfire risk assessment to be generated. This will be explored in future work.

5. Conclusions

In this paper, we presented the first study on exploring the effect of LFMC on wildfire occurrence over southwest China. The LFMC dynamics from 2007 to 2016 were first retrieved using RTM and reflectance from the MODIS MCD43A4 product and then validated using field LFMC measurements. Wildfires events were identified through the KD-tree based DBSCAN algorithm under two Koppen climate zones (Cwa and Cwb). Statistical results showed that the LFMC dynamic remarkably affects both grassland and forest fire occurrence: Forest and grassland wildfires between 2007 and 2016 were controlled by three pre-fire critical LFMC thresholds which varied slightly between different fuel classes but were similar for the two climate zones. Furthermore, regional median LFMC during a fire season with high fire activity was significantly lower than the ten year average LFMC. Therefore, this study demonstrated that LFMC dynamics have clear effects on wildfire occurrence over southwest China, allowing the possibility to develop a new operational wildfire early-warning model over this area.

Author Contributions: Study conception and design: X.Q. and B.H. Methodology refining: M.Y. Acquisition of data: K.L. and X.Q. Analysis and interpretation of data: K.L., X.Q., and B.H. The first draft of the manuscript: K.L. Critical revisions: X.Q., M.Y., and B.H., X.Q. and K.L. contributed equally to this article.

Funding: This work was supported by the National Natural Science Foundation of China (Contract No. 41671361 and 41801272) and the Fundamental Research for the Central Universities (Contract No. ZYGX2017KYQD195).

Acknowledgments: The authors wish to thank NASA for providing MODIS MCD12Q1, MCD64A1, MCD43A2, MCD43A4 and Landsat 8 OLI datasets; Google for providing a planetary-scale platform Google Earth Engine for Earth science data and analysis; Peel et al for providing the Koppen climate classification map as a raster file.

Conflicts of Interest: The authors declare no conflict of interest.

List of abbreviations

Abbreviation	Meaning
LFMC	Live Fuel Moisture Content
RTM	Radiative Transfer Model
Cwa	Subtropical Highland Zone
Cwb	Humid Subtropical Zone
BA	Burned Area
IGBP	International Geosphere–Biosphere Programme
KBDI	Keetch–Byram Drought Index
DEM	Digital Elevation Model
BRDF	Bidirectional Reflectance Distribution Function
Landsat 8 OLI	Landsat 8 Operational Land Imager
NDVI	Normalized Difference Vegetation Index
SD_{NDVI}	Standard Deviation
CV_{NDVI}	Coefficient of Variance
DBSCAN	Density-Based Spatial Clustering of Applications with Noise
CI	Confidence Interval
DOY	Day of Year

References

1. Kilgore, B.M. The ecological role of fire in Sierran conifer forests: Its application to National Park management. *Quat. Res.* **1973**, *3*, 496–513. [CrossRef]

2. Van der Werf, G.R.; Randerson, J.T.; Giglio, L.; Collatz, G.J.; Kasibhatla, P.S.; Arellano, A.F., Jr. Interannual variability in global biomass burning emissions from 1997 to 2004. *Atmos. Chem. Phys.* **2006**, *6*, 3423–3441. [CrossRef]

3. Werf, G.R.V.D.; Morton, D.C.; Defries, R.S.; Olivier, J.G.J.; Kasibhatla, P.S.; Jackson, R.B.; Collatz, G.J.; Randerson, J.T. CO_2 emissions from forest loss. *Nat. Geosci.* **2009**, *2*, 737–738. [CrossRef]

4. Boerner, R.E.J.; Jianjun, H.; Hart, S.C. Impacts of fire and fire surrogate treatments on forest soil properties: A meta-analytical approach. *Ecol. Appl.* **2009**, *19*, 338–358. [CrossRef] [PubMed]

5. Liu, X.; He, B.; Quan, X.; Yebra, M.; Qiu, S.; Yin, C.; Liao, Z.; Zhang, H. Near real-time extracting wildfire spread rate from himawari-8 satellite data. *Remote Sens.* **2018**, *10*, 1654. [CrossRef]

6. Rieman, B.E.; Gresswell, R.E.; Young, M.K.; Luce, C.H. Introduction to the effects of wildland fire on aquatic ecosystems in the Western USA. *For. Ecol. Manag.* **2003**, *178*, 1–3. [CrossRef]

7. Bradstock, R.A. A biogeographic model of fire regimes in Australia: Current and future implications. *Glob. Ecol. Biogeogr.* **2010**, *19*, 145–158. [CrossRef]

8. Meyn, A.; White, P.S.; Buhk, C.; Jentsch, A. Environmental drivers of large, infrequent wildfires: The emerging conceptual model. *Prog. Phys. Geogr.* **2007**, *3*, 287–312. [CrossRef]

9. Nolan, R.H.; Boer, M.M.; de Dios, V.R.; Caccamo, G.; Bradstock, R.A. Large-scale, dynamic transformations in fuel moisture drive wildfire activity across southeastern Australia. *Geoph. Res. Lett.* **2016**, *43*, 4229–4238. [CrossRef]

10. Chuvieco, E.; Aguado, I.; Dimitrakopoulos, A.P. Conversion of fuel moisture content values to ignition potential for integrated fire danger assessment. *Can. J. For. Res.* **2004**, *34*, 2284–2293. [CrossRef]

11. Viegas, D.X.; Viegas, M.; Ferreira, A.D. Moisture content of fine forest fuels and fire occurrence in central Portugal. *Int. J. Wildland Fire* **1992**, *2*, 69–86. [CrossRef]

12. Rossa, C.G. The effect of fuel moisture content on the spread rate of forest fires in the absence of wind or slope. *Int. J. Wildland Fire* **2017**, *26*, 24. [CrossRef]

13. Rossa, C.G.; Fernandes, P.M. Live fuel moisture content: The 'pea under the mattress' of fire spread rate modeling? *Fire* **2018**, *1*, 43. [CrossRef]

14. Davis, F.W.; Michaelsen, J. *Sensitivity of Fire Regime in Chaparral Ecosystems to Climate Change*; Springer: New York, NY, USA, 1995; pp. 435–456.

15. Dennison, P.E.; Moritz, M.A. Critical live fuel moisture in chaparral ecosystems: A threshold for fire activity and its relationship to antecedent precipitation. *Int. J. Wildland Fire* **2009**, *18*, 1021–1027. [CrossRef]

16. Agee, J.K.; Wright, C.S.; Williamson, N.; Huff, M.H. Foliar moisture content of Pacific Northwest vegetation and its relation to wildland fire behavior. *For. Ecol. Manag.* **2002**, *167*, 57–66. [CrossRef]

17. Chuvieco, E.; González, I.; Verdú, F.; Aguado, I.; Yebra, M. Prediction of fire occurrence from live fuel moisture content measurements in a Mediterranean ecosystem. *Int. J. Wildland Fire* **2009**, *18*, 430–441. [CrossRef]

18. Dimitrakopoulos, A.P.; Papaioannou, K.K. Flammability assessment of Mediterranean forest fuels. *Fire Technol.* **2001**, *37*, 143–152. [CrossRef]

19. Yebra, M.; Dennison, P.E.; Chuvieco, E.; Riaño, D.; Zylstra, P.; Hunt, E.R., Jr.; Danson, F.M.; Yi, Q.; Jurdao, S. A global review of remote sensing of live fuel moisture content for fire danger assessment: Moving towards operational products. *Remote Sens. Environ.* **2013**, *136*, 455–468. [CrossRef]

20. Schoenberg, F.P.; Peng, R.; Huang, Z.J.; Rundel, P. Detection of non-linearities in the dependence of burn area on fuel age and climatic variables. *Int. J. Wildland Fire* **2003**, *12*, 1–6. [CrossRef]

21. Dennison, P.E.; Moritz, M.A.; Taylor, R.S. Evaluating predictive models of critical live fuel moisture in the Santa Monica Mountains, California. *Int. J. Wildland Fire* **2008**, *17*, 18–27. [CrossRef]

22. Burgan, R.E. *1988 Revisions to the 1978 National Fire-Danger Rating System*; Research Paper SE-273; US Department of Agriculture, Forest Service, Southeastern Forest Experiment Station: Asheville, NC, USA, 1988; p. 144.

23. Dennison, P.E.; Roberts, D.A.; Thorgusen, S.R.; Regelbrugge, J.C.; Weise, D.; Lee, C. Modeling seasonal changes in live fuel moisture and equivalent water thickness using a cumulative water balance index. *Remote Sens. Environ.* **2003**, *88*, 442–452. [CrossRef]

24. Dimitrakopoulos, A.; Bemmerzouk, A. Predicting live herbaceous moisture content from a seasonal drought index. *Int. J. Biometeorol.* **2003**, *47*, 73–79. [PubMed]

25. Ruffault, J.; Martin-StPaul, N.; Pimont, F.; Dupuy, J.L. How well do meteorological drought indices predict live fuel moisture content (LFMC)? An assessment for wildfire research and operations in Mediterranean ecosystems. *Agric. For. Meteorol.* **2018**, *262*, 391–401. [CrossRef]

26. Viegas, D.; Piñol, J.; Viegas, M.; Ogaya, R. Estimating live fine fuels moisture content using meteorologically-based indices. *Int. J. Wildland Fire* **2001**, *10*, 223–240. [CrossRef]

27. Quan, X.; He, B.; Li, X.; Tang, Z. Estimation of grassland live fuel moisture content from ratio of canopy water content and foliage dry biomass. *IEEE Geosci. Remote Sens. Lett.* **2015**, *12*, 1903–1907. [CrossRef]

28. Quan, X.; He, B.; Li, X.; Liao, Z. Retrieval of grassland live fuel moisture content by parameterizing radiative transfer model with interval estimated LAI. *IEEE J. Sel. Top. Appl. Earth Obs. Remote Sens.* **2016**, *9*, 910–920. [CrossRef]

29. Quan, X.; He, B.; Yebra, M.; Yin, C.; Liao, Z.; Zhang, X.; Li, X. A radiative transfer model-based method for the estimation of grassland aboveground biomass. *Int. J. Appl. Earth Obs. Geoinf.* **2017**, *54*, 159–168. [CrossRef]

30. Myoung, B.; Kim, S.H.; Nghiem, S.V.; Jia, S.; Whitney, K.; Kafatos, M.C. Estimating live fuel moisture from MODIS satellite data for wildfire danger assessment in Southern California USA. *Remote Sens.* **2018**, *10*, 87. [CrossRef]

31. Jurdao, S.; Chuvieco, E.; Arevalillo, J.M. Modelling fire ignition probability from satellite estimates of live fuel moisture content. *Fire Ecol.* **2012**, *8*, 77–97. [CrossRef]

32. Nolan, R.H.; Dios, V.R.D.; Boer, M.M.; Caccamo, G.; Goulden, M.L.; Bradstock, R.A. Predicting dead fine fuel moisture at regional scales using vapour pressure deficit from MODIS and gridded weather data. *Remote Sens. Environ.* **2016**, *174*, 100–108. [CrossRef]

33. Al-Moustafa, T.; Armitage, R.P.; Danson, F.M. Mapping fuel moisture content in upland vegetation using airborne hyperspectral imagery. *Remote Sens. Environ.* **2012**, *127*, 74–83. [CrossRef]

34. Houborg, R.; Anderson, M.; Daughtry, C. Utility of an image-based canopy reflectance modeling tool for remote estimation of LAI and leaf chlorophyll content at the field scale. *Remote Sens. Environ.* **2009**, *113*, 259–274. [CrossRef]

35. Huang, J.X.; Sedano, F.; Huang, Y.B.; Ma, H.Y.; Li, X.L.; Liang, S.L.; Tian, L.Y.; Zhang, X.D.; Fan, J.L.; Wu, W.B. Assimilating a synthetic Kalman filter leaf area index series into the WOFOST model to improve regional winter wheat yield estimation. *Agric. For. Meteorol.* **2016**, *216*, 188–202. [CrossRef]

36. Quan, X.; He, B.; Yebra, M.; Yin, C.; Liao, Z.; Li, X. Retrieval of forest fuel moisture content using a coupled radiative transfer model. *J. Environ. Modell. Softw.* **2017**, *95*, 290–302. [CrossRef]

37. Yebra, M.; Chuvieco, E.; Riaño, D. Estimation of live fuel moisture content from MODIS images for fire risk assessment. *Agric. For. Meteorol.* **2008**, *148*, 523–536. [CrossRef]

38. Friedl, M.A.; Sulla-Menashe, D.; Tan, B.; Schneider, A.; Ramankutty, N.; Sibley, A.; Huang, X. MODIS collection 5 global land cover: Algorithm refinements and characterization of new datasets. *Remote Sens. Environ.* **2010**, *114*, 168–182. [CrossRef]

39. Peel, M.C.; Finlayson, B.L.; McMahon, T.A. Updated world map of the Koppen-Geiger climate classification. *Hydrol. Earth Syst. Sci.* **2007**, *11*, 1633–1644. [CrossRef]

40. Giglio, L.; Schroeder, W.; Justice, C.O. The collection 6 MODIS active fire detection algorithm and fire products. *Remote Sens. Environ.* **2016**, *178*, 31–41. [CrossRef]

41. Arganaraz, J.P.; Landi, M.A.; Scavuzzo, C.M.; Bellis, L.M. Determining fuel moisture thresholds to assess wildfire hazard: A contribution to an operational early warning system. *PLoS ONE* **2018**, *13*, e0204889. [CrossRef]

42. Quan, X.; He, B.; Li, X. A bayesian network-based method to alleviate the ill-posed inverse problem: A case study on leaf area index and canopy water content retrieval. *IEEE Trans. Geosci. Remote Sens.* **2015**, *53*, 6507–6517. [CrossRef]

43. Yebra, M.; Quan, X.; Riaño, D.; Larraondo, P.R.; Dijk, A.I.J.M.V.; Cary, G.J. A fuel moisture content and flammability monitoring methodology for continental Australia based on optical remote sensing. *Remote Sens. Environ.* **2018**, *212*, 260–272. [CrossRef]

44. Strahler, A.H. *MODIS BRDF/Albedo Product: Algorithm Theoretical Basis Document Version 5.0.* 1999. Available online: https://www.semanticscholar.org/paper/MODIS-BRDF-%2F-Albedo-Product-%3A-Algorithm-Theoretical-Strahler-Lucht/1adc54eac2199b93536c5988bb2bd1952127d74f (accessed on 9 September 2019).

45. Yebra, M.; Dijk, A.V.; Leuning, R.; Huete, A.; Guerschman, J.P. Evaluation of optical remote sensing to estimate actual evapotranspiration and canopy conductance. *Remote Sens. Environ.* **2013**, *129*, 250–261. [CrossRef]

46. Jurdao, S.; Yebra, M.; Guerschman, J.P.; Chuvieco, E. Regional estimation of woodland moisture content by inverting radiative transfer models. *Remote Sens. Environ.* **2013**, *132*, 59–70. [CrossRef]

47. Padilla, M.; Stehman, S.V.; Ramo, R.; Corti, D.; Hantson, S.; Oliva, P.; Alonso-Canas, I.; Bradley, A.V.; Tansey, K.; Mota, B.J.R.S.O.E. Comparing the accuracies of remote sensing global burned area products using stratified random sampling and estimation. *Remote Sens. Environ.* **2015**, *160*, 114–121. [CrossRef]

48. Feret, J.B.; Francois, C.; Asner, G.P.; Gitelson, A.A.; Martin, R.E.; Bidel, L.P.R.; Ustin, S.L.; le Maire, G.; Jacquemoud, S. PROSPECT-4 and 5: Advances in the leaf optical properties model separating photosynthetic pigments. *Remote Sens. Environ.* **2008**, *112*, 3030–3043. [CrossRef]

49. Verhoef, W. Light scattering by leaf layers with application to canopy reflectance modeling: The SAIL model. *Remote Sens. Environ.* **1984**, *16*, 125–141. [CrossRef]

50. Kuusk, A. *The Hot Spot Effect in Plant Canopy Reflectance*; Springer: Berlin/Heidelberg, Germany, 1991; pp. 139–159.

51. Huemmrich, K.F. The GeoSail model: A simple addition to the SAIL model to describe discontinuous canopy reflectance. *Remote Sens. Environ.* **2001**, *75*, 423–431. [CrossRef]

52. Adab, H.; Devi Kanniah, K.; Beringer, J. Estimating and up-scaling fuel moisture and leaf dry matter content of a temperate humid forest using multi resolution remote sensing data. *Remote Sens.* **2016**, *8*, 961. [CrossRef]

53. Yebra, M.; Scortechini, G.; Badi, A.; Beget, M.E.; Boer, M.M.; Bradstock, R.; Chuvieco, E.; Danson, F.M.; Dennison, P.; Resco de Dios, V.; et al. Globe-LFMC, a global plant water status database for vegetation ecophysiology and wildfire applications. *Sci. Data* **2019**, *6*, 155. [CrossRef]

54. Gorelick, N.; Hancher, M.; Dixon, M.; Ilyushchenko, S.; Thau, D.; Moore, R. Google Earth engine: Planetary-scale geospatial analysis for everyone. *Remote Sens. Environ.* **2017**, *202*. [CrossRef]

55. Hahsler, M. Density Based Clustering of Applications with Noise (DBSCAN) and Related Algorithms [R Package Dbscan Version 0.9-7]; 2016. Available online: https://rdrr.io/cran/dbscan/ (accessed on 9 September 2019).

56. Pimont, F.; Ruffault, J.; Martin-Stpaul, N.K.; Dupuy, J.L. A cautionary note regarding the use of cumulative burnt areas for the determination of fire danger index breakpoints. *Int. J. Wildland Fire* **2019**. [CrossRef]

57. Shi, Q.; He, B.; Zhe, Z.; Liao, Z.; Quan, X. Improving Fmask cloud and cloud shadow detection in mountainous area for Landsats 4–8 images. *Remote Sens. Environ.* **2017**, *199*, 107–119.

58. Zhu, Z.; Woodcock, C.E. Object-based cloud and cloud shadow detection in Landsat imagery. *Remote Sens. Environ.* **2012**, *118*, 83–94. [CrossRef]

59. Roteta, E.; Bastarrika, A.; Padilla, M.; Storm, T.; Chuvieco, E. Development of a sentinel-2 burned area algorithm: Generation of a small fire database for sub-Saharan Africa. *Remote Sens. Environ.* **2019**, *222*, 1–17. [CrossRef]

60. Asner, G.P. Cloud cover in landsat observations of the Brazilian Amazon. *Int. J. Remote Sens.* **2001**, *22*, 3855–3862. [CrossRef]

61. Tsang, L.; Kong, J.A.; Shin, R.T. *Theory of Microwave Remote Sensing*; Wiley: New York, NY, USA, 1985.

62. Fan, L.; Wigneron, J.-P.; Xiao, Q.; Al-Yaari, A.; Wen, J.; Martin-StPaul, N.; Dupuy, J.-L.; Pimont, F.; Al Bitar, A.; Fernandez-Moran, R. evaluation of microwave remote sensing for monitoring live fuel moisture content in the Mediterranean region. *Remote Sens. Environ.* **2018**, *205*, 210–223. [CrossRef]

63. Tanase, M.; Panciera, R.; Lowell, K.; Aponte, C. Monitoring live fuel moisture in semiarid environments using L-band radar data. *Int. J. Wildland Fire* **2015**, *24*, 560–572. [CrossRef]

64. Wang, L.; Quan, X.; He, B.; Yebra, M.; Xing, M.; Liu, X. Assessment of the dual polarimetric sentinel-1A data for forest fuel moisture content estimation. *Remote Sens.* **2019**, *11*, 1568. [CrossRef]

65. Jia, S.; Kim, S.H.; Nghiem, S.V.; Kafatos, M. Estimating live fuel moisture using SMAP L-band radiometer soil moisture for Southern California, USA. *Remote Sens.* **2019**, *11*, 1575. [CrossRef]

 *forests*

Article

Terpenoid Accumulation Links Plant Health and Flammability in the Cypress-Bark Canker Pathosystem

Gianni Della Rocca [1], Roberto Danti [1], Carmen Hernando [2,3], Mercedes Guijarro [2,3], Marco Michelozzi [4], Cristina Carrillo [2,5] and Javier Madrigal [2,3,*]

[1] Institute for Sustainable Plant Protection, IPSP-CNR, Via Madonna del Piano 10, I-50019 Sesto Fiorentino (FI), Italy; gianni.dellarocca@ipsp.cnr.it (G.D.R.); roberto.danti@ipsp.cnr.it (R.D.)

[2] INIA–CIFOR, Department of Forest Dynamics and Management, Ctra. A Coruña Km 7.5, 28040 Madrid, Spain; lara@inia.es (C.H.); guijarro@inia.es (M.G.); cristina.carrillo@inia.es (C.C.)

[3] iuFOR, Sustainable Forest Management Institute UVa-INIA, 34004 Palencia, Spain

[4] Istituto di Bioscenze e Biorisorse IBBR-CNR, Via Madonna del Piano 10, I-50019 Sesto Fiorentino (FI), Italy; marco.michelozzi@ibbr.cnr.it

[5] ETSI Montes, Forestal y del Medio Natural, Universidad Politécnica de Madrid (UPM), Ramiro de Maeztu s/n, 28040 Madrid, Spain

* Correspondence: incendio@inia.es

Received: 22 April 2020; Accepted: 5 June 2020; Published: 7 June 2020

Abstract: To explore the possible relationship between diseased trees and wildfires, we assessed the flammability of canker-resistant and susceptible common cypress clones that were artificially infected with *Seiridium cardinale* compared to healthy trees. This study explored the effect of terpenoids produced by the host plant in response to infection and the presence of dead plant portions on flammability. Terpenoids were extracted and quantified in foliage and bark samples by gas chromatography–mass spectrometry (GC–MS). A Mass Loss Calorimeter was used to determine the main flammability descriptors. The concentration of terpenoids in bark and leaf samples and the flammability parameters were compared using a generalized linear mixed models (GLMM) model. A partial least square (PLS) model was generated to predict flammability based on the content of terpenoid, clone response to bark canker and the disease status of the plants. The total terpenoid content drastically increased in the bark of both cypress clones after infection, with a greater (7-fold) increase observed in the resistant clone. On the contrary, levels of terpenoids in leaves did not alter after infection. The GLMM model showed that after infection, plants of the susceptible clone appeared to be much more flammable in comparison to those of resistant clones, showing higher ignitability, combustibility, sustainability and consumability. This was mainly due to the presence of dried crown parts in the susceptible clone. The resistant clone showed a slightly higher ignitability after infection, while the other flammability parameters did not change. The PLS model ($R^2Y = 56\%$) supported these findings, indicating that dead crown parts and fuel moisture content accounted for most of the variation in flammability parameters and greatly prevailed on terpenoid accumulation after infection. The results of this study suggest that a disease can increase the flammability of trees. The deployment of canker-resistant cypress clones can reduce the flammability of cypress plantations in Mediterranean areas affected by bark canker. Epidemiological data of diseased tree distribution can be an important factor in the prediction of fire risk.

Keywords: *Cupressus sempervirens*; fire risk; fuels; fuel moisture content; mass loss calorimeter; *Seiridium cardinale*; vulnerability to wildfires; disease; alien pathogen; allochthonous species; introduced fungus

1. Introduction

The relationship between non-native plant disease and the frequency of wildfires (the effect on fire regimes), and the implications for fire management has become an increasing focus of research in recent years [1,2]. Both wildfires and disease caused by invasive pathogens (and insects) are key factors in determining tree mortality in forests worldwide and are linked to the global change context [3–10]. The relationship between wildfire and forest disease depends on the host–pathogen species involved and their mutual interaction, knowledge in this field is still lacking detailed information.

Current global change scenarios in terms of the combination of climate, shift in land use, and the expansion of trade networks and volume of goods, exacerbate the seasonal drought and warming stress periods that in turn influence plant physiology, biochemical defences and disease severity, in terms of pest and disease movement and outbreaks [9,11,12]. At the stand level, the interaction between wildfire and an emerging fungal forest disease was studied in Californian and Oregon forests affected by sudden oak death (SOD) (caused by *Phytophthora ramorum*) [1,4,13]. This new disease altered the physical and biochemical characteristics of the ecosystem e.g., fuel load, increasing the surface fuel, restructuring the forest canopy, decreasing canopy continuity and increasing tree mortality. This altered the species composition and in turn affected wildfire dynamics (severity, risk of crown ignition, etc.) [4,14–17]. In Californian forests affected by SOD, the rate of standing dead trees was higher, the tanoak (*Notholithocarpus densiflorus*) mass of woody debris on the soil was tens of times greater and the depth of the fuel bed in diseased stands was four times that in disease-free forest [18]. Simulation modelling with the *BehavePlus* fire model system, indicated that flame length, fire spread rate and fireline intensity, increased several times in infected Douglas fir and Redwood stands compared to their healthy counterparts [18].

At the tree level, both climatic and biotic stress factors affect the health of trees. These stresses decrease the water content of plant organs, increase the ratio of dead to alive crown portions, and especially in conifers, influence the qualitative and quantitative amount of several plant defensive compounds, such as terpenoids [19–22]. Terpenoids are considered to be one of the most important molecules affecting forest fuel flammability [23–28]. Terpenoids are constitutive induced lines of defence in conifers; an increase in absolute amounts, changes in their proportions and de novo production of molecules (phytoalexins) have been observed after infection depending on the pathosystem [29].

Cypress canker disease (CCD) is a non-native lethal disease affecting many *Cupressaceae* (above all *Cupressus sempervirens* L., in the Mediterranean area). It is caused by the invasive fungal pathogen *Seiridium cardinale* (Wagener) Sutton et Gibson introduced in Europe (and spread across the globe) from California, USA [30–32]. This destructive disease causes the dieback of crown portions and the desiccation of twigs and branches, due to the girdling of the woody organs by the necrotrophic fungal agent [33]. An additional effect of CCD is also the de novo genesis of traumatic resin ducts (TRD) in bark tissues affected by canker [34–36], and the consequent abundant exudation of resin that flows down from the infected organs [33]. Both of these effects supposedly affect the flammability of the infected trees or their portions. A long-term genetic research program developed since the 1970s in Italy, France and Greece led to the selection of several *C. sempervirens* genotypes resistant to CCD (some of which were patented and made commercially available).

The CCD resistant clones are able to block the growth of the fungus in the infected bark within a few weeks and completely heal the lesion within a few years [37–41]. The efforts undertaken to control the CCD are justified by the high ecological, symbolic, historical and cultural value of this tree in Mediterranean countries. *Cupressus sempervirens* is used in forestry, landscaping in peri-urban and urban contexts and also as a windbreak and hedge [32]. The induction of terpenoids as part of the defensive reaction of common cypress plants (both CCD-resistant and susceptible clones and not selected for CCD resistance plants) to *Seiridium cardinale* infection is characterized by the production and accumulation of several de novo specific compounds [29].

The flammability of the live crown of plants of *C. sempervirens* has already been studied extensively [42–47]; nevertheless, the flammability descriptors (ignitability, sustainability, combustibility and consumability) of healthy and diseased cypress clones selected for CCD resistance have not yet been assessed. This work explores the links between diseased trees and wildfire, comparing the flammability of canker-resistant and susceptible common cypress clones, artificially infected by *S. cardinale*, in comparison to healthy ramets of both clones. We set out to address the following questions: (i) Is a diseased plant more flammable than a healthy one, and if so, to what extent? (ii) Is a CCD-resistant cypress clone less flammable than the CCD-non-resistant equivalent? (iii) How do terpenoids produced by the host in response to infection, and dead plant portions killed by the fungal pathogen, affect flammability?

2. Materials and Methods

2.1. Experimental Set-Up: Plant Selection, Growth Conditions and Artificial Inoculation

Twenty 3-year-old grafted ramets of *Cupressus sempervirens* of the canker-resistant (PM-322; patented cypress clone 'Bolgheri') and CCD-susceptible (10 ramets each) clones were used for this study. The plants were grown under natural field conditions in 4 litre pots (15 × 15 × 20 cm) containing a mixture of peat, compost and perlite (3:1:1, *v/v/v*) in the experimental area of the Institute for Sustainable Plant Protection (IPSP) of the Italian National Research Council (CNR) in Sesto Fiorentino, Italy (43°49′05″ N; 11°12′07″ E). During the experiment, the potted cypress plants were irrigated 2 times per week and were fertilized every 20 days with half-strength Hoagland solution.

At the beginning of June 2018, four ramets of each cypress clone were artificially inoculated with a standard isolate of *S. cardinale* (ATCC 38654) following the procedure described in Danti et al. [48], while the other ramets were left intact. A 3 mm plug of stem bark was removed with a cork borer and replaced with a plug the same size of *S. cardinale* mycelium grown on PDA in Petri dishes for 15 days at 25 °C in the dark. The inoculum was then covered with wet cotton wrapped with parafilm around the trunk for one week.

We performed 5 stem inoculations for each plant to simulate a severe CCD attack and induce severe infection symptoms. The sites of inoculation started approximately 10 cm below the top of the plant and were spaced 5 cm apart from one another, where the stem was between 0.5 and 1 cm in width, determined using a stem calliper. The duration of the study was 3 months.

Three months after the inoculation (September 2018), when the typical CCD symptoms were fully evident (development of necrotic lesions around the inoculation points, a little resin exudation from the inoculation points, with apical twigs and shoot desiccation), the 4 diseased plants and 4 unaffected plants of each clone were sampled for the flammability tests and the determination of terpenoid content. At the same time, two more intact ramets of each clone were used to determine the moisture content (FMC) and dry mass (see below).

2.2. Fuel Moisture and Biomass Determination

To determine the moisture content and biomass, two intact ramets of each clone were subdivided into three parts: upper, middle and lower third, and each portion was in turn divided into green leaves and twigs, bark tissues and xylem. The cutting and splitting of the ramets was carried out in a cold room (5 °C) as fast as possible (taking a few minutes). Each type of sample was then immediately weighed (fresh weight) with a precision balance and placed in an oven at 70 °C until a constant weight was achieved and considered to represent the dry weight. The fresh and dry mass and the moisture content (FMC) of the leaves—twigs, bark and xylem were separately determined for the three parts (upper third, middle third, lower third) of the ramets. These measures allowed us to determine the real amount of terpenoids contained in the leaves and bark of each sample used for flammability tests (see below).

2.3. Sample Splitting for the Determination of Terpenoids and Flammability Tests

For the extraction of terpenoids and to conduct the flammability tests, 16 ramets (8 per clone, 4 inoculated and 4 intact) were separately cut in 10 cm long stem portions (in a cold room as before) from the upper, lower and medium third of the ramets. From each portion, a 4 cm long stem segment (including the inoculation point, for the infected plants) and 5 g in fresh weight of leaves, randomly chosen, were sampled for the determination of terpenoids. From the 4 cm stem segments, bark tissues (from the cambium to the outer periderm) were removed and separately stored (while the xylem was discarded). For the determination of terpenoids, the leaves and bark samples were stored in falcon tubes at −20 °C until the extraction of the terpenoids. The remaining material from each of the stem portions, that were initially 10 cm in length, was placed in hermetically sealed plastic bags and immediately stored at −20 °C and shipped in dry ice (the day after with a 12 h courier) to the Forest Fire Laboratory of the INIA–CIFOR in Madrid, Spain, for the flammability tests (see below).

2.4. Terpenoids Extraction, Identification and Quantification

For the determination of terpenoids, 500 mg (fresh weight) of each foliage and bark tissues were quickly fragmented into small pieces, of about 0.5–1 cm in length, with a scalpel (in a cold room) and placed separately in sealed 230 mL vials with 1 mL of heptane as the solvent and tridecane (20 ppm) as an internal standard. The vials were then submitted to 3 sonication cycles of 10 min each at room temperature (25 °C) at a frequency of 38–40 KHz (Ultrasonic cleaner Sonica, S3 EP, Soltec, Milano, Italy) and subsequently stirred overnight (for 12 h) at 35 °C in a rotating incubator shaker (Thermoshake THO 500/1, Gerhardt, Königswinter, Germany) at 90 rpm. The vials were then centrifuged for 10 min at 20 °C at 5000 rpm (Centrifuge 5810 R, Eppendorf, Hamburg, Germany), and the supernatant pipetted in 2 mL vials sealed with a Teflon septum and crimped with an aluminum cap for the detection of the terpenoids via gas chromatography–mass spectrometry (GC–MS).

The terpenoids were analysed using a Gas Chromatograph Agilent 7820 GC-Cromatograph equipped with a 5975C MSD with EI ionisation (Agilent Tech., Palo Alto, AC, USA) as described in [27]. A 1 µL sample of the aforementioned supernatant was injected in a split/splitless injector operating in split mode with 1:10 split ratio. A Gerstel MPS2 XL autosampler equipped with liquid option was used. The analysis was carried out under the following conditions: H_2 (carrier gas) at 1.2 mL min^{-1}; the injector in splitless mode set at 260 °C, J&W innovax column (30 m, 0.25 mm i.d., 0.5 µm df); oven temperature program: initial temperature 40 °C for 1 min, then 5 °C/min until 200 °C, then 10 °C/min until 220 °C, then 30 °C/min until 260 °C, with a hold time of 3 min. The mass spectrometer was operating with an electron ionisation of 70 eV, in scan mode in the m/z range 29–330 at three scans/sec.

The deconvoluted peak spectra, obtained by Agilent MassHunter Workstation software, were matched against the NIST 11 spectral library for tentative identification. Kovats' retention indices were calculated for further compound confirmation and compared with those reported in the literature for the chromatographic column used. In addition, terpenoids (mono- and sesquiterpenoids) were identified by the comparison of the retention times with those of authentic standards (high-purity components were obtained from Fluka, Aldrich and Acros) injected under the same conditions, and also by comparison with the tridecane internal standard for those compounds for which standards were not available. The identified terpenoids (TOTterp) were grouped into four categories: monoterpenoids (MT), oxygenated monoterpenoids (MTox), sesquiterpenoids (ST) and oxygenated sesquiterpenoids (STox), as outlined in Della Rocca et al. [27]. The amount of terpenoids was expressed as µg/g dry weight (DW) of the samples.

2.5. Flammability Test at MLC

An adapted Mass Loss Calorimeter (MLC) device was used [47,49–51]. The tests were performed using the MLC arranged in the standard horizontal configuration, to determine the main flammability descriptors [52]: ignitability (time to ignition, TTI), combustibility (peak of heat release rate, PHRR), sustainability (average effective heat of combustion, AEHC) and consumability (percentage mass lost, PML). A porous holder ($10 \times 10 \times 5$ cm) was used to allow the natural diffusion of air through the samples during the MLC tests. The MLC tests were conducted at 50 kW/m^2, simulating severe fire conditions [53]. The fuel moisture content (FMC) of the live foliage was promptly determined on 8 g subsamples using a Computrac MAX R 2000XL moisture analyser (Arizona Instrument LLC). Based on their FMC values, the dry mass of the fresh samples was fixed at 10 g (to balance the variability in weight due to the differences in water content among the samples [50]). At the end of each test, the residual mass fraction was determined with a precision balance (Mettler AB104-S). All samples were stored in a refrigerated chamber (at 4 °C) and processed within 5 days.

A series of tests was carried out using the experimental design described for the extraction of terpenoids (see above). The portions cut from the 16 ramets (8 per clone), were divided into the three groups previously identified (upper, middle and lower thirds), and a total of 48 samples were obtained. From each portion, one 8 g subsample of leaves was used to obtain the FMC (see above), while the remaining samples of woody stem and foliage (10 g of dry weight) were subdivided in 3–5 subsamples to carry out the flammability tests. The MLC protocol for 'alive' samples generated a high level of variability, and non-repetitive tests must be removed from analysis to obtain at least two replicates complying with the repeatability criteria (errors less than 15% [53]). Therefore, the original set of 48 samples was reduced to 36 samples (8 replicates per condition: CCD-resistant clone inoculated (RI) or non-inoculated (RC), CCD-susceptible clone inoculated (SI) and non-inoculated (SC)). This data set ($n = 36$) was considered representative (upper, middle and lower third portion of trees), and was randomly extracted from 16 ramets (replicates), avoiding pseudo-replication. For each plant portion, the FMC and dry mass were measured and the real amount of terpenoids contained in each sample used for the flammability tests was determined, starting from the concentration of terpenoids per µg found in the leaves and the bark tissues.

2.6. Statistical Analysis

The total concentrations of terpenoids (TOTterp), as well as the subcategories of MT, MTox, ST, and STox extracted from both the bark and leaves and the flammability parameters (TTI, PHRR, AEHC, PML) were used as the response variables. Considering the hierarchical nature of the data (multiple observations on single ramets of a same clone), multilevel generalized linear mixed models (GLMM) with both fixed and random effects acting at ramet and portion (upper, middle and lower third) levels were fitted. The distribution of variables was checked for parametric requisites (skewness, kurtosis, influence points) and to select the suitable link function (Gaussian or Gamma distribution). Once the model was fitted, the assumptions of normality and homoscedasticity of the residuals were evaluated. Potential autocorrelation was controlled by residual analysis. Finally, the goodness-of-fit statistics, referring to both the marginal (not considering random effects) and conditional (including random effects) predictions were determined. A GLMM model was generated to predict variables for each clone (susceptible "S" and resistant "R") using the factor 'infection' (infected "I" vs. healthy control "C") and the covariable FMC as the fixed predictor variables. As some of the upper part of susceptible inoculated (SI) ramets died as consequence of the inoculation with the pathogen, to prevent the strong effect of dead parts in the prediction of terpenoid content and flammability, models were replicated removing those samples. An additional model was generated to detect the effect of the clone (R vs. S) in determining the terpenoid concentrations before (C) and after the treatment (I).

A partial least square (PLS) model was generated to predict the flammability parameters (TTI, PHRR, AEHC, PML) using terpenoid contents per sample (MT_tot, ST_tot, MTox_tot, STox_tot, Tot_Terp) expressed in µg and FMC as predictors, including a dummy variable "alive vs. dead samples". An additional factor predictor with 4 levels (resistant infected "RI", resistant control "RC", susceptible infected "SI" and susceptible control "SC") was included in the model. The technique prevented the problems associated with multicollinearity among the multiple initial explanatory variables (e.g., terpenoid content and FMC), and the main advantage was that a linear combination of the explanatory variables can be determined. For the determination of the optimal number of components, the cross-validation method was used by applying the Stone-Geiser Q^2 statistic. To assess the relative contribution of each independent variable in the model, the value and physical sense of the scaled coefficients were checked. The final output was a multiple linear model with a fit estimated by the R^2Y statistic, which was equivalent to the adjusted R^2 of a multiple linear model obtained by generalized least squares, and the R^2X statistic, which evaluated the collinearity between the independent variables. Statistica $10^{®}$ and SPSS $20^{®}$ packages were used to analyse the data.

3. Results

After three months, the inoculation with S. cardinale on the young stems of the S clone induced the dieback of the crowns which were partially desiccated, due to the girdling of the stem by the necrotic lesion (Figure 1).

(a)   (b)

Figure 1. *Cupressus sempervirens* L. clones 3 months after multiple artificial stem inoculation with the cypress canker disease fungal agent *S. cardinale*. (**a**) Resistant (R); (**b**) Susceptible (S). The susceptible ramets displayed the dieback of crown portions due to the effect of the bark pathogen that completely girdled the inoculated axes.

3.1. Clone and Infection Effects on Terpenoids Concentration

The differences in the concentration of terpenoids (µg/g) in the leaves and the bark samples were evaluated between the inoculated and the intact ramets of the two clones (Tables 1 and 2), as well as within the same clone, between the intact (control) and the inoculated ramets (Tables 3 and 4).

The results showed strong differences between the clones for both the intact (uninoculated) (C) and the inoculated plants (I) (Table 1). In the leaves of intact plants, clone (S) had a higher concentration ($p < 0.001$) of total terpenoids compared to clone (R) (7514 vs. 6730 µg/g) due to the higher MT and MTox (4319 vs. 3093 µg/g and 221 vs. 156 µg/g, respectively; $p < 0.001$). In contrast, a higher concentration of ST was shown by the R clone. In the bark tissues (Table 2), the total concentration of terpenoids was higher ($p < 0.001$) in the R clone than in the S (1727 vs. 1295 µg/g), despite the concentration of monoterpenoids being slightly higher (MT) or higher (MTox) in the S clone (604 vs.

549 µg/g and 42 vs. 32 µg/g, respectively); a similar pattern was also observed in the leaves. Both classes of sesquiterpenoids (ST and STox) were much higher in the R clone ($p < 0.001$).

Table 1. Generalized linear mixed models (GLMM) on the concentration of terpenoids (µg/g) in common cypress leaf samples from the control (C) and infected (I) treatments comparing the resistant (R) vs. susceptible (S) clones for the following variables: MT (monoterpenoids concentration), MTox (oxygenated monoterpenoids concentration), ST (sesquiterpenoids concentration), STox (oxygenated sesquiterpenoids concentration), TOTterp (total terpenoids concentration). The fuel moisture content (FMC) was used as the covariable. The model for the S clone was repeated removing dead samples (live fuels only)*. The average and standard deviation (in brackets) for each variable are shown (N = 12). The *p*-value shows the significance of the fixed variables (clone, R vs. S) and covariable (FMC) predictors. Significant differences (>95%) are highlighted in bold.

Treatment	Clone	MT	MTox	ST	STox	TOTterp	FMC (%)
C	R	3093 (450)	156 (35)	2129 (400)	1352 (183)	6730 (894)	149 (9)
	S	4319 (818)	221 (46)	1617 (362)	1357 (199)	7514 (1147)	147 (24)
p-value	Clone	**<0.001**	**<0.001**	**<0.001**	**0.019**	**<0.001**	
	FMC	0.205	**0.019**	**0.002**	**0.009**	**0.005**	
I	R	2894 (656)	158 (44)	2231 (514)	1204 (207)	6487 (1303)	133 (17)
	S	3134 (944)	239 (118)	1529 (535)	1100 (293)	6003 (1301)	103 (76)
p-value	Clone	**<0.001**	**<0.001**	**<0.001**	**<0.001**	**<0.001**	
	FMC	**<0.001**	**<0.001**	**<0.001**	**<0.001**	**<0.001**	
	S (live)	3671 (740)	287 (124)	1240 (448)	1172 (333)	6370 (1500)	164 (19)
p-value	Clone	**<0.001**	**<0.001**	**<0.001**	**<0.001**	**<0.001**	
	FMC	**0.003**	**0.025**	**0.030**	**0.010**	**0.009**	

Table 2. GLMM on the concentration of terpenoids (µg/g) in the common cypress bark samples from the control (C) and infected (I) treatments, comparing the resistant (R) vs. susceptible (S) clones for the following variables: MT (monoterpenoids concentration), MTox (oxygenated monoterpenoids concentration), ST (sesquiterpenoids concentration), STox (oxygenated sesquiterpenoids concentration), TOTterp (total terpenoids concentration). The average and standard deviation (sd) for each variable are shown (N = 12). The *p*-value shows the significance of the fixed variables (clone, R vs. S) and covariable (FMC) predictors. Significant differences (>95%) are highlighted in bold.

Treatment	Clone	MT	MTox	ST	STox	TOTterp
C	R	549 (360)	32 (13)	746 (266)	401 (174)	1727 (753)
	S	604 (464)	42 (18)	424 (135)	226 (72)	1295 (650)
p-value		**<0.001**	**<0.001**	**<0.001**	**<0.001**	**<0.001**
I	R	5926 (3711)	489 (561)	2623 (2034)	3504 (2508)	12,542 (8541)
	S	2190 (2225)	380 (590)	1450 (1962)	1741 (1834)	5760 (6338)
p-value		**<0.001**	**<0.001**	**<0.001**	**<0.001**	**<0.001**

The comparison of the inoculated plants of the two clones (I) showed that the clone R generated a higher total concentration of terpenoids (TOTterp) ($p < 0.001$) in both leaves (6487 vs. 6003 µg/g) (Table 1) and bark tissues (more than twice the concentration of the S one, that was 12,542 vs. 5760 µg/g) (Table 2). With regard to the intact plants C, the higher ($p < 0.001$) concentration of MT (3134 vs. 2894 µg/g) and MTox (239 vs. 158 µg/g) in the leaves was found in the S clone (Table 1), while the ST and STox were higher ($p < 0.001$) in the R clone (2231 vs. 1529 µg/g and 1204 vs. 1100 µg/g, respectively). After infection, the significant differences between the clones were maintained even when the dead leaf samples were removed: TOTterp, MT and MTox were slightly higher in the clone S (live) compared to S (live and dead) while the STs were even lower. After the fungal infection, the concentration of all terpenoid categories in the bark tissues were markedly higher ($p < 0.001$) in the R clone compared to the S clone (Table 2). The effect of the FMC on the concentration of terpenoids in the leaves was always significant, excluding for the concentration of MT in the C plants (comparing R and S clones) (Table 1). The results suggested that the effect of the infection should be analysed separately within the two clones R and S (Tables 3 and 4).

Table 3. GLMM on the concentration of terpenoids (μg/g) in the common cypress leaf samples of the resistant (R) and susceptible (S) cypress clones comparing the control (C) vs. infected (I) treatments for the following variables: MT (monoterpenoids concentration), MTox (oxygenated monoterpenoids concentration), ST (sesquiterpenoids concentration), STox (oxygenated sesquiterpenoids concentration), TOTterp (total terpenoids concentration). Fuel moisture content (FMC) is used as a covariable. Model for the SI treatment was repeated removing the dead samples *. The average and standard deviation (in brackets) for each variable are shown (N = 12). The p-value shows the significance of the fixed variables (treatment, C vs. I) and covariable (FMC) predictors. Significant differences (>95%) are highlighted in bold.

Clone	Treatment	MT	MTox	ST	STox	TOTterp
R	C	3093 (450)	156 (35)	2129 (400)	1352 (183)	6730 (894)
	I	2894 (656)	158 (44)	2231 (514)	1204 (207)	6487 (1303)
p-value	Treatment	0.224	0.999	0.993	**0.026**	0.439
	FMC	**0.052**	0.122	0.588	**0.033**	0.159
S	C	4319 (818)	221 (46)	1617 (362)	1357 (199)	7514 (1147)
	I	3134 (944)	239 (118)	1529 (535)	1100 (293)	6003 (1301)
p-value	Treatment	**0.017**	0.118	0.275	0.129	**0.025**
	FMC	**0.003**	**0.011**	0.789	**0.000**	**0.004**
S *	I	3671 (740)	287 (124)	1240 (448)	1172 (333)	6370 (1500)
(vs. C)	Treatment	0.999	0.999	**0.045**	0.999	0.08
p-value	FMC	0.998	**0.025**	0.456	**0.036**	0.117

Table 4. GLMM of the concentration of terpenoids (μg/g) in the common cypress bark samples of the resistant (R) and susceptible (S) cypress clones comparing the control (C) vs. infected (I) treatments for the following variables: MT (monoterpenoids concentration), MTox (oxygenated monoterpenoids concentration), ST (sesquiterpenoids concentration), STox (oxygenated sesquiterpenoids concentration), TOTterp (total terpenoids concentration). The average and standard deviation (sd) for each variable are shown (N = 12). The p-value shows the significance of the fixed variables (treatment, C vs. I). Significant differences (>95%) are highlighted in bold.

Clone	Treatment	MT	MTox	ST_c	STox_c	TOT_terp_c
R	C	549 (360)	32 (13)	746 (266)	401 (174)	1727 (753)
	I	5926 (3711)	489 (561)	2623 (2034)	3504 (2508)	12,542 (8541)
p-value	Treatment	**<0.001**	**<0.001**	**<0.001**	**<0.001**	**<0.001**
S	C	604 (464)	42 (18)	424 (135)	226 (72)	1295 (650)
	I	2190 (2225)	380 (590)	1450 (1962)	1741 (1834)	5760 (6338)
p-value	Treatment	**0.007**	**0.001**	**0.017**	**<0.001**	**0.002**

The results showed little differences between the control (C) and the infected (I) plants of the clone R (Table 3), with only slight decreases in the STox concentration in the leaves after infection (from 1352 to 1204 μg/g; p = 0.026) (Table 3). The susceptible (S) clone reacted to the infection in a different way, the TOTterp significantly decreased in the inoculated plants (from 7514 to 6003 μg/g, p = 0.025), mainly due to the strong reduction in the concentration of MT (from 4319 to 3134 μg/g; p = 0.017) (Table 3). Nevertheless, when the dead samples were removed from the analysis, this difference between the clones was not significant with only a difference in ST concentration occurring (decreasing to 1239 μg/g) (p = 0.045). The relationship between the FMC and terpenoids was always positive: the higher the FMC, the higher the concentration of terpenoids (Table 3). The effect of infection in both clones was strongly evident in the bark tissues (Table 4). A strong increase in TOTterp was detected in the clone S (with an almost 4-fold increase: from 1295 to 5760 μg/g) but especially in clone R in which the value in inoculated plants was more than seven times higher than in the C plants (from 1727 to 12,542 µg/g; $p < 0.001$). All the classes of terpenoids increased after the infection ($p < 0.001$), and irrespective of their concentration in the bark tissues of the intact plants, the values reached in the R clone were always higher than those in the S clone. The levels of MT and MTox in the R clone showed the highest relative increase (from 549 to 5926 µg/g, almost 11 times, for MT; from 32 to 489 µg/g, more than 15 times, for MTox) (Table 4).

3.2. Clone and Infection Effects on Flammability

Significant differences between the clones were detected for all the flammability parameters in the intact plants (C) (Table 5). The R clone showed a significantly ($p < 0.001$) higher TTI compared to the S clone (140.78 vs. 121.67 s) due to the higher FMC, AEHC ($p < 0.001$) (5.54 vs. 5.07 MJ/Kg) and PML (22.90 vs. 19%). Susceptible infected (SI) clones showed significantly higher ignitability (lower TTI, 73 vs. 116 s, $p < 0.001$), combustibility (higher PHRR, 143 vs. 69 kW/m^2, $p < 0.001$), sustainability (higher AEHC, 8.10 vs. 4.93 MJ/Kg, $p < 0.038$) and consumability (higher PML, 47.50 vs. 21.67%, $p < 0.001$) than the resistant infected (RI) clones. Excluding the dead samples of the S clone (when comparing the live R and the live S portions), many differences among the clones were markedly reduced and only the TTI remained lower in the S clone (109 vs. 116 s, $p < 0.001$) (Table 5). The effect of the FMC was always significant (p ranging from <0.001 to 0.003) except for the AEHC of the clones before the infection (Table 5).

Table 5. GLMM of the flammability parameters from the control (C) and the infected (I) treatments of the resistant (R) vs. susceptible (S) common cypress clones for the following variables: TTI (time to ignition, s), PHRR (peak heat release rate, kW/m^2), AEHC (average effective heat of combustion, MJ/Kg), PML (percentage of mass lost, %). The average and standard deviation (in brackets) for each variable and treatment are shown (N = 9). The fuel moisture content (FMC) is used as a covariable. The model for the SI treatment was repeated removing dead samples (*). The p-value indicates the significance of the fixed variables (clone, R vs. S). The factor ramet is included in the model as a random variable. Significant differences (>95%) are highlighted in bold.

Treatment	Clone	TTI	PHRR	AEHC	PML
	R	141 (79)	65 (37)	5.54 (4.46)	22.90 (25.57)
C	S	122 (77)	96 (29)	5.07 (1.63)	19.00 (6.58)
p-value	Treatment	**<0.001**	**<0.001**	**<0.001**	**<0.001**
	FMC	**<0.001**	**0.003**	0.955	**<0.001**
	R	116 (73)	69 (32)	4.93 (2.95)	21.67 (17.83)
I	S	73 (51)	143 (106)	8.10 (5.65)	47.50 (36.10)
	S *	109 (27)	64 (27)	3.94 (2.15)	23.75 (25.54)
p-value	Treatment	**<0.001**	**<0.001**	**0.038**	**<0.001**
	FMC	**<0.001**	**<0.001**	**<0.001**	**<0.001**
SC vs. SI (live)	Treatment	**<0.001**	0.332	0.925	0.168
	FMC	**<0.001**	**<0.001**	0.267	0.139

The comparison of infection effects within the R and S clones showed that the main differences were related to ignitability (TTI) and combustibility (PHRR) (Table 6). In fact, in the R clone the TTI was the only parameter that changed (decreased) in the inoculated plants (from 141 to 116 s, $p < 0.001$) (Table 6) which showed a higher ignitability. The effect of the inoculation in the S clone was even stronger and involved all of the flammability parameters: ignitability (TTI dropped from 122 to 73 s, $p = 0.002$), combustibility (PHRR, $p < 0.001$), sustainability (AEHC, $p = 0.005$) and consumability (PML, $p < 0.001$) all increased. Nevertheless, when the dead samples were removed, only the TTI remained significantly lower than in the control (Table 6). The effect on the FMC was significant for all of the flammability parameters in the S clone, whereas in the R clone, it was only the TTI and PHRR that were significantly affected ($p < 0.001$). In the S clone, when the dead parts were removed from the computation, the effect of the infection on the FMC was still significant ($p < 0.001$) (Table 6).

Table 6. GLMM of the flammability parameters of the resistant (R) and susceptible (S) common cypress clones comparing the control (C) and infected (I) treatments for the following variables: TTI (time to ignition, s, PHRR (peak heat release rate, kW/m^2), AEHC (average effective heat of combustion, MJ/Kg), PML (percentage of mass lost, %). The average and standard deviation (in brackets) for each variable and treatment are shown (N = 9). The fuel moisture content (FMC) is used as a covariable. The model for the SI treatment was repeated removing the dead samples (*). The *p*-value indicates the significance of the fixed variables (treatment, C vs. I). Significant differences (>95%) are highlighted in bold.

Clone	Treatment	TTI	PHRR	AEHC	PML
R	C	141 (79)	65 (37)	5.54 (4.46)	77.10 (25.57)
	I	116 (73)	69 (32)	4.93 (2.95)	78.33 (17.83)
p-value	Treatment	**<0.001**	0.459	0.177	0.413
	FMC	**<0.001**	**<0.001**	0.387	0.958
S	C	122 (77)	96 (29)	5.07 (1.63)	81.00 (6.58)
	I	73 (51)	143 (106)	8.10 (5.65)	87.50 (36.10)
	I *	109 (27)	94 (27)	3.94 (2.15)	76.25 (25.54)
p-value	Treatment	**0.002**	**<0.001**	**0.005**	**<0.001**
	FMC	**<0.001**	**<0.001**	**<0.001**	**<0.001**
RI vs. SI (live)	Treatment	**0.001**	0.657	0.170	0.257
	FMC	**<0.001**	0.241	0.962	0.513

3.3. Linking Disease and Flammability

The PLS model indicated that 56% of the variation in the flammability can be explained by the following selected variables: clone x treatment (RC, RI, SC, SI), fuel moisture content (FMC), the dummy variable 'alive vs. dead' samples, and terpenoid concentration for each class. The fitted parameters (Table 7) show that the flammability components were explained by the predictive variables with different fits between 24% and 77%. Therefore, the model accounts for the combustibility ($R^2Y = 77\%$ for PHRR) and consumability ($R^2Y = 66\%$ for PML) more effectively than ignitability ($R^2Y = 24\%$ for TTI) or sustainability ($R^2Y = 52\%$ for AEHC). The PLS model showed that the FMC and 'alive vs. dead' (dummy variable) explain most of the variability in the data (higher scaled coefficients) (Figure 2). With regard to the amount of terpenoids, ST showed the highest scaled coefficients and a positive relationship with flammability (higher PHRR, AEHC and PML) (Figure 2). In terms of the treatment effects, the scaled coefficients in the model showed that the most important parameter related to flammability was the dead portions of plant instead of the changes in the content of terpenoids (Figure 2). When the dead samples were included (Figure 2), the susceptible infected clone (SI) showed the highest flammability.

Table 7. Fitted parameters for the partial least square (PLS) model to predict the flammability (TTI, PHRR, AEHC, PML) using: the amount of terpenoids (MT_tot, ST_tot, MTox_tot, STox_tot, Tot_Terp) contained in each sample (µg), the FMC, the treatment x clone (4 levels: resistant control, RC, resistant infected, RI, susceptible control, SC, susceptible infected, SI) as predictors. The model selects two components and includes the dummy variable 'alive vs. dead' samples to highlight the importance of the dead samples (4 samples belonging to the SI treatment) in the prediction of flammability (N = 36). The total model fit (R^2Y) and the partial model fit for the predicted flammability variables are highlighted in bold.

PLS Model	R^2X	R^2X (Cumul.)	Eigenvalues	R^2Y	R^2Y (Cumul.)	Q^2	Q^2 (Cumul.)	R^2Y for TTI	R^2Y for PHRR	R^2Y for AEHC	R^2Y for PML
Component1	0.31	0.31	3.624263	0.53	0.53	0.462919	0.462919	**0.23**	**0.68**	**0.41**	**0.64**
Component2	0.16	0.47	1.616667	0.03	0.56	−0.24175	0.333077	**0.24**	**0.77**	**0.53**	**0.66**

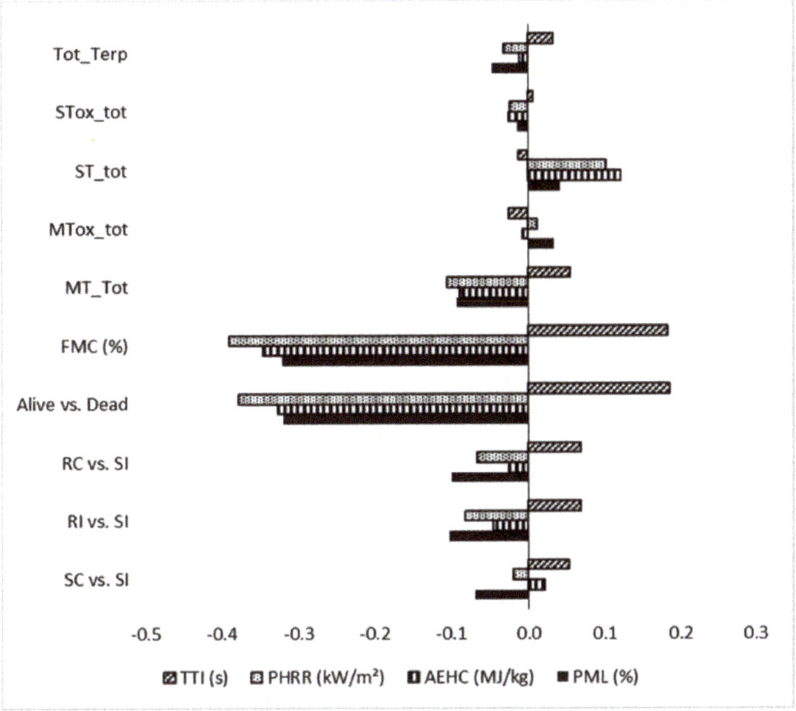

Figure 2. Scaled coefficients for the predictive variables of the flammability parameters: TTI (ignitability), PHRR (combustibility), AEHC (sustainability) and PML (consumability). The scaled coefficients show the relative importance of each variable in the structure of the PLS model (N = 36).

4. Discussion

Volatile terpenoids are secondary plant metabolites that undertake many ecological functions and roles [54,55]. As part of resins, terpenoids play an essential role in the plant defence against microbes, especially in conifers [22,56–59]. The expected increase in the content of terpenoids after infection with S. cardinale was observed in both the cypress clones in the bark tissue (although it was much stronger in the resistant clone). The accumulation of terpenoids did not occur in the leaves, in accordance with the biology of the fungal pathogen acting at the cortical level [30,33]. This result was consistent with the known de novo production and the accumulation of all the classes of terpenoids as a reaction to the attack of fungal pathogens in many species of conifers [20,29,59–63]. In addition, it was recently reported that a S. cardinale inoculation on the cypress stems or branches induced a reaction in the host which consisted in the production of traumatic resin ducts in the phloem [36] and the consequent accumulation of terpenoids in the bark tissues near the site of infection. This defence reaction was observed to be stronger in canker resistant cypress genotypes [29], confirming that the production of resin terpenoids was an important and effective response of cypress to CCD. In unaffected plants (C), the differences among the clones mainly concerned STs that were higher than in the R clone, indicating a possible deployment of a constitutive chemical barrier of quantitatively 'minor' terpenoids with a higher biological efficacy, instead of MTs that have less antifungal activity against S. cardinale [29].

Low weight terpenoids (volatiles terpenoids), such as monoterpenoids (C_{10}) and sesquiterpenoids (C_{15}), possess relatively low boiling and flash points, and as a consequence high flammability [26,27]. The role played by the plant volatile terpenoids in driving the flammability of vegetation is now widely accepted [26,64], though the quantitative effect is still debated [27]. In other words, to what extent does the terpenoid content influence combustion, and the wildfires on the scale of a forest fire, is not yet well defined. Few laboratory studies have attempted to evaluate the effect of these compounds on the variability of flammability parameters in different tree species. This is particularly relevant given the amount of terpenoids contained in the leaf tissues or plant twigs [27].

Inoculation with S. cardinale on the young cypress plants induced the dieback of the upper part of the crowns of the S clone after three months (Figure 1). This has two implications: firstly, the presence of dried plant material, and secondly, the accumulation of terpenoids in bark tissues around the necrotic lesions. It is well known that cypress canker disease may cause large bark lesions, inducing serious diebacks of cypress crowns [33] and causing the copious exudation of resin which flows down the affected trunks in severely cankered stems. In this study, the tissues around the inoculated points on the stem did not show the exudation of resin outside the cankered lesion. This could be due to the relatively short time between the inoculations and the collection of samples (3 months), the young age of the plants or the relatively small diameters of the inoculated stems and their subsequent early death.

To our knowledge this is one of the first studies to evaluate the effect of the estimated amount of terpenoids contained in plant tissues (μg) on flammability tests. The GLMM model showed that the SI (susceptible–infected) clone samples appeared to be significantly more flammable in comparison to the RI (resistant–infected) samples, showing higher ignitability, combustibility, sustainability and consumability. This is mainly due to the presence of dead crown portions, as when the dead samples of SI were removed from the computation most of the differences between the treatments disappeared. This finding was also supported by the values of the R clone, which showed a slightly higher ignitability in infected plants, while the other parameters did not change as a consequence of the S. cardinale infection. In contrast, the ignitability of the S clone was higher than that of the R clone, when only the living tissues were considered.

The outcomes from the GLMM were supported by the PLS model results, which indicated that the FMC and the 'alive vs. dead' accounted for most of the data variability. Moreover, concerning the treatment, the model showed that the most important parameter related to flammability was the presence of dead plant portions rather than changes in the content of terpenoids (μg). As accounted for above, this could be partially explained by the absence of a copious amount of resin flowing on the stems, and by the relatively small proportion of bark tissues (where the highest increase in terpenoids was observed) when compared to the sample as a whole. In fact, the scaled coefficients showed a negative correlation between both 'non-dead' plant material and FMC vs. all of the tested flammability parameters. An interesting positive relationship was found between the total amount of ST and combustibility and sustainability, confirming the role of terpenoids in flammability, as already hypothesized in Della Rocca et al. [27]. In contrast, a surprising negative relationship between the total amount of MTs and all the flammability parameters was found. This might be explained by the observed strong reduction in MT concentration in the dead tissues of the susceptible clone, while in the live tissues of the same inoculated clone the MT concentration increased. These results suggested a possible MT leak as the diseased crown portions dry, which requires further investigation to fully account for the possible relationship between the tissues' water content and the terpenoids loss.

5. Conclusions

In response to the experimental questions outlined above: (i) in the common cypress—cypress bark canker pathosystem (at least for young plants such as those considered in this study), the diseased plants were more ignitable and showed increased combustibility and enhanced sustainability; (ii) the CCD-resistant cypress clone appeared less flammable than the susceptible clone when infected; and (iii) the proportion of the dried crown parts (as a consequence of the disease) was a stronger factor in determining the overall flammability than the terpenoid accumulation.

The selection of CCD-resistant cypress genotypes [65,66] for their use in plantations to replace trees killed or compromised by the disease, and sanitation to remove the heavily infected portions of trees [67] not only improves the health, aesthetic and recreational attributes of plantations [32,48] but also reduces their flammability. The results of this work suggest some general considerations for the phytosanitary management of woodlands, plantations and hedges. First, a disease can strongly increase the flammability of a tree, especially conifers, because it can cause the dehydration or desiccation of crown portions, the retention of dead material in the crowns and the accumulation of flammable compounds, such as terpenoids, in living bark tissues as a reaction to the disease. Secondly, the flammability of a tree species can be highly genotype dependent (i.e., genotypes resistant to a disease or more tolerant to drought stress can be less flammable than their susceptible or less tolerant counterparts). Finally, the spread of disease at the tree and stand level can be an important factor in the prediction of fire risk and should be actively monitored.

Author Contributions: Conceptualization, G.D.R. and J.M.; methodology, G.D.R., R.D. and J.M.; formal analysis, G.D.R., M.M., C.C. and J.M.; investigation, G.D.R. and J.M.; data curation, G.D.R. and J.M.; writing—original draft, G.D.R., R.D. and J.M.; writing—review & editing, C.H., M.G. and C.C.; funding acquisition, C.H. and J.M. All authors have read and agreed to the published version of the manuscript.

Funding: This study was supported by the VIS4FIRE (Integrated Vulnerability of Forest Systems to Wildfire: Implications on Forest Management Tools), Spanish R&D project (RTA2017-00042-C05-01). VIS4FIRE is co-funded by the EU through the FEDER program. This study was also co-financed by INIA (FPI-SGIT 2018) and a European Social Fund grant awarded to Cristina Carrillo.

Acknowledgments: We thank Carmen Díez from INIA–CIFOR for the assistance during the flammability experiment, Vincenzo Di Lonardo for growing the cypress plants and Matthew Haworth for the revision of the manuscript.

Conflicts of Interest: The authors declare no conflict of interest.

References

1. Chen, G.; He, Y.; De Santis, A.; Li, G.; Cobb, R.; Meentemeyer, R.K. Assessing the impact of emerging forest disease on wildfire using Landsat and KOMPSAT-2 data. *Remote. Sens. Environ.* **2017**, *195*, 218–229. [CrossRef]

2. Metz, M.; Varner, J.M.; Frangioso, K.M.; Meentemeyer, R.K.; Rizzo, D.M. Unexpected redwood mortality from synergies between wildfire and an emerging infectious disease. *Ecology* **2013**, *94*, 2152–2159. [CrossRef] [PubMed]

3. Killi, D.; Bussotti, F.; Gottardini, E.; Pollastrini, M.; Mori, J.; Tani, C.; Papini, A.; Ferrini, F.; Fini, A. Photosynthetic and morphological responses of oak species to temperature and [CO2] increased to levels predicted for 2050. *Urban For. Urban Green.* **2018**, *31*, 26–37. [CrossRef]

4. Metz, M.; Varner, J.M.; Simler-Williamson, A.; Frangioso, K.M.; Rizzo, D.M. Implications of sudden oak death for wildland fire management. *For. Phytophthoras* **2017**, *7*, 30–44. [CrossRef]

5. Boyd, I.L.; Freer-Smith, P.H.; Gilligan, C.A.; Godfray, H.C.J. The Consequence of Tree Pests and Diseases for Ecosystem Services. *Science* **2013**, *342*, 1235773. [CrossRef] [PubMed]

6. Stephens, S.L.; Agee, J.K.; Fulé, P.Z.; North, M.P.; Romme, W.H.; Swetnam, T.W.; Turner, M.G. Managing Forests and Fire in Changing Climates. *Science* **2013**, *342*, 41–42. [CrossRef]

7. Ruffault, J.; Curt, T.; Martin-StPaul, N.K.; Moron, V.; Trigo, R.M. Extreme wildfire events are linked to global-change-type droughts in the northern Mediterranean. *Nat. Hazards Earth Syst. Sci.* **2018**, *18*, 847–856. [CrossRef]

8. Huang, Y.; Wu, S.; Kaplan, J. Sensitivity of global wildfire occurrences to various factors in the context of global change. *Atmos. Environ.* **2015**, *121*, 86–92. [CrossRef]

9. Ramsfield, T.; Bentz, B.; Faccoli, M.; Jactel, H.; Brockerhoff, E.G. Forest health in a changing world: Effects of globalization and climate change on forest insect and pathogen impacts. *Forestry* **2016**, *89*, 245–252. [CrossRef]

10. Loehman, R.; Keane, R.; Holsinger, L.M.; Wu, Z. Interactions of landscape disturbances and climate change dictate ecological pattern and process: Spatial modeling of wildfire, insect, and disease dynamics under future climates. *Landsc. Ecol.* **2016**, *32*, 1447–1459. [CrossRef]

11. Pautasso, M.; Schlegel, M.; Holdenrieder, O. Forest Health in a Changing World. *Microb. Ecol.* **2014**, *69*, 826–842. [CrossRef] [PubMed]

12. Hossain, M.; Veneklaas, E.J.; Hardy, G.E.S.J.; Poot, P. Tree host–pathogen interactions as influenced by drought timing: Linking physiological performance, biochemical defence and disease severity. *Tree Physiol.* **2018**, *39*, 6–18. [CrossRef] [PubMed]

13. Metz, M.; Frangioso, K.M.; Meentemeyer, R.K.; Rizzo, D.M. Interacting disturbances: Wildfire severity affected by stage of forest disease invasion. *Ecol. Appl.* **2011**, *21*, 313–320. [CrossRef] [PubMed]

14. Meentemeyer, R.K.; Rank, N.E.; Shoemaker, D.A.; Oneal, C.B.; Wickland, A.C.; Frangioso, K.M.; Rizzo, D.M. Impact of sudden oak death on tree mortality in the Big Sur ecoregion of California. *Boil. Invasions* **2007**, *10*, 1243–1255. [CrossRef]

15. Cobb, R.C.; Filipe, J.A.N.; Meentemeyer, R.K.; Gilligan, C.A.; Rizzo, D.M. Ecosystem transformation by emerging infectious disease: Loss of large tanoak from California forests. *J. Ecol.* **2012**, *100*, 712–722. [CrossRef]

16. Cobb, R.C.; Rizzo, D.M. Decomposition and N cycling changes in redwood forests caused by sudden oak death. In Proceedings of the Coast Redwood Forests in a Changing California: A Symposium for Scientists and Managers, Albany, CA, USA, 21–23 June 2011; 2012; pp. 357–362.

17. Varner, J.M.; Kuljian, H.G.; Kreye, J.K. Fires without tanoak: The effects of a non-native disease on future community flammability. *Boil. Invasions* **2017**, *19*, 2307–2317. [CrossRef]

18. Forrestel, A.B.; Ramage, B.; Moody, T.; Moritz, M.A.; Stephens, S.L. Disease, fuels and potential fire behavior: Impacts of Sudden Oak Death in two coastal California forest types. *For. Ecol. Manag.* **2015**, *348*, 23–30. [CrossRef]

19. Trapp, S.; Croteau, R. Defensive resin biosynthesis in conifers. *Annu. Rev. Plant Physiol. Plant Mol. Biol.* **2001**, *52*, 689–724. [CrossRef]

20. Franceschi, V.R.; Krokene, P.; Christiansen, E.; Krekling, T. Anatomical and chemical defenses of conifer bark against bark beetles and other pests. *New Phytol.* **2005**, *167*, 353–376. [CrossRef]

21. Bakkali, F.; Averbeck, S.; Averbeck, D.; Idaomar, M. Biological effects of essential oils—A review. *Food Chem. Toxicol.* **2008**, *46*, 446–475. [CrossRef]

22. Boulogne, I.; Petit, P.; Ozier-Lafontaine, H.; Desfontaines, L.; Loranger-Merciris, G. Insecticidal and antifungal chemicals produced by plants: A review. *Environ. Chem. Lett.* **2012**, *10*, 325–347. [CrossRef]

23. Chetehouna, K.; Barboni, T.; Zarguili, I.; Leoni, E.; Simeoni, A.; Fernandez-Pello, A.C. Investigation on the Emission of Volatile Organic Compounds from Heated Vegetation and Their Potential to Cause an Accelerating Forest Fire. *Combust. Sci. Technol.* **2009**, *181*, 1273–1288. [CrossRef]

24. Chetehouna, K.; Courty, L.; Garo, J.P.; Viegas, D.; Fernandez-Pello, C. Flammability limits of biogenic volatile organic compounds emitted by fire-heated vegetation (Rosmarinus officinalis) and their potential link with accelerating forest fires in canyons: A Froude-scaling approach. *J. Fire Sci.* **2013**, *32*, 316–327. [CrossRef]

25. Courty, L.; Chetehouna, K.; Lemee, L.; Mounaïm-Rousselle, C.; Halter, F.; Garo, J. Pinus pinea emissions and combustion characteristics of limonene potentially involved in accelerating forest fires. *Int. J. Therm. Sci.* **2012**, *57*, 92–97. [CrossRef]

26. Pausas, J.G.; Alessio, G.A.; Moreira, B.; Segarra-Moragues, J.G. Secondary compounds enhance flammability in a Mediterranean plant. *Oecologia* **2015**, *180*, 103–110. [CrossRef] [PubMed]

27. Della Rocca, G.; Madrigal, J.; Marchi, E.; Michelozzi, M.; Moya, B.; Danti, R. Relevance of terpenoids on flammability of Mediterranean species: An experimental approach at a low radiant heat flux. *iFor. Biogeosci. For.* **2017**, *10*, 766–775. [CrossRef]

28. Fares, S.; Bajocco, S.; Salvati, L.; Camarretta, N.; Dupuy, J.-L.; Xanthopoulos, G.; Guijarro, M.; Madrigal, J.; Hernando, C.; Corona, P. Characterizing potential wildland fire fuel in live vegetation in the Mediterranean region. *Ann. For. Sci.* **2017**, *74*, 1–14. [CrossRef]

29. Achotegui-Castells, A.; Della Rocca, G.; Llusià, J.; Danti, R.; Barberini, S.; Bouneb, M.; Simoni, S.; Michelozzi, M.; Penuelas, J. Terpene arms race in the Seiridium cardinale—Cupressus sempervirens pathosystem. *Sci. Rep.* **2016**, *6*, 18954. [CrossRef]

30. Graniti, A. CYPRESS CANKER: A Pandemic in Progress. *Annu. Rev. Phytopathol.* **1998**, *36*, 91–114. [CrossRef]

31. Della Rocca, G.; Osmundson, T.; Danti, R.; Doulis, A.G.; Pecchioli, A.; Donnarumma, F.; Casalone, E.; Garbelotto, M. AFLP analyses of California and Mediterranean populations ofSeiridium cardinaleprovide insights on its origin, biology and spread pathways. *For. Pathol.* **2013**, *43*, 211–221. [CrossRef]

32. Danti, R.; Della Rocca, G. Epidemiological History of Cypress Canker Disease in Source and Invasion Sites. *Forests* **2017**, *8*, 121. [CrossRef]

33. Danti, R.; Della Rocca, G.; Panconesi, A. Chapter 17 Cypress canker. In *Infectious Forest Disease*; Nicolotti, G., Gonthier, P., Eds.; CAB International: Wallingford, UK; Boston, MA, USA, 2013; pp. 358–371.

34. Ponchet, J.; Andreoli, C. Histopathologie du cancre du cyprès à Seiridium cardinale. *Eur. J. For. Path.* **1989**, *19*, 212–221.

35. Spanos, B.K.A.; Pirrie, A.; Woodward, S.; Xenopoulos, S. Responses in the bark of Cupressus sempervirens clones artificially inoculated with Seiridium cardinale under field conditions. *Eur. J. For. Pathol.* **1999**, *29*, 135–142. [CrossRef]

36. Papini, A.; Moricca, S.; Danti, R.; Tani, C.; Posarelli, I.; Falsini, S.; Della Rocca, G. Ultrastructure of traumatic resin duct formation in Cupressus sempervirens L. in response to the attack of the fungus Seiridium cardinale (Wag.) Sutton & Gibson. In Proceedings of the 14th Multinational Congress on Microscopy, Belgrade, Serbia, 15–20 September 2019; pp. 310–311.

37. Raddi, P.; Panconesi, A. Pathogenicity of some isolates of Seiridium (Coryneum) cardinale, agent of cypress canker disease. *For. Pathol.* **1984**, *14*, 348–354. [CrossRef]

38. Santini, A.; Donardo, V. Genetic variability of the 'bark canker resistance' character in several natural provenances of Cupressus sempervirens. *For. Pathol.* **2000**, *30*, 87–96. [CrossRef]

39. Danti, R.; Panconesi, A.; Di Lonardo, V.; Della Rocca, G.; Raddi, P. *Italico*'and Mediterraneo': Two Seiridium cardinale Canker-Resistant Cypress Cultivars of Cupressus sempervirens. *Am. Soc. Hort. Sci.* **2006**, *41*, 1357–1359. [CrossRef]

40. Danti, R.; Di Lonardo, V.; Pecchioli, A.; Della Rocca, G. 'Le Crete 1' and 'Le Crete 2': Two newly patentedSeiridium cardinalecanker-resistant cultivars ofCupressus sempervirens. *For. Pathol.* **2012**, *43*, 204–210. [CrossRef]

41. Danti, R.; Rotordam, M.G.; Emiliani, G.; Giovannelli, A.; Papini, A.; Tani, C.; Barberini, S.; Della Rocca, G. Different clonal responses to cypress canker disease based on transcription of suberin-related genes and bark carbohydrates' content. *Trees* **2018**, *32*, 1707–1722. [CrossRef]

42. Valette, J.-C. Inflammabilités des espèces forestières méditerranéennes. Conséquences sur la combustibilité des formations forestières. *Rev. For. Française* **1990**, *42*, 76–92. [CrossRef]

43. Dimitrakopoulos, A.; Papaioannou, K.K. Flammability Assessment of Mediterranean Forest Fuels. *Fire Technol.* **2001**, *37*, 143–152. [CrossRef]

44. Liodakis, S.; Bakirtzis, D.; Lois, E. TG and autoignition studies on forest fuels. *J. Therm. Anal. Calorim* **2002**, *69*, 519–528. [CrossRef]

45. Neyisci, T.; Intini, M. The use of cypress barriers for limiting fires in Mediterranean countries. In Proceedings of the Il Cipresso e Gli Incendi, Valencia, Spain, 14–16 June 2006; Arsia Toscana: Firenze, Italy, 2006; pp. 3–18.

46. Ganteaume, A.; Jappiot, M.; Lampin, C.; Guijarro, M.; Hernando, C. Flammability of Some Ornamental Species in Wildland–Urban Interfaces in Southeastern France: Laboratory Assessment at Particle Level. *Environ. Manag.* **2013**, *52*, 467–480. [CrossRef] [PubMed]

47. Della Rocca, G.; Hernando, C.; Madrigal, J.; Danti, R.; Moya, J.; Guijarro, M.; Pecchioli, A.; Moya, B. Possible land management uses of common cypress to reduce wildfire initiation risk: A laboratory study. *J. Environ. Manag.* **2015**, *159*, 68–77. [CrossRef] [PubMed]

48. Danti, R.; Barberini, S.; Pecchioli, A.; Di Lonardo, V.; Della Rocca, G. The Epidemic Spread of Seiridium cardinaleon Leyland Cypress Severely Limits Its Use in the Mediterranean. *Plant Dis.* **2014**, *98*, 1081–1087. [CrossRef] [PubMed]

49. Madrigal, J.; Hernando, C.; Guijarro, M.; Díez, C.; Marino, E.; De Castro, A.J. Evaluation of Forest Fuel Flammability and Combustion Properties with an Adapted Mass Loss Calorimeter Device. *J. Fire Sci.* **2009**, *27*, 323–342. [CrossRef]

50. Madrigal, J.; Hernando, C.; Guijarro, M. A new bench-scale methodology for evaluating the flammability of live forest fuels. *J. Fire Sci.* **2012**, *31*, 131–142. [CrossRef]

51. Della Rocca, G.; Danti, R.; Hernando, C.; Guijarro, M.; Madrigal, J. Flammability of Two Mediterranean Mixed Forests: Study of the Non-additive Effect of Fuel Mixtures in Laboratory. *Front. Plant Sci.* **2018**, *9*, 825. [CrossRef]

52. White, R.H.; Zipperer, W.C. Testing and classification of individual plants for fire behaviour: Plant selection for the wildland—urban interface. *Int. J. Wildland Fire* **2010**, *19*, 213–227. [CrossRef]

53. Cruz, M.G.; Butler, B.W.; Viegas, D.; Palheiro, P. Characterization of flame radiosity in shrubland fires. *Combust. Flame* **2011**, *158*, 1970–1976. [CrossRef]

54. Tholl, D. Biosynthesis and Biological Functions of Terpenoids in Plants. *Plant Cells* **2015**, *148*, 63–106. [CrossRef]

55. Yazaki, K.; Arimura, G.; Ohnishi, T. 'Hidden' Terpenoids in Plants: Their Biosynthesis, Localization and Ecological Roles. *Plant Cell Physiol.* **2017**, *58*, 1615–1621. [CrossRef] [PubMed]

56. Freeman, K. An Overview of Plant Defenses against Pathogens and Herbivores. *Plant Heal. Instr.* **2008**, 149. [CrossRef]

57. Selim, S.A.; E Adam, M.E.; Hassan, S.M.; AlBalawi, A.R. Chemical composition, antimicrobial and antibiofilm activity of the essential oil and methanol extract of the Mediterranean cypress (Cupressus sempervirens L.). *BMC Complement. Altern. Med.* **2014**, *14*, 179. [CrossRef] [PubMed]

58. Zulak, K.G.; Bohlmann, J. Terpenoid Biosynthesis and Specialized Vascular Cells of Conifer Defense. *J. Integr. Plant Boil.* **2010**, *52*, 86–97. [CrossRef]

59. Celedon, J.M.; Bohlmann, J. Oleoresin defenses in conifers: Chemical diversity, terpene synthases and limitations of oleoresin defense under climate change. *New Phytol.* **2019**, *224*, 1444–1463. [CrossRef]

60. Krokene, P.; Nagy, N.E.; Krekling, T. Chapter 7 Traumatic resin ducts and polyphenolic parenchyma cells in conifers. In *Induced Plant Resistance to Herbivory*; Springer: Berlin/Heidelberg, Germany, 2008; pp. 147–169.

61. Bonello, P.; Capretti, P.; Luchi, N.; Martini, V.; Michelozzi, M. Systemic effects of Heterobasidion annosum s.s. infection on severity of Diplodia pinea tip blight and terpenoid metabolism in Italian stone pine (Pinus pinea). *Tree Physiol.* **2008**, *28*, 1653–1660. [CrossRef]

62. Pepori, A.L.; Michelozzi, M.; Santini, A.; Cencetti, G.; Bonello, P.; Gonthier, P.; Sebastiani, F.; Luchi, N. Comparative transcriptional and metabolic responses of Pinus pinea to a native and a non-native Heterobasidion species. *Tree Physiol.* **2018**, *39*, 31–44. [CrossRef]

63. Whitehill, J.; Yuen, M.M.S.; Henderson, H.; Madilao, L.L.; Kshatriya, K.; Bryan, J.; Jaquish, B.; Bohlmann, J. Functions of stone cells and oleoresin terpenes in the conifer defense syndrome. *New Phytol.* **2018**, *221*, 1503–1517. [CrossRef]

64. Ormeño, E.; Céspedes, B.; Sánchez, I.A.; García, A.V.; Moreno, J.M.; Fernandez, C.; Baldy, V. The relationship between terpenes and flammability of leaf litter. *For. Ecol. Manag.* **2009**, *257*, 471–482. [CrossRef]

65. Raddi, P.; Panconesi, A.; Xenopoulos, S.; Ferrandes, P.; Andreoli, C. Genetic improvement for resistance to cypress canker. In *Agrimed Reserach Programme, Progress in EEC Research on Cypress Disease*; Ponchet, J., Ed.; Report EU 12493 EN; Commission of the European Communities: Brussels, Belgium; Luxembourg, 1990; pp. 127–134.

66. Danti, R.; Della Rocca, G.; Di Lonardo, V.; Pecchioli, A.; Raddi, P. Genetic improvement program of cypress: Results and outlook. In *Status of the Experimental Network of Mediterranean Forest Genetic Resources*; Besacier, C., Ducci, F., Malagnoux, M., Souvannavong, O., Eds.; CRA SAL, Arezzo and FAO Silva Mediterranea: Rome, Italy, 2011; pp. 88–96.

67. Panconesi, A.; Danti, R. Esperienze técnico-scientifiche nella bonifica del cipresso. In Proceedings of the Il Recupero Del Cipresso Nel Paesaggio E Nel Giardino Storico' Collodi, Pistoia, Italy, 15 March 1995; Regione Toscana Giunta Regionale Dipartimento Agricoltura e Foreste: Florence, Italy, 1995; pp. 9–21.

 *forests*

MDPI

Article

The Effect of Antecedent Fire Severity on Reburn Severity and Fuel Structure in a Resprouting Eucalypt Forest in Victoria, Australia

Luke Collins [1,2,3,*], Adele Hunter [1], Sarah McColl-Gausden [4], Trent D. Penman [4] and Philip Zylstra [5]

1   Department of Ecology, Environment & Evolution, La Trobe University, Bundoora, VIC 3086, Australia; adeleh.18@gmail.com
2   Arthur Rylah Institute for Environmental Research, Department of Environment, Land, Water and Planning, P.O. Box 137, Heidelberg, VIC 3084, Australia
3   Research Centre for Future Landscapes, La Trobe University, Bundoora, VIC 3086, Australia
4   School of Ecosystem and Forest Sciences, University of Melbourne, Creswick, VIC 3363, Australia; mccoll.s@unimelb.edu.au (S.M.-G.); trent.penman@unimelb.edu.au (T.D.P.)
5   School of Molecular and Life Sciences, Curtin University, Kent Street, Bentley, WA 6102, Australia; philip.zylstra@curtin.edu.au
*   Correspondence: l.collins3@latrobe.edu.au

check for **updates**

Citation: Collins, L.; Hunter, A.; McColl-Gausden, S.; Penman, T.D.; Zylstra, P. The Effect of Antecedent Fire Severity on Reburn Severity and Fuel Structure in a Resprouting Eucalypt Forest in Victoria, Australia. *Forests* 2021, 12, 450. https://doi.org/10.3390/f12040450

Academic Editors: Joaquim S. Silva and Paulo M. Fernandes

Received: 16 December 2020
Accepted: 3 April 2021
Published: 8 April 2021

Publisher's Note: MDPI stays neutral with regard to jurisdictional claims in published maps and institutional affiliations.

Abstract: Research highlights—Feedbacks between fire severity, vegetation structure and ecosystem flammability are understudied in highly fire-tolerant forests that are dominated by epicormic resprouters. We examined the relationships between the severity of two overlapping fires in a resprouting eucalypt forest and the subsequent effect of fire severity on fuel structure. We found that the likelihood of a canopy fire was the highest in areas that had previously been exposed to a high level of canopy scorch or consumption. Fuel structure was sensitive to the time since the previous canopy fire, but not the number of canopy fires. Background and Objectives—Feedbacks between fire and vegetation may constrain or amplify the effect of climate change on future wildfire behaviour. Such feedbacks have been poorly studied in forests dominated by highly fire-tolerant epicormic resprouters. Here, we conducted a case study based on two overlapping fires within a eucalypt forest that was dominated by epicormic resprouters to examine (1) whether past wildfire severity affects future wildfire severity, and (2) how combinations of understorey fire and canopy fire within reburnt areas affect fuel properties. Materials and Methods—The study focused on ≈77,000 ha of forest in south-eastern Australia that was burnt by a wildfire in 2007 and reburnt in 2013. The study system was dominated by eucalyptus trees that can resprout epicormically following fires that substantially scorch or consume foliage in the canopy layer. We used satellite-derived mapping to assess whether the severity of the 2013 fire was affected by the severity of the 2007 fire. Five levels of fire severity were considered (lowest to highest): unburnt, low canopy scorch, moderate canopy scorch, high canopy scorch and canopy consumption. Field surveys were then used to assess whether combinations of understorey fire (<80% canopy scorch) and canopy fire (>90% canopy consumption) recorded over the 2007 and 2013 fires caused differences in fuel structure. Results—Reburn severity was influenced by antecedent fire severity under severe fire weather, with the likelihood of canopy-consuming fire increasing with increasing antecedent fire severity up to those classes causing a high degree of canopy disturbance (i.e., high canopy scorch or canopy consumption). The increased occurrence of canopy-consuming fire largely came at the expense of the moderate and high canopy scorch classes, suggesting that there was a shift from crown scorch to crown consumption. Antecedent fire severity had little effect on the severity patterns of the 2013 fire under nonsevere fire weather. Areas affected by canopy fire in 2007 and/or 2013 had greater vertical connectivity of fuels than sites that were reburnt by understorey fires, though we found no evidence that repeated canopy fires were having compounding effects on fuel structure. Conclusions—Our case study suggests that exposure to canopy-defoliating fires has the potential to increase the severity of subsequent fires in resprouting eucalypt forests in the short term. We propose that the increased vertical connectivity of fuels caused by resprouting and seedling recruitment were responsible for the elevated fire severity. The effect of antecedent fire severity on reburn severity will likely be constrained by a range of factors, such as fire weather.

163

Keywords: epicormic resprouter; eucalyptus; fire severity; flammability feedbacks; temperate forest

1. Introduction

Wildfire size, the annual area burned and the extent of high severity fire are increasing across many forested regions worldwide [1]. Much of the change in fire activity has been attributed to recent warming and drying trends associated with anthropogenic climate change [2,3], owing to the dominant influence of fuel aridity and fire weather on forest fire behaviour [4–6]. The frequency and duration of climatic conditions that are conducive to large wildfires are projected to increase across many forested regions in the future [7–9], with models forecasting greater exposure of forests to large and severe wildfires [10,11]. Feedbacks between fire and vegetation have the potential to either constrain or increase the effect of climate on future wildfire behaviour through the modification of fuel properties and ecosystem flammability [11,12]. However, these feedbacks remain poorly understood [13,14], and as such, are rarely incorporated into projections of the effects of climate change on ecosystem flammability and fire regimes [11].

Flammability is a multidimensional trait of fuels that encompasses their probability of ignition, rate of combustion and amount of heat released [15]. The flammability of a fuel particle is determined by its physical properties (e.g., size, shape, moisture content and calorific value) and the exogenous conditions under which the fuel ignites and burns [16,17]. Fire behaviour at the stand or ecosystem scale, which provides a contextual measure of flammability [18], is determined by the flammability traits of individual fuel particles and their vertical and horizontal arrangement within a plant, fuel stratum and stand [19]. Exogenous factors that affect fuel moisture (i.e., soil moisture content, temperature and relative humidity) and flame propagation (i.e., wind) will strongly affect fire behaviour, as they influence the likelihood that fire will bridge the gaps between fuel particles within and between fuel strata [19]. Past exposure of ecosystems to fire may alter leaf flammability traits and the spatial arrangement of fuels [20], as well as the exogenous conditions affecting fuel moisture [21], though these effects will depend on the immediate and longer-term response of the vegetation community to fire [14,22].

Fire severity is a measure that is used to quantify the immediate impact of fire on ecosystems [23]. Metrics used to quantify fire severity vary, though most tend to focus on the degree of change to canopy and understorey foliage (e.g., stem mortality, foliage scorch and consumption), as these changes are readily detected using remote sensing [24–26]. In the context of this paper, we describe fire severity based on the degree of scorch and consumption of foliage, providing a measure of the immediate impacts of fire on vegetation [23,27]. In this schema, fires that result in the complete consumption of canopy foliage will represent the upper extreme of the fire severity spectrum within a community, whereas fires causing little or no impact to the canopy or understorey foliage represent the lower end of the spectrum [27–29]. Fire causing intermediate levels of consumption and scorch to foliage are typically considered to be of intermediate severity [24,27].

An ecosystem's response to a fire will depend on the capacity of the vegetation community to resist the direct effects of fire, termed "fire resistance", or recover following a fire via resprouting or seedling recruitment, termed "fire resilience" [30]. Fires or fire regimes that exceed the resistance or resilience of the dominant tree species will cause substantial changes to vegetation composition and structure [31–33], where in extreme cases, this leads to the conversion of forests to nonforested states [34]. These structural and compositional changes often increase the amount of live and dead fuel close to the forest floor (e.g., [14,35]). This is particularly evident following high-severity fires occurring at short-intervals, which can instigate the conversion of forests dominated by fire-sensitive obligate seeder tree species to highly flammable nonforest states [30,32,36]. In more fire-tolerant forests dominated by resprouters, high-severity fire can cause substantial changes to canopy structure, reducing the gaps between the tree canopy and litter fuels on the forest floor [31,33]. Such changes to the biomass and arrangement of fuels are often inferred as evidence of positive fire feedbacks (e.g., [32]), though empirical tests over multiple fire cycles are lacking for many ecosystems.

The eucalypt forests of south-eastern Australia are highly fire-tolerant communities that are primarily composed of plant species that display some resilience to fire (i.e., >95% of species [37]). Most of these forests are dominated by eucalypt trees (i.e., *Eucalyptus*, *Corymbia* and *Angophora* spp.) [38] that are capable of resprouting from buds on the stem and branches in response to fires that impact the canopy foliage, which is a trait that is referred to as "epicormic resprouting" [39,40]. Canopy species in these forests show a high degree of resistance to understorey fires that primarily burn the surface litter, herbaceous and shrub layers, and result in a low-to-moderate degree of canopy scorch [27,31,41,42]. The resistance of these canopy species decreases as the degree of scorch or consumption of canopy foliage increases, with high rates of branch and stem mortality being observed following fires that consume most of the canopy leaves (i.e., canopy fires) [31,42,43]. Mass seedling recruitment and vigorous resprouting along surviving defoliated stems and branches often occur following fires that cause extensive leaf scorch or consumption [31,44]. It has been proposed that these structural changes to vegetation increase the flammability of eucalypt forests [22,35], resulting in positive feedbacks between canopy fires (e.g., [45]), though research linking the fire severity–vegetation structure–flammability feedback is currently lacking.

Here, we looked for evidence that fire severity feedbacks had occurred in a forest dominated by epicormic resprouters using a case study located in south-eastern Australia. Our study focused on a large area (≈77,000 ha) of eucalypt forest that was burnt by two successive wildfires six years apart, with the first occurring during the 2007 fire season and the second during the 2013 season. We address two questions in this study: (1) Are patterns in reburn severity influenced by antecedent fire severity? (2) Has fuel structure changed in response to combinations of canopy fire and understorey fire within areas of reburnt forest? We used remotely sensed measures of wildfire severity to test whether the severity of the 2013 fire was affected by the severity of the 2007 fire. We conducted field surveys in areas burnt by different combinations of understorey fire (i.e., <80% canopy scorch) and canopy fire (i.e., >90% canopy consumption) over the two successive wildfires to assess how fuel properties varied in response to combinations of fire types.

2. Materials and Methods

2.1. Study Area

The study took place in the West Gippsland region of Victoria, Australia, approximately 140 km east of Melbourne (Figure 1). The study area was impacted by two major wildfires prior to the commencement of the study. These fires were the "Great Divide fire", which burnt approximately one million hectares between December 2006 and February 2007 (referred to hereon as the "2007 fire"), and the "Aberfeldy fire", which burnt 87,000 ha between January and February 2013 (referred to hereon as the "2013 fire"). We note that a third fire, the "Walhalla fire", burnt ≈8700 ha within the study area in February 2019, reburning areas impacted by both the 2007 and 2013 wildfires. The study was confined to the footprint of the 2013 fire and focused on the impact of the 2007 and 2013 fires.

The study was conducted within forest communities dominated by eucalypt species that possess the capacity to resprout epicormically [31,43,46]. Open forest communities dominate the exposed ridges and slopes, whereas tall open eucalypt forest occupies the mesic sheltered topographic locations (i.e., poleward aspects, lower slopes and gullies) [47]. Canopy cover ranges between 30% and 70% across both forest types [48], except after high-severity wildfire, when it is substantially reduced [49]. Canopy height is generally less than 30 m in the open forests but exceeds 30 m in the tall open forest communities [48]. The forests are comprised of a diverse mix of eucalypt species, including *Eucalyptus consid... niana*, *Eucalyptus cypellocarpa*, *Eucalyptus dives*, *Eucalyptus muelleriana*, *Eucalyptus obliqua*, *Eucalyptus radiata*, *Eucalyptus sieberi* and *Eucalyptus tricarpa* [31,46]. Across both forest types, the understorey is characterised by a well-developed shrub and herb layer. The composition and structure of the understorey vegetation is strongly affected by site productivity and fire history [47–49].

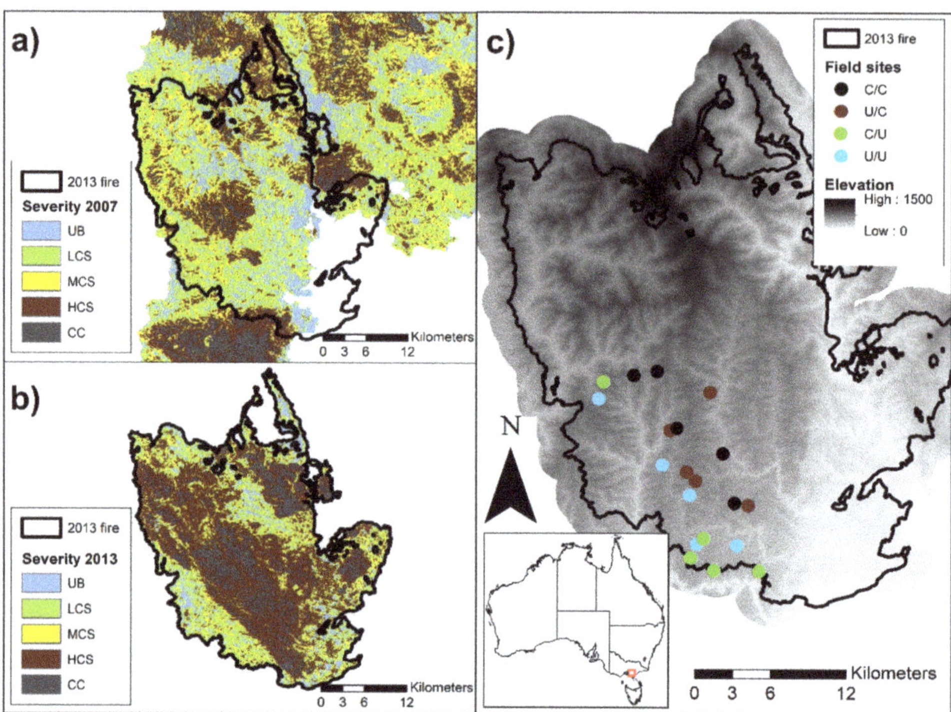

Figure 1. Severity maps for the (**a**) 2007 Great Divide wildfire and (**b**) 2013 Aberfeldy wildfire. Panel (**c**) shows the location of the study area (red square in the inset) and the field sites used to assess the effect of the fire combinations on fuel properties. The footprint of the 2013 Aberfeldy wildfire has been included in each panel to provide a reference. The fire severity classes presented in panels (**a,b**) are unburnt (UB), low canopy scorch (LCS), moderate canopy scorch (MCS), high canopy scorch (HCS) and canopy consumption (CC). Descriptions of these severity classes are provided in Section 2.2. The fire severity combinations presented in panel (**c**), which describe the severity of the 2007 and 2013 fires, are understorey fire followed by an understorey fire (U/U), canopy fire then an understorey fire (C/U), understorey fire then a canopy fire (U/C) and canopy fire then a canopy fire (C/C). Descriptions of these severity combinations are provided in Section 2.4.

The climate across the study region is temperate, with the average monthly maximum temperature for the region ranging between 13.7 °C (July) and 26.7 °C (December), and the average minimum ranging between 3.7 °C (July) and 12.9 °C (December) (station 85280; www.bom.gov.au, accessed on 14 May 2019). The average annual rainfall was 735.5 mm, with no strong seasonal trend occurring throughout the year (station 85280; www.bom.gov.au, accessed on 14 May 2019).

2.2. Fire Severity Mapping

Fire severity mapping was derived for the 2007 and 2013 wildfires using Landsat imagery and a random forest classification method described in Collins et al. [50]. This classification scheme targets the degree of scorch and consumption of the canopy foliage, providing a severity metric that is correlated with flame dimensions within structurally similar communities [23,27]. The classification method identified five fire severity classes, including (i) unburnt vegetation (UB; <10% of the understorey burnt), (ii) low canopy scorch (LCS; <20% canopy scorch), (iii) moderate canopy scorch (MCS; 20–80% canopy scorch), (iv) high canopy scorch (HCS; >80% canopy scorch) and (v) canopy consumption (CC; canopy mostly consumed) [24,50]. The classification method was shown to have a very high classification accuracy across eucalypt forests (88% global accuracy) when independently cross-validated on fires not included in the random forest training dataset [50]. Fire severity maps were produced in the Google Earth Engine platform [28,51].

2.3. Relationships between the Severity Patterns of the 2007 and 2013 Fires

We used fire severity mapping to examine whether the severity patterns of the 2013 fire (Figure 1a) were influenced by the severity patterns of the 2007 fire (Figure 1b). Several environmental covariates were also considered to account for the effects of fire weather, terrain and vegetation type on the severity patterns during the 2013 fire. These variables are described below.

2.3.1. Environmental Datasets

Fire weather conditions in eucalypt forests are operationally classified using the McArthur Forest Fire Danger Index (FFDI). The FFDI provides a single index of fire danger (0 to >100) that is calculated using temperature, relative humidity, wind speed and antecedent precipitation [52]. The index is related to the likelihood of a fire starting, the fire intensity, the rate of spread and the suppression difficulty [53]. Six categories of FFDI are recognised for operational purposes: low (0–12), high (13–25), very high (26–49), severe (50–74), extreme (75–99) and catastrophic ($\geq$100) [53]. Fires burning under severe, extreme or catastrophic fire weather (i.e., FFDI > 50) are predominantly weather-driven fires that are characterised by rapid rates of spread and large areas of canopy-consuming fire [53–55]. The effect of topography and fuels on fire behaviour typically increases when FFDI falls below $\approx$50 [53].

Fire progression data and weather observations were used to assign fire weather conditions across the 2013 fire extent. Progression data recorded by the Victorian Department of Environment, Land, Water and Planning (DELWP) was used in combination with hotspot data derived from the Moderate Resolution Imaging Spectroradiometer (MODIS) (http://sentinel.ga.gov.au, accessed on 17 January 2019) to assign a day of burn across areas of the 2013 fire. The weather data were obtained from the closest Bureau of Meteorology weather station (station no. 85280, www.bom.gov.au, accessed on 26 April 2016). Fire weather conditions within the fire ground can deviate considerably from conditions experienced at the weather station, owing to topographic effects on the wind and the effect of the fire on local weather conditions. Consequently, we used information from the nearest weather station, coupled with observed daily rates of fire spread, to inform our classification of the fire weather conditions (see [6]). The FFDI was calculated at 30 min intervals and the maximum daily FFDI was extracted. We defined "severe" weather (SEV) as those days where the maximum FFDI exceeded 49 or the average rate of forest fire spread over a 24 h period exceeded 1000 m h^{-1}. We defined "nonsevere" weather (NSEV) as conditions when the maximum FFDI was less than 35 with average rates of spread that were less than 1000 m h^{-1}.

A digital elevation model (30 m resolution) that was generated from the Shuttle Radar Topography Mission [56] was used to derive spatial layers of slope, aspect and topographic position, as these topographic variables have previously been found to influence fire behaviour patterns in eucalypt forests [6,45,57]. Elevation, slope and aspect were calculated and extracted using the Google Earth Engine platform [51]. Aspect was adjusted such that the values were relativised with respect to north. This was done by calculating the absolute difference between 360° and those aspects greater than 180°. Values approaching 0° represent north-facing aspects, while values approaching 180° represent south-facing aspects. A topographic position index (TPI) was calculated as the difference between the elevation of a focal pixel and the average elevation of pixels within a 500 m radius. Positive values of the TPI represent exposed topographic positions (i.e., ridges and upper slopes), negative values represent sheltered positions (i.e., gullies and lower slopes) and values close to zero represent flat areas or mid-slopes. Vegetation mapping acquired from DELWP [47] was used to assign areas as an open forest or a tall open forest. Other vegetation communities were excluded.

2.3.2. Sampling Design

Fire severity and environmental datasets were sampled using a points-based approach following protocols developed for the analysis of wildfire severity patterns in eucalypt forests [54,57]. The spacing between sample points is an important consideration when sampling fire severity data, as fire behaviour displays spatial dependence within forest ecosystems [57,58]. Topographic position imposes a strong influence on spatial patterns of fire severity across the study landscape, with ridges and upper slopes typically burning at a higher severity than gullies and lower slopes [54,58]. We determined that a minimum spacing of 400 m between sample points was suitable for our study as this is the typical distance between ridges and gullies across the study region [6]. A grid of points with 400 m spacing was established across the extent of the 2013 fire. Fire severity, environmental data and spatial coordinates were extracted from each point (n = 3736) and used for the analysis. The data extraction and calculation of the TPI were undertaken in ArcGIS v10.7.1 (Environmental Systems Research Institute, Redlands, CA, USA).

2.3.3. Statistical Analysis

Ordinal regression was used to analyse the effects of past fire severity, fire weather, terrain and vegetation on the likelihood of occurrence for each fire severity class. The effect of past fire severity and vegetation community were fitted as two-way interactions with fire weather. The interaction between the vegetation community and past severity was also included to account for the different fire responses between communities. Topographic variables were included as additive effects. Interacting smooth terms for longitude and latitude were also included in the model to account for spatial autocorrelation. Bayesian regression was used as it provided a robust means for fitting ordinal regression with a complex model structure. We lacked sufficient a priori information to define meaningful priors; therefore, uninformed priors were used when fitting the models. The models were fitted using Markov chain Monte Carlo (MCMC) as follows: four Markov chains were sampled, each consisting of 5000 iterations, with a 1000-iteration warm-up period, resulting in 16,000 iterations to derive the posterior distributions for the model parameters. Model convergence was assessed using the Gelman–Rubin diagnostic [59]. Plots of the median and 95% credible intervals were used to visualise the effect of model parameters on the likelihood of each fire severity class. Statistical analysis was conducted in the R statistical package v4.0.2 [60]. Bayesian models were fitted using the "brms" package [61].

2.4. The Effect of Fire Severity on Fuel Properties

2.4.1. Field Study Design

The field study examining the effect of fire severity on fuel structure targeted areas of open forest that were burnt by both the 2007 and 2013 fires. We focused on open forests, as they typically occur on ridges and upper slopes in the study area and are therefore more accessible than the tall open forests occurring on steep slopes and in gullies. We targeted areas that were affected by fires that either predominantly burnt in the understorey (i.e., understorey fires) or burnt extensively in both the understorey and canopy (i.e., canopy fires), as they are known to trigger contrasting responses from the dominant eucalypt species. Areas that had experienced low-to-moderate canopy scorch (<80% canopy scorch) were considered as having experienced an understorey fire, as field-based assessments have found that this degree of canopy scorch is associated with fires burning in the understorey fuels in eucalypt forests [27,62,63]. Areas that had experienced a high degree of canopy consumption (>90% of the canopy foliage) were considered to have experienced a canopy fire. Four fire severity combinations were identified across the 2007 and 2013 fires, with these being consecutive understorey fires (U/U), a canopy fire followed by an understorey fire (C/U), an understorey fire followed by a canopy fire (U/C) and consecutive canopy fires (C/C) (Figure 1c).

Large patches (>150 m diameter) of each fire severity combination were initially identified using both fire severity maps and high-resolution (<35 cm) post-fire aerial photographs (see [31]). Five replicate sites were selected for each fire severity combination ($n = 20$). Most (90%) of the instances of understorey fire had less than 20% canopy scorch. Fire severity classes were corroborated in the field where possible through observations of the scorch heights on tree trunks and the presence of dead and resprouting branches on trees. The sites were located on ridges and upper slopes with a north-facing aspect (270–360° and 0–90°) in areas that had no record of timber harvesting in the past 30 years to control for these potentially confounding factors. Site centroids were located at least 50 m from the severity patch edge and more than 50 m from roads or major clearings to avoid the influence of edge effects on the vegetation structure. A minimum spacing of 500 m was imposed between sites with different fire severity combinations, with a minimum spacing of 1000 m between sites with the same fire severity combination to ensure the independence of sites [31]. Field sampling took place between November 2018 and April 2019, approximately six years following the 2013 wildfire.

2.4.2. Fuel Surveys

Four vegetation strata were identified within the open forest community targeted in our study: near-surface (grasses and herbs), elevated (shrubs and regrowth), mid-storey (large shrubs and intermediate trees) and canopy (canopy trees) (Figure 2). These strata have been identified as being influential in determining fire behaviour in eucalypt forests [64,65]. The assignment of plants to each strata was based on the vegetation aggregation at the site, rather than predetermined heights, to facilitate differences in vegetation structure driven by environmental factors, such as fire severity (see Figure 2) [65]. Plants were assigned to the strata that encompassed most of their foliage (Figure 2).

Measurements targeted the dominant species in each of the four strata, with up to three species being surveyed per stratum. Plots were established at each site centroid to derive a representative sample of the dominant species. The plot size varied such that up to 15 individuals were surveyed per dominant species, resulting in plot radii ranging from 1 m to a predefined maximum of 25 m. Measurements of the crown base height, top height and width of the plant in two directions were made for each individual. If fewer than 15 individuals were found within a 25-m-radius plot, then no further measurements were made. If more than 15 individuals occurred within a plot, irrespective of the plot size, then these additional individuals were counted to obtain an overall density of the species. Species that were not dominant were tallied and added to the counts for the dominant species that had the most similar traits. The average base height and top height were calculated for each dominant species within a stratum. Crown cover was estimated for each stratum from the number of plants per area multiplied by the species-weighted area of each plant crown, where crowns were treated as circular. Cover values could exceed 100% as crowns within a stratum can overlap and crown dimensions were not always symmetrical.

Surface fine fuels were measured using destructive sampling. Surface fine fuels were classified as any dead flammable material <0.6 mm thick, including leaves, twigs and bark on the ground. Four samples were collected using a circular fuel ring of 0.1 m^2 at each site. These samples were collected 5 m from the centre of the site running perpendicular and parallel to the slope. Surface fine fuel samples were placed in a 70 °C oven for a minimum of 72 h or until a constant weight was achieved. Fuels were weighted and the mass (tonnes) of fuel per hectare was calculated.

2.4.3. Statistical Analysis

The analysis of vegetation structure focused on plant top and base heights and crown cover. Linear mixed-effect models were used to assess the effect of the fire severity combinations on the top and base heights of the plants within each fuel strata. Individual plants were treated as the unit of replication. The models included the interaction between the severity of the 2007 (SEV07) and 2013 (SEV13) fires, with the site identifier included as a random effect to account for plants being nested within sites. Analysis of variance was used to analyse the effect of the interaction between SEV07 and SEV13 on the estimated canopy cover. Model residuals were visually assessed for all models to see whether they met the assumptions of homogeneity of variance and normality of residuals. A log transformation (ln) was performed when these assumptions were not met. Parameters with $p < 0.05$ were considered statistically significant. Model predictions were used for the graphical interpretation of the model effects. Confidence intervals (95%) were generated using bootstrapping ($n = 1000$ replicates). Statistical analysis was conducted in the R statistical package v4.0.2 [60]. Linear mixed models were fitted using the "lme4" package [66]. Bootstrapping was conducted using the "boot" package [67].

Figure 2. Graphic depiction of the fuel layers measured in the study. The panels show the delineation of the fuel strata for open forests that are either (**a**) long unburnt or recently burnt by understorey fire and (**b**) recently burnt (i.e., ≈6 years post-fire) by canopy fire. Photos (**c,d**) were taken six years following an understorey fire and a canopy fire, respectively. The fuel type depicted in panel (**a**) corresponds to the photo in panel (**c**) and the fuel type in panel (**b**) corresponds to the photo in panel (**d**). The broken horizontal lines show the boundaries between fuel strata. The fuel strata are near-surface (NS), which includes grasses and small shrubs (typically <50 cm tall); elevated (E), which includes medium-sized shrubs and tree saplings (typically 50–200 cm tall); mid-storey (M), which includes tall shrubs and subcanopy trees (typically >200 cm); canopy (C), which is the uppermost tree stratum. Plants were assigned to the strata that encompassed most of their foliage. For example, the tree that is resprouting from the trunk and branches in panel (**b**) would be assigned to the canopy stratum, whereas the tree resprouting from the base would be assigned to the mid-storey stratum.

3. Results

3.1. Relationships between the Severity Patterns of the 2007 and 2013 Fires

The severity patterns of the 2013 fire were influenced by the two-way interactions between fire weather, antecedent fire severity and vegetation community, and the additive effects of aspect and topographic position (Table S1, Figure 3). Fire weather and antecedent fire severity exerted a strong influence on the severity patterns of the 2013 fire (Figure 3). The likelihood of a point remaining unburnt (UB) or experiencing an understorey fire (i.e., LCS and MCS) was typically lower during severe fire weather (i.e., SEV) relative to nonsevere (i.e., NSEV) fire weather conditions, with the opposite trend being observed for fires that scorched or consumed most of the canopy foliage (i.e., HCS and CC) (Figure 3). Under SEV weather, the likelihood of CC was greater in areas that were previously exposed to fires that consumed or scorched most of the canopy foliage (i.e., HCS and CC) relative to areas that were affected by LCS or were UB (Figure 3). In the open forest communities, the increase in CC came at the expense of the HCS class (Figure 3a, SEV), whereas in the tall open forest communities, the increase in CC came at the expense of the MCS and HCS classes (Figure 3b, SEV). Under NSEV weather, the effect of past fire severity was muted, with small ($\Delta P < 0.1$) changes in the probability of fire severity classes being observed for both open forest and tall open forest types (Figure 3).

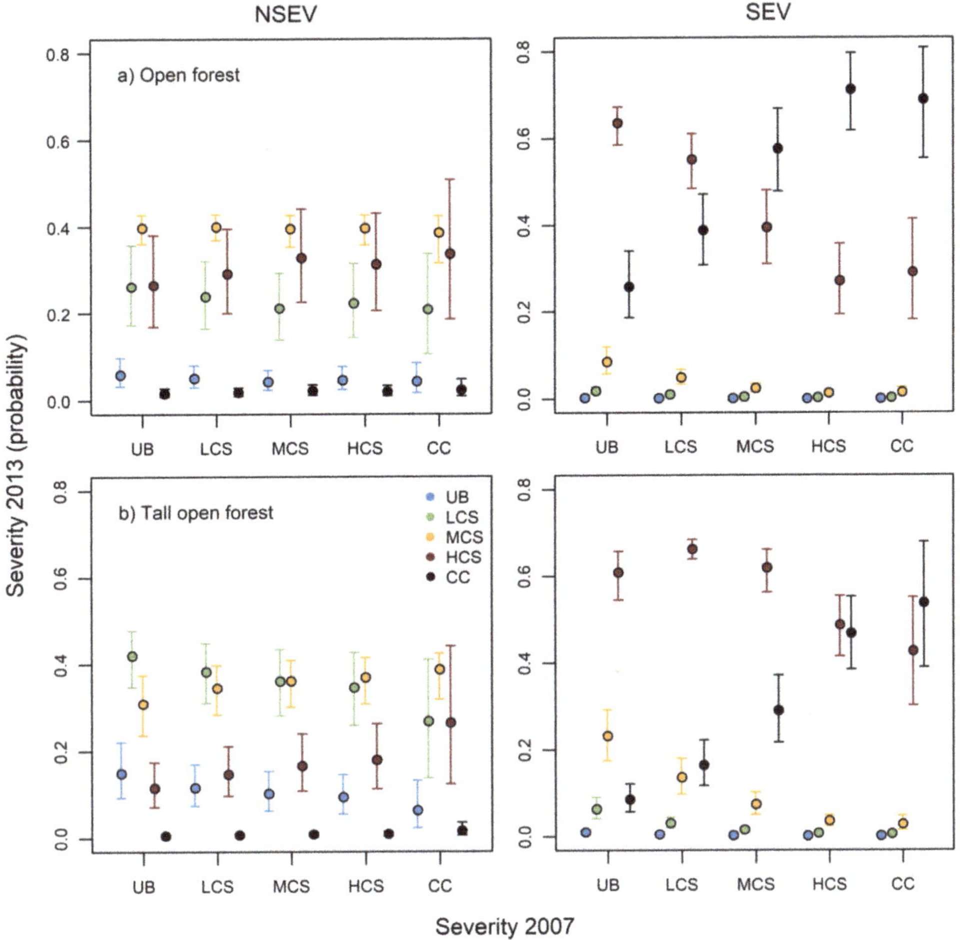

Figure 3. The effect of the severity of the 2007 fire (*x*-axis) on the probability of each severity class occurring during the 2013 fire (*y*-axis) for (**a**) open forest and (**b**) tall open forest types under nonsevere (NSEV) and severe (SEV) fire weather. Points show the mean and error bars are the 95% credible intervals. The fire severity codes are unburnt (UB), low canopy scorch (LCS), moderate canopy scorch (MCS), high canopy scorch (HCS) and canopy consumption (CC). Slope, aspect, topographic position (TPI), latitude and longitude have been held constant at their mean values.

Fire severity was typically lower on poleward-facing aspects compared to equatorial facing aspects and protected topographic locations (i.e., gullies) compared to exposed locations (i.e., ridges) (Figures S1 and S2).

3.2. The Effect of Fire Severity on Fuel Properties

Canopy height and cover were affected by the severity of the 2013 wildfire (SEV13), but not the preceding fire in 2007 (SEV07) (Table 1). Canopy top height and base height were significantly shorter at sites experiencing canopy fire in 2013 (Figure 4). This was particularly evident for canopy base height, which displayed substantial differences (mean $\pm$ S.E.) in areas impacted by understorey fire (8.92 $\pm$ 0.16 m) and canopy fire (2.51 $\pm$ 0.14 m) (Figure 4). The canopy cover at sites affected by a canopy fire in 2013 (57 $\pm$ 11%) was one-third of that recorded at sites affected by an understorey fire in 2013 (177 $\pm$ 32%). Crown height (base and top height) of the mid-storey layer was affected by the interaction between SEV07 and SEV13 (Table 1), with taller crowns being observed at sites experiencing a sequence of a canopy fire then an understorey fire (i.e., C/U) compared to the other fire severity combinations (Figure 4). The cover of the mid-storey layer was not affected by fire severity (Table 1). The fire severity classes did not affect the crown properties or the cover of plants in the elevated and near-surface layers (Table 1).

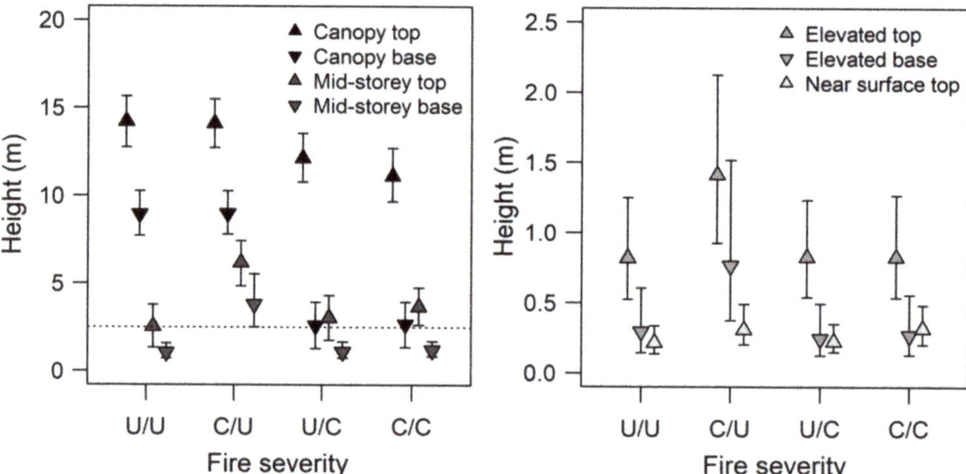

Figure 4. Model predictions for the plant crown top and base heights across the four fuel strata. The left panel shows the canopy and mid-storey layers. The right panel shows the elevated and near-surface layers. Upward-facing triangles represent the average crown top height and downward-facing triangles represent the average crown base height. Error bars show the 95% confidence intervals. The broken line in the left panel corresponds to the maximum height (2.5 m) in the right panel. The fire severity combinations are understorey fire followed by an understorey fire (U/U), canopy fire then an understorey fire (C/U), understorey fire then a canopy fire (U/C) and canopy fire then a canopy fire (C/C).

The surface fine-fuel load was affected by SEV13 but not SEV07 (Table 1). The average fine-fuel load on sites burnt by a canopy fire in 2013 (2.45 t ha^{-1}) was less than half that recorded at sites burnt by an understorey fire (5.90 t ha^{-1}).

Table 1. Summary of the models that were used for testing the effects of fire severity combinations on fuel properties. Five fuel strata were considered. The effect of the fire severity combinations on crown top height, base height and cover were assessed for each vegetation stratum (excluding the base height for near-surface fuels). Fuel biomass was assessed for surface fuels. Values highlighted in bold are statistically significant ($p < 0.05$).

Stratum	Response	SEV07		SEV13		SEV07 × SEV13	
		F	*p*-Value	*F*	*p*-Value	*F*	*p*-Value
Canopy	Top height	0.561	0.465	11.958	**0.003**	0.391	0.541
	Base height	0.008	0.931	93.426	**<0.001**	0.001	0.972
	Cover	0.443	0.515	20.957	**<0.001**	0.836	0.374
Mid-storey	Top height	11.041	**0.006**	2.990	0.107	5.854	**0.031**
	Base height	10.825	**0.006**	8.669	0.011	9.023	**0.010**
	Cover	0.159	0.696	0.342	0.569	3.366	0.090
Elevated	Top height	1.535	0.233	1.422	0.250	1.536	0.233
	Base height	2.065	0.170	2.743	0.117	1.402	0.254
	Cover	0.116	0.737	0.776	0.391	0.009	0.924
Near-surface	Top height	2.400	0.141	0.022	0.884	0.001	0.972
	Cover	0.499	0.490	2.229	0.155	0.729	0.406
Surface	Biomass	0.582	0.457	11.768	**0.003**	0.215	0.650

4. Discussion

Our study found evidence that the reburn patterns of the 2013 Aberfeldy wildfire were influenced by the severity patterns of the 2007 Great Divide wildfire, though these effects were strongly constrained by fire weather conditions. Under conditions of severe weather, the likelihood of canopy consumption (CC) during the 2013 fire was the highest in areas where the preceding fire had a substantial impact on the canopy structure (i.e., HCS or CC) and the lowest in areas where the canopy was minimally affected (LCS) or did not burn (UB). Reburn severity exhibited a shift from moderate or high canopy scorch to canopy consumption as the prior fire severity, and hence canopy disturbance, increased. However, under conditions of nonsevere fire weather, the antecedent fire severity had little effect on the reburn severity. These results provide some agreement with previous research that has found evidence of positive relationships between reburn severity and antecedent fire severity in resprouting eucalypt forests across the Sydney basin bioregion of south-eastern Australia [45] and conifer forests of the western United States (e.g., [68,69]).

Fuels were most sensitive to the severity of the 2013 fire at the time of sampling, which took place approximately six years post fire. There were three key changes to fuels in response to the severity of the 2013 fire, with sites impacted by canopy fires having (i) smaller gaps between the canopy fuels and understorey fuels (i.e., surface, near-surface, elevated and mid-storey), (ii) reduced tree canopy cover and (iii) less biomass of surface fine fuels. We found no evidence that repeated exposure to canopy fire resulted in a transition to a different fuel state. The resilience of these forests to repeated canopy fire contrasts with forests dominated by obligate seeder eucalypts, which display transitions to alternative fuel states (e.g., *Acacia*-dominated communities) with elevated flammability [32,70].

Changes to the structural properties of vegetation were primarily driven by the fire response of the mid-storey and canopy species. Resprouting eucalypts display high rates of survival following canopy fires (typically >95%), with aerial resprouting being common amongst large stems (>30 cm diameter at breast height), and basal resprouting being prevalent amongst smaller stems [31,46]. Vigorous epicormic and basal resprouting following canopy fires increased the vertical connectivity of fuels, creating a ladder structure from the surface to the canopy (Figure 2d). In contrast, substantial gaps between the understorey and canopy fuels were present at sites that had only been exposed to understorey fires (Figure 2c). The observed differences in the vertical connectivity of fuels likely explain the greater propensity for canopy consumption during the 2013 fire in areas that had previously experienced a canopy-defoliating fire (i.e., HCS and CC), as increasing fuel hazard in strata close to the surface fuel layer often leads to taller flame heights in eucalypt forests [64], while a reduction in the spacing between the canopy and surface fuel strata should increase the likelihood of canopy fire initiation [71]. The loss of canopy cover following a canopy fire may also contribute to these trends by increasing in-stand windspeed, the rate of fire spread, and consequently, flame height [64]. Although our study was not specifically designed to assess the temporal changes in fuel following canopy-defoliating fires, our results suggest that by 12 years post canopy fire (i.e., C/U), gaps between the canopy-stratum and mid-storey fuels were considerably smaller than at sites that had experienced two understorey fires (i.e., 2.8 m vs. 6.4 m; Figure 4). The rapid growth rates of seedlings and basal resprouts, coupled with the mortality of epicormic shoots on the lower stems of trees (LC, personal observation) were likely driving these temporal changes in fuel connectivity following a high severity fire.

Research examining the effect of fire severity on ecosystem flammability has predominantly focused on forests dominated by obligate seeder and basal resprouter canopy species, where a high-severity fire typically causes complete mortality or topkill of canopy species (e.g., [14,32]). The transition towards shrub-like structure, driven by the mass recruitment and regeneration of shrubs and trees following a high-severity fire, typically increases fuel hazard and flammability in these systems [14,32]. Our study suggests that an alternative mechanism, namely, resprouting from the tree stems, may have driven the increased propensity for canopy fires in eucalypt forests dominated by epicormic resprouters that have been recently burnt (i.e., 6 years since the fire) by canopy-defoliating fires. However, we do note that shrub cover was relatively sparse across the forests examined in our study, possibly reflecting the combined effects of low rainfall, shallow rocky soils and recent short interfire intervals [36,48]. Therefore, post-fire shrub recruitment may be a more influential determinant of fire severity in eucalypt forests dominated by epicormic resprouters elsewhere (see [22,35,45]). Furthermore, shrubs will likely be important in determining fire severity patterns at longer interfire intervals (e.g., 10–30 years), owing to the timing of shrub maturation and senescence [45,47,54]. Knowledge of the distribution of community fire response traits will be an important requirement for predicting changes to ecosystem flammability in response to fire severity.

The fuel assessments conducted in our study were limited to two contrasting fire types (understorey vs. canopy consuming fires), representing the upper and lower ends of the fire severity spectrum [27]. However, vegetation and fuels may display a range of responses, depending on the degree of scorch and the consumption of canopy foliage. For example, fire severity classes involving extensive canopy scorch will produce different patterns in epicormic resprouting compared to canopy consumption, owing to the greater likelihood of branch and stem mortality following canopy consuming fires [31,72]. Epicormic resprouting will generally occur higher along the stem and branch profile following canopy scorch, leading to a more rapid recovery of canopy height and cover compared to canopy consumption, though tree characteristics (e.g., size, bark thickness and type) will also be influential in this regard [31,72–74]. Further research examining the response of the structural properties of fuels to a broader range of fire severity classes is warranted as this would facilitate a better understanding of fire severity feedbacks between fires.

Fire weather was found to be an important driver of the occurrence of the two highest fire severity classes (i.e., HCS and CC) in the fire examined in our study. This finding is in agreement with a large body of work that has identified top-down drivers (e.g., wind, temperature, humidity and drought) as the primary determinants of fire occurrence, size and severity in forests, with bottom-up factors (e.g., vegetation structure and terrain) having secondary effects (e.g., [4,6,58,75]). The effect of fuel properties (i.e., biomass and structure) on fire behaviour is typically assumed to decrease as the fire weather becomes more severe [53]. However, we found that antecedent fire severity, a proxy for vertical connectivity of fuels, had a greater influence on reburn severity under severe fire weather as opposed to nonsevere fire weather conditions. While high-severity fire reduces the vertical spacing between the surface and canopy fuel strata, the propagation of fire into the canopy is strongly dependent on the occurrence of severe weather conditions. The ignition of canopy fuels is influenced by the exogenous factors affecting the flammability of fuel particles (e.g., moisture content), although the spacing of these fuels will place limits on whether ignition is possible [19,76].

It has been proposed that high-severity fires in resprouting eucalypt forests could potentially generate a "runaway positive feedback", whereby high-severity fire becomes self-sustaining [45]. While our findings demonstrate a positive association between antecedent canopy defoliating fires and the occurrence of a subsequent canopy fire, it was evident that fire weather and landscape factors (e.g., terrain, vegetation community) imposed major constraints on fire severity. For example, the likelihood of repeated canopy-consuming fires decreased considerably as the weather conditions transitioned from severe to nonsevere, with canopy-consuming fire being largely absent under nonsevere weather (Figure 3). Fire weather has been found to impose similar constraints on the occurrence of canopy fire more broadly across the eucalypt forests of south-eastern Australia (e.g., [54,58,77,78]). Topographic factors (e.g., aspect and TPI) further moderate fire behaviour in eucalypt forests, with the likelihood of a canopy fire typically decreasing from exposed aspects (equatorial-facing upper slopes) to sheltered aspects (poleward-facing lower slopes) (Figures S1 and S2) [45,54,58]. Therefore, self-sustaining fire severity feedbacks will likely be localised in eucalypt forests that are dominated by epicormic resprouters given the spatial and temporal constraints imposed by fire weather, terrain and the rapid regeneration of fire-resilient plant species, as highlighted for ecosystems elsewhere (e.g., [69]).

Fuel hazard management across the eucalypt forests of southern Australia primarily focuses on the time since fire, with little consideration of fire severity patterns (e.g., [20,79]). This partly reflects the historic emphasis on the importance of fine-fuel biomass in models that are used to predict fire behaviour (e.g., [80,81]), with little regard for the arrangement of fine-fuels. Accessibility to reliable fire severity mapping has been another limitation, though this has been resolved through recent advances in fire severity classification techniques [28,82]. Research aimed at quantifying the effect of fire severity on fuel hazard in eucalypt forests over time should be a priority for fire management agencies. Understanding the patterns in the temporal development of fuel hazard following low- and high-severity fires will be critical for quantifying the risk fire poses to both environmental and built assets. Further development of models that can incorporate the effect of fire severity on vegetation structure and composition (e.g., [19]) should be a priority for fire risk modelling research.

5. Conclusions

Climate change is increasing the frequency and severity of conditions that are conducive to large wildfires across forested regions worldwide [2,9,10]. Feedbacks between fire severity and vegetation structure have the potential to accelerate or constrain the effect of climate change on extreme wildfire events [11,12]. Our results demonstrate that canopy-defoliating fires can increase the vertical connectivity of fuels in resprouting eucalypt forests in the short term (e.g., 6–12 years post fire), increasing the likelihood of future canopy fires under severe fire weather conditions. These findings are consistent with a growing body of evidence showing that high-severity canopy-disturbing fires cause transitions towards fuel states with greater ignitability and propensity to burn at a high severity [14,30,68]. There is evidence that the increasing frequency of severe fire weather has already driven the contraction of interfire intervals and increased the area affected by high severity fire across large areas of south-eastern Australia [83–85]. These changes to fire weather and fire regimes, coupled with increased canopy fuel connectivity resulting from exposure to high-severity fire, have likely increased the propensity for high-severity fires across areas of southern Australia.

Supplementary Materials: The following are available online at https://www.mdpi.com/article/10.3390/f12040450/s1, Figure S1: The effect of slope, aspect relative to north (ASPN) and TPI on the probability of each fire severity class in open forests, Figure S2: The effect of slope, aspect relative to north (ASPN) and TPI on the probability of each fire severity class in tall open forests, Table S1: Summary of model coefficients and 95% credible intervals from the ordinal regression model testing the effect of the severity patterns of the 2007 fire on the severity of the 2013 fire.

Author Contributions: L.C. conceived the project; L.C., A.H., S.M.-G., T.D.P. and P.Z. designed the methodology; A.H., L.C. and P.Z. collected the data; L.C. analysed the data; L.C. wrote the paper in consultation with A.H., S.M.-G., T.D.P. and P.Z.; L.C. supervised all phases of research and administrated the project. All authors have read and agreed to the published version of the manuscript.

Funding: La Trobe University and the Arthur Rylah Institute provided financial support to conduct the field surveys. L.C. was supported by a postdoctoral fellowship at La Trobe University and the Arthur Rylah Institute.

Data Availability Statement: The data used in this manuscript is available via Figshare https://doi.org/10.6084/m9.figshare.14379104 (accessed on 7 April 2021).

Acknowledgments: We are grateful to several volunteers who assisted with the field data collection.

Conflicts of Interest: The authors declare no conflict of interest.

References

1. Bowman, D.M.J.S.; Kolden, C.A.; Abatzoglou, J.T.; Johnston, F.H.; van der Werf, G.R.; Flannigan, M. Vegetation fires in the Anthropocene. *Nat. Rev. Earth Environ.* **2020**, *1*, 500–515. [CrossRef]
2. Abatzoglou, J.T.; Williams, A.P. Impact of anthropogenic climate change on wildfire across western US forests. *Proc. Natl. Acad. Sci. USA* **2016**, *113*, 11770–11775. [CrossRef] [PubMed]
3. Parks, S.A.; Abatzoglou, J.T. Warmer and Drier Fire Seasons Contribute to Increases in Area Burned at High Severity in Western US Forests From 1985 to 2017. *Geophys. Res. Lett.* **2020**, *47*, e2020GL089858. [CrossRef]
4. Abatzoglou, J.T.; Williams, A.P.; Boschetti, L.; Zubkova, M.; Kolden, C.A. Global patterns of interannual climate–fire relationships. *Glob. Chang. Biol.* **2018**, *24*, 5164–5175. [CrossRef] [PubMed]
5. Nolan, R.H.; Boer, M.M.; Resco de Dios, V.; Caccamo, G.; Bradstock, R.A. Large-scale, dynamic transformations in fuel moisture drive wildfire activity across southeastern Australia. *Geophys. Res. Lett.* **2016**, *43*, 4229–4238. [CrossRef]
6. Collins, L.; Bennett, A.F.; Leonard, S.W.J.; Penman, T.D. Wildfire refugia in forests: Severe fire weather and drought mute the influence of topography and fuel age. *Glob. Chang. Biol.* **2019**, *25*, 3829–3843. [CrossRef] [PubMed]
7. Clarke, H.; Evans, J.P. Exploring the future change space for fire weather in southeast Australia. *Theor. Appl. Climatol.* **2018**, 1–13. [CrossRef]
8. Goss, M.; Swain, D.L.; Abatzoglou, J.T.; Sarhadi, A.; Kolden, C.A.; Williams, A.P.; Diffenbaugh, N.S. Climate change is increasing the likelihood of extreme autumn wildfire conditions across California. *Environ. Res. Lett.* **2020**, *15*, 094016. [CrossRef]
9. Abram, N.J.; Henley, B.J.; Sen Gupta, A.; Lippmann, T.J.R.; Clarke, H.; Dowdy, A.J.; Sharples, J.J.; Nolan, R.H.; Zhang, T.; Wooster, M.J.; et al. Connections of climate change and variability to large and extreme forest fires in southeast Australia. *Commun. Earth Environ.* **2021**, *2*, 8. [CrossRef]
10. Turco, M.; Rosa-Cánovas, J.J.; Bedia, J.; Jerez, S.; Montávez, J.P.; Llasat, M.C.; Provenzale, A. Exacerbated fires in Mediterranean Europe due to anthropogenic warming projected with non-stationary climate-fire models. *Nat. Commun.* **2018**, *9*, 3821. [CrossRef]
11. Hurteau, M.D.; Liang, S.; Westerling, A.L.; Wiedinmyer, C. Vegetation-fire feedback reduces projected area burned under climate change. *Sci. Rep.* **2019**, *9*, 2838. [CrossRef]
12. Parks, S.A.; Miller, C.; Abatzoglou, J.T.; Holsinger, L.M.; Parisien, M.-A.; Dobrowski, S.Z. How will climate change affect wildland fire severity in the western US? *Environ. Res. Lett.* **2016**, *11*, 035002. [CrossRef]
13. McColl-Gausden, S.C.; Penman, T.D. Pathways of change: Predicting the effects of fire on flammability. *J. Environ. Manag.* **2019**, *232*, 243–253. [CrossRef]
14. Landesmann, J.B.; Tiribelli, F.; Paritsis, J.; Veblen, T.T.; Kitzberger, T. Increased fire severity triggers positive feedbacks of greater vegetation flammability and favors plant community-type conversions. *J. Veg. Sci.* **2020**, *32*. [CrossRef]
15. Pausas, J.G.; Keeley, J.E.; Schwilk, D.W. Flammability as an ecological and evolutionary driver. *J. Ecol.* **2017**, *105*, 289–297. [CrossRef]
16. Grootemaat, S.; Wright, I.J.; van Bodegom, P.M.; Cornelissen, J.H.C. Scaling up flammability from individual leaves to fuel beds. *Oikos* **2017**, *126*, 1428–1438. [CrossRef]
17. Gill, A.M.; Zylstra, P. Flammability of Australian forests. *Aust. For.* **2005**, *68*, 87–93. [CrossRef]
18. Zylstra, P.J. Flammability dynamics in the Australian Alps. *Aust. Ecol.* **2018**, *43*, 578–591. [CrossRef]

19. Zylstra, P.; Bradstock, R.A.; Bedward, M.; Penman, T.D.; Doherty, M.D.; Weber, R.O.; Gill, A.M.; Cary, G.J. Biophysical mechanistic modelling quantifies the effects of plant traits on fire severity: Species, not surface fuel loads, determine flame dimensions in eucalypt forests. *PLoS ONE* **2016**, *11*, e0160715. [CrossRef]

20. McColl-Gausden, S.C.; Bennett, L.T.; Duff, T.J.; Cawson, J.G.; Penman, T.D. Climatic and edaphic gradients predict variation in wildland fuel hazard in south-eastern Australia. *Ecography* **2020**, *43*, 443–455. [CrossRef]

21. Cawson, J.G.; Duff, T.J.; Tolhurst, K.G.; Baillie, C.C.; Penman, T.D. Fuel moisture in Mountain Ash forests with contrasting fire histories. *For. Ecol. Manag.* **2017**, *400*, 568–577. [CrossRef]

22. Gordon, C.E.; Price, O.F.; Tasker, E.M.; Denham, A.J. Acacia shrubs respond positively to high severity wildfire: Implications for conservation and fuel hazard management. *Sci. Total Environ.* **2017**, *575*, 858–868. [CrossRef]

23. Keeley, J.E. Fire intensity, fire severity and burn severity: A brief review and suggested usage. *Int. J. Wildland Fire* **2009**, *18*, 116–126. [CrossRef]

24. McCarthy, G.; Moon, K.; Smith, L. Mapping fire severity and fire extent in forest in Victoria for ecological and fuel outcomes. *Ecol. Manag. Restor.* **2017**, *18*, 54–65. [CrossRef]

25. Key, C.H.; Benson, N.C. *Landscape Assessment (LA)*; Department of Agriculture, Forest Service, Rocky Mountain Research Station: Fort Collins, CO, USA, 2006.

26. Fernández-Alonso, J.M.; Vega, J.A.; Jiménez, E.; Ruiz-González, A.D.; Álvarez-González, J.G. Spatially modeling wildland fire severity in pine forests of Galicia, Spain. *Eur. J. For. Res.* **2017**, *136*, 105–121. [CrossRef]

27. Hammill, K.A.; Bradstock, R.A. Remote sensing of fire severity in the Blue Mountains: Influence of vegetation type and inferring fire intensity. *Int. J. Wildland Fire* **2006**, *15*, 213–226. [CrossRef]

28. Collins, L.; Griffioen, P.; Newell, G.; Mellor, A. The utility of Random Forests for wildfire severity mapping. *Remote Sens. Environ.* **2018**, *216*, 374–384. [CrossRef]

29. Chafer, C.J.; Noonan, M.; Macnaught, E. The post-fire measurement of fire severity and intensity in the Christmas 2001 Sydney wildfires. *Int. J. Wildland Fire* **2004**, *13*, 227–240. [CrossRef]

30. Coop, J.D.; Parks, S.A.; McClernan, S.R.; Holsinger, L.M. Influences of prior wildfires on vegetation response to subsequent fire in a reburned Southwestern landscape. *Ecol. Apps.* **2016**, *26*, 346–354. [CrossRef]

31. Collins, L. Eucalypt forests dominated by epicormic resprouters are resilient to repeated canopy fires. *J. Ecol.* **2020**, *108*, 310–324. [CrossRef]

32. Bowman, D.M.J.S.; Murphy, B.P.; Neyland, D.L.J.; Williamson, G.J.; Prior, L.D. Abrupt fire regime change may cause landscape-wide loss of mature obligate seeder forests. *Glob. Chang. Biol.* **2014**, *20*, 1008–1015. [CrossRef] [PubMed]

33. Fairman, T.A.; Bennett, L.T.; Tupper, S.; Nitschke, C.R. Frequent wildfires erode tree persistence and alter stand structure and initial composition of a fire-tolerant sub-alpine forest. *J. Veg. Sci.* **2017**, *28*, 1151–1165. [CrossRef]

34. Bowman, D.M.J.S.; Murphy, B.P.; Boer, M.M.; Bradstock, R.A.; Cary, G.J.; Cochrane, M.A.; Fensham, R.J.; Krawchuk, M.A.; Price, O.F.; Williams, R.J. Forest fire management, climate change, and the risk of catastrophic carbon losses. *Front. Ecol. Environ.* **2013**, *11*, 66–67. [CrossRef]

35. Bennett, L.T.; Bruce, M.J.; Machunter, J.; Kohout, M.; Krishnaraj, S.J.; Aponte, C. Assessing fire impacts on the carbon stability of fire-tolerant forests. *Ecol. Appl.* **2017**, *27*, 2497–2513. [CrossRef]

36. Enright, N.J.; Fontaine, J.B.; Bowman, D.M.; Bradstock, R.A.; Williams, R.J. Interval squeeze: Altered fire regimes and demographic responses interact to threaten woody species persistence as climate changes. *Front. Ecol. Environ.* **2015**, *13*, 265–272. [CrossRef]

37. Clarke, P.J.; Lawes, M.J.; Murphy, B.P.; Russell-Smith, J.; Nano, C.E.M.; Bradstock, R.; Enright, N.J.; Fontaine, J.B.; Gosper, C.R.; Radford, I.; et al. A synthesis of postfire recovery traits of woody plants in Australian ecosystems. *Sci. Total Environ.* **2015**, *534*, 31–42. [CrossRef]

38. Boland, D.J.; Brooker, M.I.H.; Chippendale, G.M.; Hall, N.; Hyland, B.P.M.; Johnston, R.D.; Kleinig, D.A.; McDonald, M.W.; Turner, J.D. *Forest Trees of Australia*; CSIRO Publishing: Collingwood, VIC, Australia, 2006; Volume 5, p. 736.

39. Clarke, P.J.; Lawes, M.J.; Midgley, J.J.; Lamont, B.B.; Ojeda, F.; Burrows, G.E.; Enright, N.J.; Knox, K.J.E. Resprouting as a key functional trait: How buds, protection and resources drive persistence after fire. *New Phytol.* **2013**, *197*, 19–35. [CrossRef]

40. Burrows, G.E. Buds, bushfires and resprouting in the eucalypts. *Aust. J. Bot.* **2013**, *61*, 331–349. [CrossRef]

41. Watson, G.M.; French, K.; Collins, L. Timber harvest and frequent prescribed burning interact to affect the demography of Eucalypt species. *For. Ecol. Manag.* **2020**, *475*, 118463. [CrossRef]

42. Benyon, R.G.; Lane, P.N.J. Ground and satellite-based assessments of wet eucalypt forest survival and regeneration for predicting long-term hydrological responses to a large wildfire. *For. Ecol. Manag.* **2013**, *294*, 197–207. [CrossRef]

43. Bennett, L.T.; Bruce, M.J.; MacHunter, J.; Kohout, M.; Tanase, M.A.; Aponte, C. Mortality and recruitment of fire-tolerant eucalypts as influenced by wildfire severity and recent prescribed fire. *For. Ecol. Manag.* **2016**, *380*, 107–117. [CrossRef]

44. Pausas, J.G.; Keeley, J.E. Epicormic resprouting in fire-prone ecosystems. *Trends Plant Sci.* **2017**, *22*, 1008–1015. [CrossRef]

45. Barker, J.W.; Price, O.F. Positive severity feedback between consecutive fires in dry eucalypt forests of southern Australia. *Ecosphere* **2018**, *9*, e02110. [CrossRef]

46. Fairman, T.A.; Bennett, L.T.; Nitschke, C.R. Short-interval wildfires increase likelihood of resprouting failure in fire-tolerant trees. *J. Environ. Manag.* **2019**, *231*, 59–65. [CrossRef]

47. Cheal, D. *Growth Stages and Tolerable Fire Intervals for Victoria's Native Vegetation Data Sets*; Fire and Adaptive Management Report No. 84; Department of Sustainability and Environment: East Melbourne, VIC, Australia, 2010.

48. Gill, A.M.; Catling, P.C. Fire regimes and biodiversity of forested landscapes of southern Australia. In *Flammable Australia: Fire Regimes and Biodiversity of a Continent*; Bradstock, R.A., Williams, J.E., Gill, A.M., Eds.; Cambridge University Press: Cambridge, UK, 2002; pp. 351–369.

49. Haslem, A.; Leonard, S.W.J.; Bruce, M.J.; Christie, F.; Holland, G.J.; Kelly, L.T.; MacHunter, J.; Bennett, A.F.; Clarke, M.F.; York, A. Do multiple fires interact to affect vegetation structure in temperate eucalypt forests? *Ecol. Apps.* **2016**, *26*, 2414–2423. [CrossRef]

50. Collins, L.; McCarthy, G.; Mellor, A.; Newell, G.; Smith, L. Training data requirements for fire severity mapping using Landsat imagery and random forest. *Remote Sens. Environ.* **2020**, *245*, 111839. [CrossRef]

51. Gorelick, N.; Hancher, M.; Dixon, M.; Ilyushchenko, S.; Thau, D.; Moore, R. Google Earth Engine: Planetary-scale geospatial analysis for everyone. *Remote Sens. Environ.* **2017**, *202*, 18–27. [CrossRef]

52. Gill, A.M.; Christian, K.R.; Moore, P.H.R.; Forrester, R.I. Bushfire incidence, fire hazard and fuel reduction burning. *Aust. J. Ecol.* **1987**, *12*, 299–306. [CrossRef]

53. Tolhurst, K.G.; McCarthy, G. Effect of prescribed burning on wildfire severity: A landscape-scale case study from the 2003 fires in Victoria. *Aust. For.* **2016**, *79*, 1–14. [CrossRef]

54. Storey, M.; Price, O.; Tasker, E. The role of weather, past fire and topography in crown fire occurrence in eastern Australia. *Int. J. Wildland Fire* **2016**, *25*, 1048–1060. [CrossRef]

55. Penman, T.D.; Collins, L.; Price, O.F.; Bradstock, R.A.; Metcalf, S.; Chong, D.M.O. Examining the relative effects of fire weather, suppression and fuel treatment on fire behaviour—A simulation study. *J. Environ. Manag.* **2013**, *131*, 325–333. [CrossRef] [PubMed]

56. Farr, T.G.; Rosen, P.A.; Caro, E.; Crippen, R.; Duren, R.; Hensley, S.; Kobrick, M.; Paller, M.; Rodriguez, E.; Roth, L.; et al. The Shuttle Radar Topography Mission. *Rev. Geophys.* **2007**, *45*. [CrossRef]

57. Price, O.F.; Bradstock, R.A. The efficacy of fuel treatment in mitigating property loss during wildfires: Insights from analysis of the severity of the catastrophic fires in 2009 in Victoria, Australia. *J. Environ. Manag.* **2012**, *113*, 146–157. [CrossRef] [PubMed]

58. Bradstock, R.A.; Hammill, K.A.; Collins, L.; Price, O. Effects of weather, fuel and terrain on fire severity in topographically diverse landscapes of south-eastern Australia. *Landsc. Ecol.* **2010**, *25*, 607–619. [CrossRef]

59. Gelman, A.; Rubin, D.B. Inference from iterative simulation using multiple sequences. *Stat. Sci.* **1992**, *7*, 457–511. [CrossRef]

60. R Foundation for Statistical Computing. *R: A Language and Environment for Statistical Computing*; R Foundation for Statistical Computing: Vienna, Austria, 2020.

61. Bürkner, P. Brms: An R Package for Bayesian Multilevel Models Using Stan. *J. Stat. Softw.* **2017**, *80*, 1–28. [CrossRef]

62. Burrows, N. Predicting canopy scorch height in jarrah forests. *CALM Sci. Suppl.* **1997**, *2*, 267–274.

63. Gould, J.S. Evaluation of McArthur's control burning guide in regrowth Eucalyptus sieberi forest. *Aust. For.* **1994**, *57*, 86–93. [CrossRef]

64. Cheney, N.P.; Gould, J.S.; McCaw, W.L.; Anderson, W.R. Predicting fire behaviour in dry eucalypt forest in southern Australia. *For. Ecol. Manag.* **2012**, *280*, 120–131. [CrossRef]

65. Hines, F.; Tolhurst, K.G.; Wilson, A.A.G.; McCarthy, G.J. *Overall Fuel Hazard Assessment Guide*; Fire and Adaptive Management, Report No. 82; Department of Sustainability and Environment: East Melbourne, VIC, Australia, 2010; p. 41.

66. Bates, D.; Maechler, M.; Bolker, B.; Walker, S. Fitting linear mixed-effects models using lme4. *J. Stat. Softw.* **2015**, 1–48. [CrossRef]

67. Canty, A.; Ripley, B. Boot: Bootstrap R (S-Plus) Functions; R Package Version 1.3-27. 2021. Available online: https://cran.r-project.org/web/packages/boot/boot.pdf (accessed on 7 April 2021).

68. Thompson, J.R.; Spies, T.A.; Ganio, L.M. Reburn severity in managed and unmanaged vegetation in a large wildfire. *Proc. Natl. Acad. Sci. USA* **2007**, *104*, 10743–10748. [CrossRef]

69. Harvey, B.J.; Donato, D.C.; Turner, M.G. Burn me twice, shame on who? Interactions between successive forest fires across a temperate mountain region. *Ecology* **2016**, *97*, 2272–2282. [CrossRef]

70. Lindenmayer, D.B.; Hobbs, R.J.; Likens, G.E.; Krebs, C.J.; Banks, S.C. Newly discovered landscape traps produce regime shifts in wet forests. *Proc. Natl. Acad. Sci. USA* **2011**, *108*, 15887–15891. [CrossRef]

71. Cruz, M.G.; Alexander, M.E.; Wakimoto, R.H. Modeling the Likelihood of Crown Fire Occurrence in Conifer Forest Stands. *For. Sci.* **2004**, *50*, 640–658. [CrossRef]

72. Strasser, M.J.; Pausas, J.G.; Noble, I.R. Modelling the response of eucalypts to fire, Brindabella Ranges, ACT. *Aust. J. Ecol.* **1996**, *21*, 341–344. [CrossRef]

73. Moreira, F.; Catry, F.; Duarte, I.; Acácio, V.; Silva, J.S. A conceptual model of sprouting responses in relation to fire damage: An example with cork oak (*Quercus suber* L.) trees in Southern Portugal. *Plant Ecol.* **2009**, *201*, 77–85. [CrossRef]

74. Catry, F.X.; Moreira, F.; Duarte, I.; Acácio, V. Factors affecting post-fire crown regeneration in cork oak (*Quercus suber* L.) trees. *Eur. J. For. Res.* **2009**, *128*, 231–240. [CrossRef]

75. Clarke, H.; Penman, T.; Boer, M.; Cary, G.J.; Fontaine, J.B.; Price, O.; Bradstock, R. The Proximal Drivers of Large Fires: A Pyrogeographic Study. *Front. Earth Sci.* **2020**, *8*. [CrossRef]

76. Van Wagner, C.E. Conditions for the start and spread of crown fire. *Can. J. For. Res.* **1977**, *7*, 23–34. [CrossRef]

77. Collins, L.; Bradstock, R.A.; Penman, T.D. Can precipitation influence landscape controls on wildfire severity? A case study within temperate eucalypt forests of south-eastern Australia. *Int. J. Wildland Fire* **2014**, *23*, 9–20. [CrossRef]

78. Ndalila, M.N.; Williamson, G.J.; Bowman, D.M.J.S. Geographic Patterns of Fire Severity Following an Extreme Eucalyptus Forest Fire in Southern Australia: 2013 Forcett-Dunalley Fire. *Fire* **2018**, *1*, 40. [CrossRef]

79. Penman, T.D.; Collins, L.; Duff, T.D.; Price, O.F.; Cary, G.J. Scientific evidence regarding effectiveness of prescribed burning. In *Prescribed Burning in Australia: The Science and Politics of Burning the Bush*; Leavesley, A., Wouters, M., Thornton, R., Eds.; Australasian Fire and Emergency Service Authorities Council: East Melbourne, Australia, 2020; pp. 99–111.

80. Rothermel, R.C. *A Mathematical Model for Predicting Fire Spread in Wildland Fuels*; United States Department of Agriculture, Forest Service Research Paper INT-115; Intermountain Forest and Range Experiment Station: Ogden, UT, USA, 1972; p. 40.

81. Noble, I.R.; Bary, G.A.V.; Gill, A.M. McArthur's fire-danger meters expressed as equations. *Aust. J. Ecol.* **1980**, *5*, 201–203. [CrossRef]

82. Gibson, R.; Danaher, T.; Hehir, W.; Collins, L. A remote sensing approach to mapping fire severity in south-eastern Australia using Sentinel 2 and random forest. *Remote Sens. Environ.* **2020**, *240*, 111702. [CrossRef]

83. Lindenmayer, D.B.; Taylor, C. New spatial analyses of Australian wildfires highlight the need for new fire, resource, and conservation policies. *Proc. Natl. Acad. Sci. USA* **2020**, *117*, 12481–12485. [CrossRef]

84. Tran, B.N.; Tanase, M.A.; Bennett, L.T.; Aponte, C. High-severity wildfires in temperate Australian forests have increased in extent and aggregation in recent decades. *PLoS ONE* **2020**, *15*, e0242484. [CrossRef]

85. Collins, L.; Bradstock, R.A.; Clarke, H.; Clarke, M.F.; Nolan, R.H.; Penman, T.D. The 2019/2020 mega-fires exposed Australian ecosystems to an unprecedented extent of high-severity fire. *Environ. Res. Lett.* **2021**. [CrossRef]

 *forests*

Article

Study on the Diurnal Dynamic Changes and Prediction Models of the Moisture Contents of Two Litters

Yunlin Zhang [1] and **Ping Sun** [2,*]

[1] School of Biological Sciences, Guizhou Education University, Gaoxin St. 115, Guiyang 550018, Guizhou, China; zhangyunlin@gznc.edu.cn

[2] College of Forestry, Northeast Forestry University, Hexing St. 26, Harbin 150040, Heilongjiang, China

* Correspondence: sunping0202@nefu.edu.cn

Received: 14 December 2019; Accepted: 10 January 2020; Published: 12 January 2020

Abstract: The occurrence and behavior of forest fires are mainly affected by litter moisture content, which is very important for fire risk forecasting. Errors in models of litter moisture content prediction mainly stem from the neglect of diurnal variation. Consequently, it is essential to determine the diurnal variation of litter moisture content and establish a high-precision prediction model. In this study, the moisture contents of litters of Mongolian oak (*Quercus mongolica*) and Korean pine (*Pinus koraiensis*) were monitored at 1 h time steps to obtain the diurnal variations of moisture content, and two direct estimation (Nelson and Simard) methods as well as one meteorological factor regression method were selected to establish prediction models at 1 h time steps. The moisture contents of the two litter types showed obvious diurnal variation, and the changes were significantly correlated with the air temperature and relative humidity. The wind speed had no significant effect on the change within 1 h. The mean absolute error (MAE) values of the three prediction models of Mongolian oak were 1.02%, 1.03%, and 1.46%, and those of Korean pine were 0.50%, 0.50%, and 1.95%, respectively. Similarly, the mean relative error (MRE) values of the three prediction models of oak litter were 4.76%, 4.73%, and 6.65%, and those of pine were 3.53%, 3.59%, and 13.26%, respectively. These results indicated that the accuracy of the Nelson and Simard methods was similar, and both met the requirements for the forecasting of forest fire risk. Therefore, the direct estimation method was selected to predict the moisture contents of two litter types in this area.

Keywords: direct estimation; meteorological factor regression; moisture content; time lag

1. Introduction

The goal of forecasting forest fires is to calculate the probability of fire occurrence and potential damage [1,2]. With global warming and the increasing occurrence of forest fires, the study of forest fire forecasting is becoming increasingly important [3–5]. High-accuracy forest fire forecasting can reduce the losses caused by forest fires through the formulation of firefighting plans according to the possibility of forest fires and potential damage. Whether suppression can be achieved and appropriate firefighting plans formulated mainly depends on whether the forest fire forecast is accurate, and whether the accuracy can meet the needs for prevention and extinguishing of the fire. Therefore, one of the main tasks of forest fire forecasting is to improve the accuracy of predictions of fire occurrence and potential damage, which can provide accurate technical support for forest fire management decision-makers.

Research on forest fire forecasting must explore information related to meteorological conditions, fuel characteristics, geographical conditions, and human activities [6,7]. Among these, forest fuel is the carrier of forest fire and it is an important variable in forest fire forecasting [8–12]. The moisture content of fuel determines the possibility of burning and fire behaviors [13–15], and thus forms the basis of forest fire prediction. As an important part of forest fuel, surface litter is an important fuel contributing to forest fires, and its moisture content value determines the probability and initial spread rate of forest fires. Therefore, it is of great significance to establish a high-accuracy prediction model for litter moisture content for forecasting forest fires [16].

The prediction methods used to obtain the moisture content of litters mainly include the meteorological factor, physical process, and direct estimation methods. Although the physical process method can help to understand the phenomenon of litter moisture change, it is difficult to apply to fire risk systems due to its complex structure and numerous model parameters [17–20]. The principal methods employed to predict the moisture content of litter in a fire risk system are the regression method of meteorological factors and the direct estimation method, both of which still result in some errors [18,21,22]. The meteorological factor method is a statistical method with a large error of 15% or more [23,24]. The direct estimation method is a combination of physical processes, the meteorological factor method, and the principal prediction equation derived from the physical diffuse equation, and it can be used in the equilibrium moisture content response equation based on the form of physical derivation or other statistical models [25,26]. The parameters of the direct estimation method are estimated using observation data and, therefore, it has the advantages of universality and reliability of the physical method and simplicity of the statistical method [27–29].

The small errors of the direct estimation method compared with the meteorological factor method [30–32], can lead to a 1- or 2-fold difference in fire risk in some cases [32,33]. Two main reasons for the errors are as follows: one source of error is the oversimplification of the diurnal variation of litter moisture content [17,18], while the other is extrapolation and poor applicability [34,35]. Therefore, there are two mainstream methods for improving the prediction accuracy of dynamic changes in moisture content, the first of which makes predictions based on an hour (h) or shorter scales. For a specific forest type, the shorter the prediction scale, the higher the prediction accuracy [17–19]. The other method is to establish corresponding prediction models according to different stand conditions and fuel types. Because of the strong spatial heterogeneity of litter bed moisture content [36], if the second method is selected, it will be time-consuming and laborious, and thus it is difficult to apply in practice. Therefore, this study chose to shorten the prediction scale to improve the moisture content prediction accuracy, which is of great significance to understanding the diurnal dynamic changes in litter moisture content.

Meteorological factors have significant impacts on litter moisture content. The air temperature and relative humidity have a stable diurnal variation process for non-rain days, so the moisture content of litter has a strong daily variation process [17,18,24]. In addition, stochastic wind and rainfall further complicate the process. In the past, meteorological factor observation techniques have been limited, and it has been difficult to provide detailed information that can reflect the daily changes in these weather phenomena. The value of litter moisture content can be calculated using only meteorological observations at a certain time of day. The existing fire risk system basically follows the traditional method; thus, the prediction of litter moisture content is still based on 24 h as the calculation step. For example, in the Canadian Fire Risk Forecasting System, the noon temperature, humidity, wind speed, and first 24 h of rainfall are used as predictors to calculate the litter moisture content. In the United States, meteorological factors such as the daily maximum and minimum temperature are used. In other models, the meteorological factors at a certain time are directly used as regression models [37,38], which produce errors due to simplification of the daily variation in moisture content of the surface litter. It seems the main reason for this is that the influence of diurnal variation of meteorological factors upon changes in litter moisture content is not taken into consideration.

Due to technical progress, meteorological factors and litter moisture content can now be monitored in steps of 1 h or less. Forest fire researchers also recognize that fire occurrence and fire behavior are very sensitive to litter moisture content responses, and accurate forest fire forecasting requires litter moisture content data at a shorter time scale than that of an entire day to be used to establish a prediction model using steps of 1 h or less. Therefore, some studies have investigated the diurnal variation in moisture. Van Wagner [39] proposed a method for predicting the moisture content of litter at a 1 h step. Additionally, Lawson et al. [40] used the moisture content at 4 pm to calculate moisture content data at other times. Furthermore, Catchpole et al. [26] studied the diurnal variation process of moisture content and established a prediction model using the direct estimation method, with errors from 2.3% to 5.2%. Sun et al. [24] selected the direct estimation and meteorological factor regression methods to predict the diurnal variation in litter moisture content and found that direct estimation was better than the regression method, but the study was only conducted during the daytime, indicating that the applicability of the diurnal variation prediction model has limitations. Nelson [17] used the 10 h humidity bar as the object and obtained a diurnal variation model of moisture content. However, the driving mechanism of moisture content change between the humidity bar and surface litter layer was quite different, and the error was large and not suitable for forest fire forecasting when applied to the surface litter layer. Among these studies, the direct estimation and meteorological factor regression methods are the two most commonly used methods. In summary, although the diurnal variation of surface litter moisture content has been studied and relevant prediction models have been established, the diurnal variation in the moisture content of litter layer has not been fully revealed due to the limitations of previous research, including the structure of the litter layer, research duration, and research objects. Therefore, in this study, the litter moisture content was monitored hourly in the field both during daytime and nighttime, and the driving factors of moisture content change were analyzed. The direct estimation method and the meteorological factor method were selected to establish the 1 h step prediction model. Which method is more suitable for the prediction of diurnal changes in moisture content was also assessed.

Heilongjiang Province is the most serious fire-risk region in China [24]. We selected two important types of forest, Mongolian oak (*Quercus mongolica*) and Korean pine (*Pinus koraiensis*), which are widely distributed in this region. The leaf litter of Mongolian oak is large and easily collapses, making it easy to ignite; thus, oak forest often carries a higher fire risk [41]. Korean pine plantations also face a high risk of fire due to their high lipid content. Therefore, we selected the surface litter layers of oak and pine forests as the research object, and used a non-destructive sampling method to monitor the diurnal variation in litter moisture content in 1 h steps. Meteorological data were simultaneously recorded and used to analyze the driving factors of diurnal variation in litter moisture content. The direct estimation and meteorological factor regression methods were used to establish a prediction model with 1 h steps and to quantify the accuracy of the model and the difference between forest types, which will be of great significance for improving the prediction accuracy of moisture content prediction and forest fire forecasting in the future.

2. Methods

2.1. Study Area

The study area was located at the Maoer Mountain Experimental Forest Farm of the Northeast Forestry University, Shangzhi City, Heilongjiang Province, China (45°24′–45°25′ N, 127°34′–127°40′ E), 30 km from north to south and 26 km from east to west, with an area of approximately 260 km^2 (Figure 1). In this area, the forest coverage is 85% and the total forest stock is 20,500 km^2, with an elevation of 200–805 m. The climate is a temperate continental monsoon climate. The annual average temperature is 2.8 °C, the lowest temperature in January is –31.9 °C, and the highest temperature in July is 26.1 °C. The annual rainfall is 720 mm and is mainly concentrated in July and August, which accounts for more than half of the total value for the year. The vegetation mainly includes Mongolian oak (*Quercus mongolica*), Korean pine (*Pinus koraiensis*), walnut (*Juglans mandshurica*), ash (*Fraxinus mandchurica* Rupr.), and poplar (*Populus davidiana* Dode).

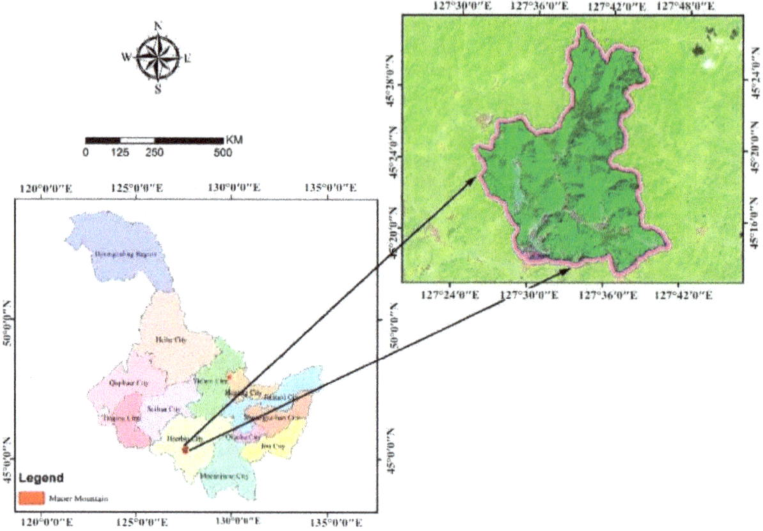

Figure 1. Location of the study site.

2.2. Field Experiment

Although the slope, aspect, and canopy density influence the dynamics of litter moisture content, taking these factors into consideration when studying the diurnal process of moisture content and establishing prediction models would result in large uncertainties that are unsuitable in practice. Therefore, this study selected only one plot for each type of forest, and plots of 20 × 20 m were set up in both the Mongolian oak forest and Korean pine plantations (see Table 1 for plot conditions). The moisture content of the surface litter was monitored from 18 May 2017 to 24 May 2017 (spring fire prevention period).

Table 1. Basic information of the two sample plots.

Forest Type	Location	Elevation (m)	Mean Height (m)	Mean Diameter at Breast Height (cm)	Canopy Density	Litter Load (t·ha⁻¹)
Mongolian oak (*Quercus mongolica*)	Upper slope	544	12	23	0.45	3.68
Korean pine (*Pinus koraiensis*)	Middle slope	382	15	21	0.55	6.22

In this study, a non-destructive sampling method was adopted to monitor moisture content. To ensure that the structure of the surface litter layer was consistent with the original state and that water exchange could be carried out in a wild state, plastic baskets were selected for holding litter. The length, width, and height of the plastic baskets were 300, 300, and 45 mm, respectively, and the baskets were covered with nylon mesh (1 × 1 mm mesh size) at the bottom and around the plastic basket. Three plastic baskets were arranged for each plot. The samples were weighed once every 1 h for a total of 7 days. After the experiment, the baskets were dried at 105 °C for 24 h, and the dry mass of the litter was recorded. The surface litter bed moisture content was then obtained according to the formula for calculating the moisture content (the ratio of the water to the dry mass of the litter). A total of 24 × 7 = 168 records of data were obtained for each basket and a total of 168 × 3 = 504 sets of data were obtained for each plot. The arithmetic means of the moisture content of the three baskets of each plot were calculated as the moisture content value of the plot.

A HOBO meteorological station, developed by Onset Company (US) that can automatically record weather data in the field and was installed in each plot. The sensors were placed on the ground and the temperature, humidity, wind speed, and rainfall meteorological factors were recorded every 1 h. The recording of meteorological factors occurred earlier than the moisture content monitoring start time, although the recording of meteorological factors was synchronized with the moisture content monitoring when they overlapped.

2.3. Model Description

2.3.1. Direct Estimation Method

Catchpole [26] proposed a direct estimation method to predict litter moisture content. Based on the equilibrium moisture content, the field litter moisture content data and meteorological data were used to directly predict litter moisture content. This method can directly use meteorological data to predict the moisture content, so has a better applicability and higher accuracy than an indoor simulation [42]. For the response equation of the equilibrium moisture content in the direct estimation method, the Nelson model [27] was selected based on semiphysical conditions, and the Simard model [43] was chosen based on statistics to achieve better prediction results [44]. Therefore, when the direct estimation method was selected to establish the prediction model, the Nelson and Simard equilibrium moisture content prediction models were used to calculate the equilibrium moisture content. Hereinafter, they are simply referred to as the Nelson method and the Simard method.

The main equation of the direct estimation method was based on the differential equation of litter moisture content proposed by Byram [42]. The form of the equation is shown in Equation (1):

$$\frac{dm}{dt} = \frac{M - E}{\tau} \tag{1}$$

where M indicates the litter moisture content (%), E is the equilibrium moisture content (%), and τ represents the time lag (h); the same applies to the equations below.

The equilibrium moisture content prediction model in Equation (1) was calculated by the Nelson and Simard models, respectively. The forms of the Nelson model [27] and the Simard model [43] for the equilibrium moisture content are shown in Equations (2) and (3):

$$E = \alpha + \beta log(-\frac{RT}{m}logH) \tag{2}$$

where R is the universal gas constant (8.314 J·K^{-1}·mol^{-1}); T and H indicate the air temperature (K) and relative humidity (%), respectively; m indicates the molecular mass of water (18.0153 g·mol^{-1}); and α and β are empirical parameters.

$$E = \begin{cases} 0.03 + 0.2626H - 0.00104HT \ if \ H < 10 \\ 1.76 + 0.1601H - 0.0266T \ if \ 10 \leq H \leq 50 \\ 21.06 - 0.4944H + 0.005565H^2 - 0.00063HT \ if \ H > 50 \end{cases} \tag{3}$$

where T indicates the air temperature (°C) and H indicates the relative humidity (%).

When the direct estimation method is used to predict the moisture content of litter, the time lag τ of the litter must be fixed [45]. Because the monitoring step of litter moisture content was 1 h, Δt was equal to 1 h. Discretization of the differential Equation (1) for moisture calculation yielded a discrete equation for the moisture content calculation (Equation (4)).

$$M_i = \lambda^2 M_{i-1} + \lambda(1 - \lambda)E_{i-1} + (1 - \lambda)E_i \tag{4}$$

where M_i indicates the moisture content of the litter at time i; M_{i-1} is the moisture content of the litter at time $i - 1$; and E_i and E_{i-1} are the equilibrium moisture contents at time i and $i - 1$, respectively.

The Nelson and Simard methods were separately selected for moisture content prediction. The ordinary least squares method was used to obtain the parameters.

2.3.2. Meteorological Factor Regression Method

With the litter moisture content as the dependent variable, the meteorological factors with significant influence were used as the independent variables in the stepwise regression method. The equation employed for this has the form of a multivariate linear equation, as shown in Equation (5).

$$M = \sum\nolimits_{i=0}^{n} X_i b_i \tag{5}$$

where M is the litter moisture content; X_i indicates the weather variable, such as the current or n-hour earlier ($n < 10$ h) temperature, humidity, wind speed, or rainfall; and b_i is the estimated parameter.

2.4. Data Analysis

When the litter moisture content is higher than 35%, it is difficult for it to ignite and cause forest fires [46]. Therefore, all the data in this paper were analyzed for a moisture content of less than 35%. For the oak litter, a total of 133 data records were collected from 18 May 2017 at 9:40 to 23 May 2017 at 21:40, and for the pine litter, a total of 125 data records were collected from 18 May 2017 at 17:40 to 23 May 2017 at 21:40.

2.4.1. Basic Statistics

Statistical analysis was performed on the surface litter moisture content data to calculate the minimum, maximum, and average changes in the monitoring interval of litter moisture contents in the oak and pine plots. Using the sampling sequence as the abscissa and the moisture contents of the litter bed as the coordinates, the daily variation in the moisture contents of the two types of litter bed was plotted.

2.4.2. Correlation Analysis

Changes in litter moisture content are driven by meteorological factors, and the moisture content responds differently to different meteorological factors. Therefore, it was necessary to analyze the correlation between the litter moisture content and meteorological factors. This study specified that the meteorological factor before n ($\leq$10) hours was represented by the angle n, that is, T_5 indicates the air temperature 5 h before. Pearson correlation analysis was carried out, and the correlation coefficient was plotted as a function of time.

2.4.3. Model

In this study, the methods of direct estimation and meteorological factor regression were used to establish the prediction model of diurnal variation in litter moisture content (see Section 2.3 for the methods). N-fold cross-validation was selected to test the accuracy of the model, and the mean absolute error (MAE) and mean relative error (MRE) of the prediction model were then calculated. The calculation formulas are shown in Equations (6) and (7), and error bars were drawn to compare the errors of the prediction model.

$$MAE = \frac{1}{n} \sum\nolimits_{i=1}^{n} \left| M_i - \hat{M}_i \right| \tag{6}$$

$$MRE = \frac{1}{n} \sum\nolimits_{i=1}^{n} \frac{\left| M_i - \hat{M}_i \right|}{M_i} \tag{7}$$

where M_i indicates the observed moisture content (%) and $\hat{M}_i$ indicates the predicted moisture content (%).

The forecasting performances of the three models at different time periods of the day were compared by drawing polyline maps. A 1:1 scatter plot of the three methods was drawn by taking the measured values as the abscissa and the predicted values as the vertical coordinates. The prediction effects of the three prediction models were then compared according to different sections of litter bed moisture contents.

3. Results

3.1. Diurnal Variation in the Litter Moisture Content

Table 2 shows the changes in the litter moisture contents for the two forest types. The minimum, maximum, and mean of the litter moisture content for oak were 17.91%, 32.37%, and 22.02%, respectively. The range of moisture content of the pine plantation was greater than that of the oak forest, with an average of 15.35%. The change in the litter moisture contents was calculated at 1 h steps. The minimum change in the litter moisture content of the oak and pine was 0. The maximum change for oak was 11.49%, and the mean was 1.06%. Additionally, the maximum change for pine was only 2.94%, and the average change value was 0.58%.

Table 2. Basic statistics of the moisture content of surface litter.

Litter Type	N	Mean (%)	Minimum (%)	Maximum (%)	Mean Change in 1 h (%)	Minimum Change in 1 h (%)	Maximum Change in 1 h (%)
Mongolian oak	134	22.02	17.91	32.37	1.06	0.00	11.49
Korean pine	125	15.35	9.28	33.89	0.58	0.00	2.94

The litter moisture contents exhibited obvious diurnal variations. Because there was rainfall (8 mm in total) before the beginning of the monitoring period, the litter moisture contents of the two forests began to decline on the first day. The litter moisture content of the oak showed obvious changes from the second to sixth days, and the pine planation showed a similar trend from the third to the sixth day; that is, the litter moisture content increased from early morning to a peak at approximately 8:40 AM, and then began to decline. The litter moisture content was lowest from 13:40 to 15:40, and then began to increase again. In this process, due to the effect of wind speed, the changes in litter moisture content in different time periods of the day displayed some fluctuations, but the overall trend was similar (Figure 2).

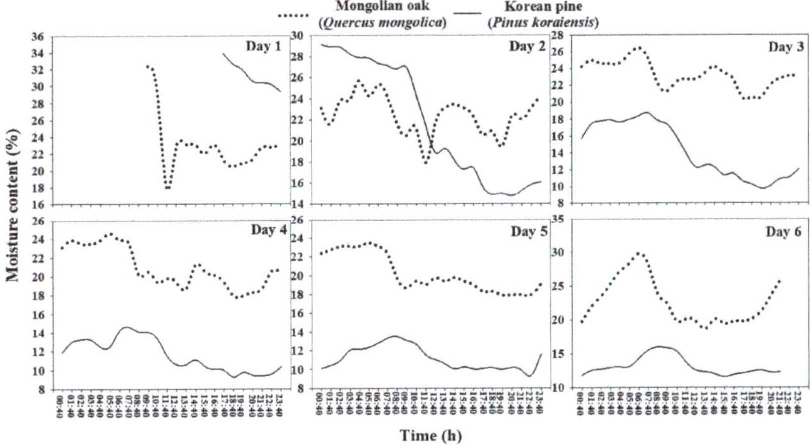

Figure 2. Diurnal variation of litter moisture contents of the Mongolian oak forest and Korean pine plantation.

3.2. Correlation Analysis

The change in the moisture content of the oak litter as a response to meteorological factors was stronger than that of the pine. The litter moisture content was negatively correlated with the air temperature and positively correlated with the relative humidity. The correlation coefficient first increased and then decreased with lagging time. For the oak, the temperature and humidity 3 h before had the most significant effect on the litter moisture content. For the pine, the temperature and humidity 5 h before had the most significant effect on the litter moisture content (Figure 3). Wind speed had no significant effect on the change in litter moisture content.

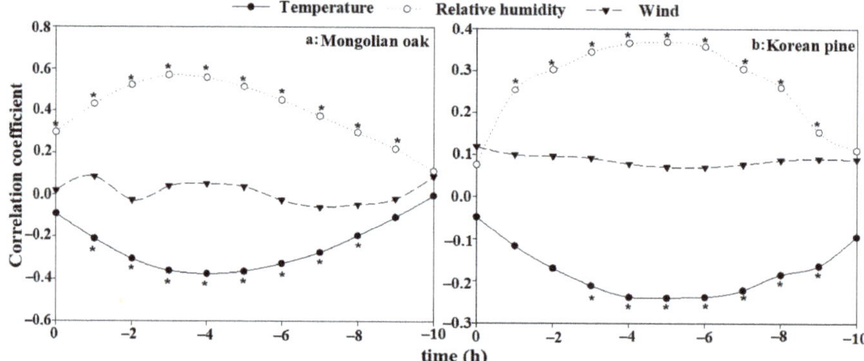

Figure 3. The coefficient of cross correlation between meteorological factors and the moisture content lagged with different times. * Indicates significant correlation.

3.3. Model

3.3.1. Model Parameters

Table 3 shows the estimated parameters and errors of the litter moisture content prediction model. The time lags of oak obtained by the Nelson and Simard methods were 2.0261 and 16.0968 h, respectively, and those of the pine litter were 12.2335 and 18.4463 h, respectively, indicating that the time lag of oak was shorter than that of pine. For the same litter type, the Simard method obtained a higher time lag than the Nelson method. In the meteorological factor regression method, the relative humidity 3 h before and the air temperature 4 h before were selected as the meteorological factors for oak litter, and the relative humidity values at 3, 4, and 5 h before were selected as the meteorological factors for pine litter moisture content.

Table 3. Estimated parameters and errors of all the models established using three different methods. MAE and MRE indicates the mean absolute error and mean relative error, respectively).

Method	Parameters	Mongolian Oak	MAE (%)	MRE (%)	Korean Pine	MAE (%)	MRE (%)
Nelson method	α	0.2458			0.0039		
	β	−0.0261	1.02	4.76	−0.0023	0.50	3.53
	λ	0.7813			0.9600		
	τ	2.0261			12.2335		
Simard method	λ	0.9694	1.03	4.73	0.9732	0.50	3.59
	τ	16.0968			18.4463		
Direction regression method	b_0	19.7130			15.009		
	b_1	0.0630	1.46	6.65	−08400	1.95	13.26
	b_2	−0.1211			0.4130		
	b_3	-			0.079		

3.3.2. Model Error Comparison

For Mongolian oak, the MAEs based on the Nelson, Simard, and meteorological factor regression methods were 1.02%, 1.03%, and 1.46%, respectively, with corresponding MREs of 4.76%, 4.73%, and 6.65%. The MAEs of the three prediction models for the diurnal variation in litter moisture content of the pine litters were 0.50%, 0.50%, and 1.95%, respectively, and the corresponding MREs were 3.53%, 3.59%, and 13.26%. For the direct estimation method, the prediction performance of the litter moisture content model was better for pine than for oak, while the opposite performance trend was observed in the case of the meteorological factor regression method for the two forest types. The direct estimation method was better than the regression method for both forest types (Figure 4).

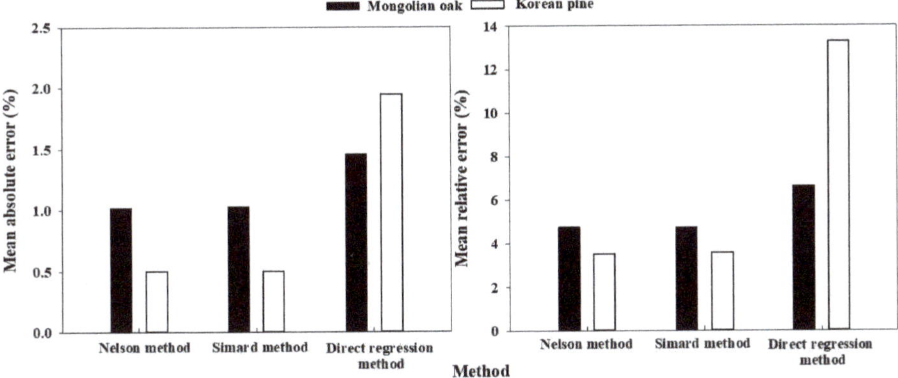

Figure 4. Comparison of litter moisture content model errors for the Mongolian oak and Korean pine.

3.3.3. Comparison of Measured and Predicted Values of Diurnal Variation

For the prediction model of diurnal variation in litter moisture content of oak, the Nelson and Simard methods produced similar predictions, but the prediction effect of the two methods depended on the daily stage, and the meteorological factor regression method had the worst prediction performance. When the peak litter moisture content (maximum and minimum) of the oak began to change, the Nelson and Simard methods had the largest error, although the Simard method was slightly better and its error was lower than that of the Nelson method. The predicted value of the meteorological factor regression method always had a certain lag for the measurements (Figure 5).

The performance of the prediction of litter moisture content model for pine was similar to that for oak. The meteorological factor regression method had the worst prediction effect but also showed a certain lag. The predicted results of the Nelson and Simard methods were also similar to each other, but the errors between the predicted values and the measured values at different stages of the day showed a certain regularity; that is, from approximately 9:40 to 17:40, the measured values were slightly higher than the predicted values, while in the remaining stages of the day, the measured values were lower than the predicted values. The error of the direct estimation method mainly occurred when the litter moisture content reached its peak value, and the model performance was good for all other times (Figure 6).

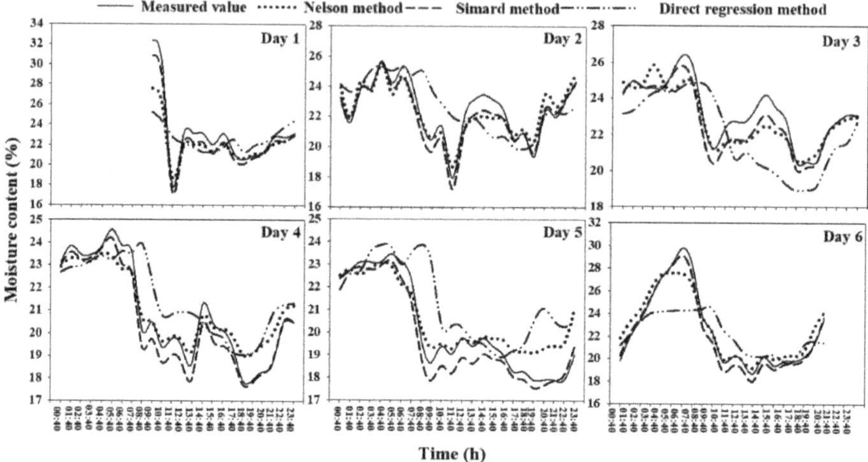

Figure 5. Measured and predicted values of the diurnal variation in litter moisture content of Mongolian oak.

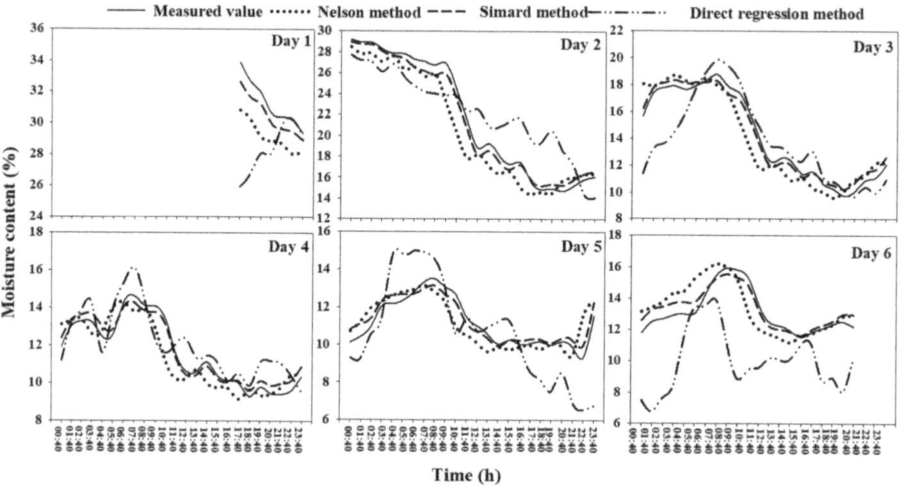

Figure 6. Measured and predicted values of the diurnal variation in litter moisture content of Korean pine.

3.3.4. 1:1 Comparison of Measured and Predicted Values

As seen in Figure 7, variation in the litter moisture content range resulted in differences between the predicted and measured values for the different methods. For the litter moisture content of oak, the Simard method value was closer to the measured value and could be evenly distributed on both sides of the 1:1 line, particularly when the moisture content was low. When the moisture content increased, the Simard method slightly underestimated the litter moisture content. When the litter moisture content was low, the Nelson method values were almost uniformly distributed near the 1:1 line. When the moisture content exceeded 30%, the litter moisture content was substantially underestimated. By contrast, the predicted and measured values of the meteorological factor regression method were irregularly distributed near the 1:1 line, and the model overestimated when the litter moisture content was lower, but underestimated when the litter moisture content was higher than 25%. The model performance for pine was similar to that for oak. With different moisture content intervals, the Simard method had the best prediction effect, followed by the Nelson method, and the meteorological factor regression method had the worst prediction effect. Regardless of the method, the prediction model performance for pine with a low moisture content was better than that with a high moisture content.

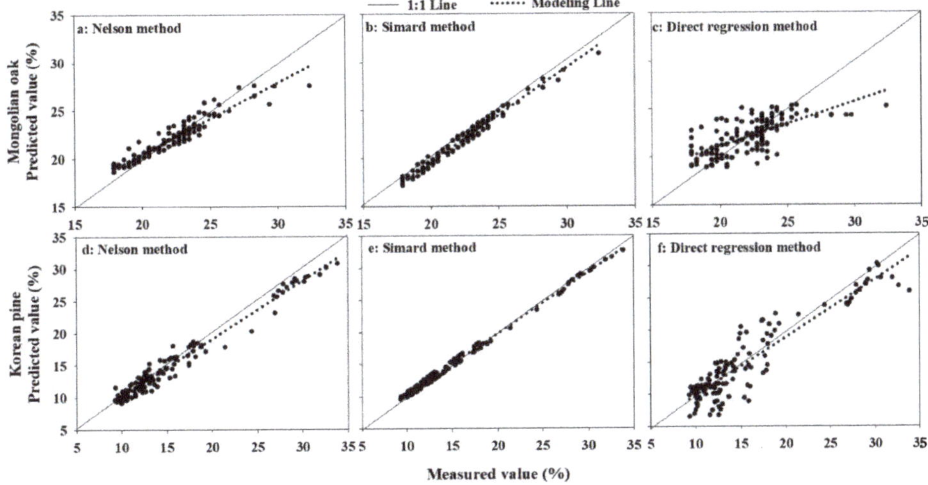

Figure 7. Comparisons of measured and predicted values. (**a–c**) indicate the comparisons of measure and predicted values of Mongolian oak by Nelson, Simard and Direct regression methods, respectively; (**d–f**) indicates the comparisons of measure and predicted values of Korean pine by Nelson, Simard and Direct regression methods, respectively.

4. Discussion

4.1. Correlation Analysis

The diurnal variation in the litter moisture contents of oak and pine was significantly correlated with the air temperature and relative humidity, and was not related to the wind speed. As the collection time of the meteorological factor progressed, its influence on litter moisture content first increased and then decreased, indicating that the litter moisture content had a specific lag associated with the meteorological factor, which is in line with the results of numerous studies [26,47–49]. The wind speed had no significant effect on the diurnal variation of litter moisture contents, which is different to the research results published by Zhang et al. [50]. This difference may have been because the influence of wind speed on the moisture content of the litter needed to be analyzed over a longer time interval.

4.2. Model Parameters

The litter moisture contents of oak and pine were established using the Nelson method with an hourly step prediction model, and the α values were 0.2458 and 0.0039, respectively. Catchpole et al. [26] found that the α value ranged from 0.26 to 0.37, whilst Slijepcevic et al. [51] found that the α value ranged from 0.28 to 0.41. Additionally, Sun et al. [24] obtained an α value of 0.087 to 0.594 for Dahurian larch forests. These different research results may be related to the litter type, test conditions, and setting time period. In the Nelson method, β reflects the degree of response of the equilibrium moisture content to the air temperature and relative humidity. The larger the absolute value of β, the more sensitive the response of the litter moisture content to temperature and humidity [27]. The absolute value of β of the oak (0.0261) was higher than that of the pine (0.0023), indicating that the water holding capacity of the broad-leaf layer was lower than that of the needle layer, and that the change in moisture content was more sensitive to temperature and humidity for the former. The packing ratio and physical properties of the litter layer have a significant impact on the dynamic change of litter moisture content [52]. Consequently, the differences in the estimated β and water holding capacity between the two forest types may have been caused by differences in the structure and physical properties of the litter layer.

4.3. Time Lag

The observed time lags for the oak and pine litter beds were 2.0261 and 12.2335 h, respectively, as calculated using the Nelson method. Sun et al. [24] studied Dahurain larch litter and obtained a time lag from 2.375 to 4.180 h. Yu et al. [53,54] obtained a time lag from 1 to 8 h for a broad-leaved bed, and that of the needle bed they studied was between 1 and 4 h. Furthermore, Lu [55] used the Nelson method to obtain time lags of broad-leaved and needle beds of 29.1 and 15.6 h, respectively. The results of this study were different to those of previous investigations. In addition to the differences in litter structure and morphology, there were also differences associated with the research monitoring time intervals. In this paper, the moisture content was monitored continuously for 7 days at 1 h steps. Sun et al. [24] only monitored the moisture content data from 8:00 to 16:00, i.e., during the daytime, but the water vapor exchange at night is different from that in the daytime. Therefore, the time lag may be different between studies. The time lag of the broad-leaved Mongolian oak was shorter than that of the pine needle, which may also have been related to the packing ratio of the litter layer. The average packing ratios of the litter layer of the oak and pine stands were 0.0138 and 0.0236, respectively, which suggests that the water vapor exchange between litter and air was faster, and the response to meteorological factors was stronger for oak litter. Therefore, the time lag of oak litter layer was shorter than that of pine litter layer.

The time lags of the oak and pine litter layer obtained using the Simard method (16.0968 and 18.4463 h, respectively) were higher than those obtained using the Nelson method. This difference occurred mainly because the equilibrium moisture content value obtained by the Simard equilibrium moisture content formula was approximately 4% higher than that obtained using the Nelson method at the same temperature and humidity. Sun et al. [24] concluded that the time lag of Dahurian larch litter was 5.705 to 18.880 h using the Simard method, and the time lags of the broad-leaved and needle specimens obtained by Liu [56] using fixed indoor temperature and humidity were 9.35 and 25.3 h, respectively—similar to the Simard results presented in this paper. This results showed that the time lag value obtained using the Simard method was more reliable than that obtained using the Nelson method.

4.4. Model Accuracy

Using the Nelson and Simard methods, the accuracies in litter bed moisture content daily variation prediction models for oak were 1.02% and 1.03%, respectively, and those for pine were 0.5% and 0.5%, respectively. The results were similar to the 0.41%–1.30% reported by Sun et al. [24] and the 0.8%–1.9% reported by Catchpole et al. [26], indicating that a direct estimation method can be used in prediction models for the diurnal variation of oak and pine litter moisture content in this region. Its prediction accuracy meets the accuracy requirements for fire behavior prediction and fire risk forecasting [16]. The MAEs of the prediction model of the oak and pine obtained by meteorological factor regression were 6.65% and 13.26%, respectively, which did not meet the forecast accuracy requirements.

5. Conclusions

This study showed that the moisture contents of two types of litter had strong diurnal variation. The diurnal variation of the litter moisture content was significantly correlated with the air temperature and relative humidity at 1 h intervals, but not with the wind speed. The time lags of the oak litter bed obtained using the Nelson and Simard methods were 2.0261 and 16.0968 h, respectively, and the time lags for the pine were 12.2335 and 18.4463 h, respectively. The moisture content of the Mongolian oak litter bed was more sensitive to meteorological factors than that of pine. The accuracy of the two kinds of litter moisture content diurnal variation prediction models established by the direct estimation method met the fire behavior prediction and forest fire risk forecast accuracy requirements, but the prediction model obtained by the meteorological factor regression method did not meet the requirements.

This study had certain limitations. For example, specific litter types, different stand characteristics, bed packing ratios, slopes, and aspects have significant impacts on litter moisture content [57], but this research did not consider the impacts of stand characteristics. The use of the equilibrium moisture content prediction model using a direct estimation method is limited, i.e., to only the spring fire season research. In future, the diurnal variation of litter moisture content in various stand and topographic conditions should be considered comprehensively, and the time scale should be expanded, which will be of great significance for improving the accuracy and practical application of litter moisture content prediction models.

Author Contributions: Y.Z. analyzed the data, data collection, literature search and wrote manuscript; P.S. conceived and designed experiment. All authors have read and agreed to the published version of the manuscript.

Funding: This research was funded by [Research results of Guizhou Education University's PhD Project in 2019] grant number [2019BS005] and the APC was funded by [National Natural Science Foundation of China] grant number [31370656].

Acknowledgments: We are indebted to Ziyuan Man, Fangce Liu, and Yuan Li for their support in field work. We are also grateful to two anonymous reviewers for their valuable comments.

Conflicts of Interest: The authors declare no conflict of interest.

References

1. Chuvieco, E.; Aguado, I.; Yebra, M.; Nieto, H.; Salas, J.; Martín, M.P.; Vilar, L.; Martínez, J.; Martín, S.; Ibarra, P.; et al. Development of a framework for fire risk assessment using remote sensing and geographic information system technologies. *Ecol. Model.* **2010**, *221*, 46–58. [CrossRef]
2. Finney, M.A. The challenge of quantitative risk analysis for wildland fire. *For. Ecol. Manag.* **2005**, *211*, 97–108. [CrossRef]
3. Shu, L.F.; Tian, X.R. Current Status and Prospects of Forest Fire Prevention work abroad. *World For. Res.* **1997**, *2*, 28–36.
4. Ager, A.; Vaillant, N.; Finney, M. A comparison of landscape fuel treatment strategies to mitigate wildland fire risk in the urban interface and preserve old forest structure. *For. Ecol. Manag.* **2010**, *259*, 1556–1570. [CrossRef]
5. Turco, M.; Marcos-Matamoros, R.; Castro, X.; Canyameras, E.; Llasat, M. Seasonal prediction of climate-driven fi re risk for decision-making and operational applications in a Mediterranean region. *Sci. Total Environ.* **2019**, *676*, 577–583. [CrossRef]
6. Chowdhury, E.H.; Hassan, Q.K. Operational Perspective of Remote Sensing-based Forest Fire Danger Forecasting Systems. *ISPRS J. Photogramm. Remote Sens.* **2015**, *104*, 224–236. [CrossRef]
7. Sokolova, G.V.; Makogonov, S.V. Development of the Forest Fire Forecast Method (a Case Study for the Far East). *Russ. Meteorol. Hydrol.* **2013**, *38*, 222–226. [CrossRef]
8. Deeeming, J.E.; Burgan, R.E.; Cohen, J.D. The Nation Fire Danger Rating System-1978. *USDA For. Serv. GTR* **1978**. [CrossRef]
9. Bradshaw, L.S.; Deeming, J.E.; Burgan, R.E. *The 1978 National Fire-Danger Rating System: Technical documentation*; General Technical Report INT-169; United States Department of Agriculture, Forest Service, Intermountain Research Station: Ogden, UT, USA, 1983; p. 44.
10. Van Wagner, C.E. *Development and Structure of the Canadian Forest Fire Weather Index System*; Forestry Technical Report; Canadian Forestry Service: Ottawa, ON, Canada, 1987; pp. 35–37.
11. Hu, H.Q. *Forest Ecology and Management*; China Forestry Press: Beijing, China, 2005.
12. Sun, P.; Zhang, Y.L. A Probabilistic Method Predicting Forest Fire Occurrence Combining Firebrands and the Weather-Fuel Complex in the Northern Part of the Daxinganling Region, China. *Forests* **2018**, *9*, 428. [CrossRef]

13. Rothermel, R.C.; Wilson, R.A.J.; Morris, G.A.; Scatett, S.S. Modeling Moisture Content of Fine Dead Wildland Fuels: Input to the BEHAVE Fire Prediction System. *Usda For. Serv. Intermt. Res. Stn. Res. Pap.* **1986**, *11*, 1–61. [CrossRef]

14. Marsdensmedley, J.B.; Catchpole, W.R.; Marsdensmedley, J.B.; Catchpole, W.R. Fire Modelling in Tasmanian Buttongrass Moorlands. III. Dead Fuel Moisture. *Int. J. Wildland Fire* **2001**, *10*, 241–253. [CrossRef]

15. Chuvieco, E.; Aguado, I.; Dimitrakopoulos, A.P. Conversion of Fuel Moisture Content Values to Ignition Potential for Integrated Fire Danger Assessment. *Can. J. For. Res.* **2004**, *34*, 2284–2293. [CrossRef]

16. Trevitt, A.C.F. Weather parameters and fuel moisture content: Standards for fire model inputs. In Proceedings of the Conference on Bushfire Modeling and Fire Danger Rating Systems, Canberra, ACT, Australia, 11–12 July 1988.

17. Nelson, R.M.J. Prediction of Diurnal Change in 10-h Fuel Stick Moisture Content. *Can. J. For. Res.* **2000**, *30*, 1071–1087. [CrossRef]

18. Matthews, S.; Mccaw, W.L. A Next-generation Fuel Moisture Content Model for Fire Behavior Prediction. *For. Ecol. Manag.* **2006**, *234*, s27–s30. [CrossRef]

19. Matthews, S.; Mccaw, W.L.; Neal, J.E.; Smith, R.H. Testing a Process-based Fine Fuel Moisture Model in Two Forest Types. *Can. J. For. Res.* **2007**, *37*, 23–35. [CrossRef]

20. Matthews, S.; Gould, J.; Mccaw, L. Simple models for predicting dead fuel moisture in eucalyptus forests. *Int. J. Wildland Fire* **2010**, *19*, 459–467. [CrossRef]

21. Wang, C.; Gao, H.Z.; Cheng, S.; Lu, D.L.; Chang, W.Q. Study on water content of Forest Fuel and Forest Fire Risk Prediction in Saihanba Forest Area. *For. Sci. Technol. Dev.* **2009**, *23*, 59–61. [CrossRef]

22. Pook, E.W.; Gill, A.M. Variation of live and dead fine fuel moisture in *Pinus radiata* plantations of the Autralian Capital Territory. *Int. J. Wildland Fire* **1993**, *3*, 155–168. [CrossRef]

23. Li, S.Y.; Shu, Q.T.; Ma, A.L.; Zhang, Q.R.; Liu, Y.; Wang, X.Q. Modeling on Moisture Content Predicting of Surface Fuel of Litter- fall Layer in Pinus armandii Plantations. *For. Resour. Manag.* **2009**, *1*, 84–88. [CrossRef]

24. Sun, P.; Yu, H.Z.; Jin, S. Predicting hourly litter moisture content of larch stands in daxinganling region, china using three vapour-exchange methods. *Int. J. Wildland Fire* **2015**, *24*, 114–119. [CrossRef]

25. Jin, S.; Li, L. Validation of the Method for Direct Estimation of Timelag and Equilibrium Moisture Content of Forest Fuel. *Sci. Silvae Sin.* **2010**, *46*, 98–105.

26. Catchpole, E.A.; Catchpole, W.R.; Viney, N.R.; McCaw, W.L.; Marsden-Smedley, J.B. Estimating Fuel Response Time and Predicting Fuel Moisture Content From Field Data. *Int. J. Wildland Fire* **2001**, *10*, 215–222. [CrossRef]

27. Nelson, R.M. A method for describing equilibrium moisture content of forest fuels. *Can. J. For. Res.* **1984**, *14*, 597–600. [CrossRef]

28. Ruiz, A.D.; Maseda, C.M.; Lourido, C.; Viegas, D.X. Possibilities of dead fine fuels moisture prediction in Pinus pinaster Ait. stands at "Cordal de Ferreiros" (Lugo, north-western of Spain). In Proceedings of the Forest Fire Research & Wildland Fire Safety: IV International Conference on Forest Fire Research Wildland Fire Safety Summit, Coimbra, Portugal, 18–23 November 2002.

29. Slijepcevic, A.; Anderson, W. Hourly variation in fine fuel moisture in eucalypt forests in tasmania. *For. Ecol. Manag.* **2006**, *234*. [CrossRef]

30. Simard, A.J.; Main, W.A. Comparing methods of predicting jack pine slash moisture. *Can. J. For. Res.* **1982**, *12*, 793–802. [CrossRef]

31. Simard, A.J.; Eenigenburg, J.E.; Blank, R.W. Predicting fuel moisture in jack pine slash: A test of two systems. *Can. J. For. Res.* **1984**, *14*, 68–76. [CrossRef]

32. Groot, W.J.; Wardati Wang, Y. Calibrating the fine fuel moisture code for grass ignition potential in sumatra, indonesia. *Int. J. Wildland Fire* **2005**, *14*, 161–168. [CrossRef]

33. Wang, H.Y.; Li, L.; Liu, Y.; Jin, S. Suitability of Canadian Forest Fire Danger Rating System in Tahe Forestry Bureau, Heilongjiang Province. *J. Northeast For. Univ.* **2008**, *36*, 47–49.

34. Wotton, B.M.; Stocks, B.J.; Martell, D.L. An index for tracking sheltered forest floor moisture within the canadian forest fire weather index system. *Int. J. Wildland Fire* **2005**, *14*, 169–182. [CrossRef]

35. Wotton, B.K.; Beverly, J.L. Stand-specific litter moisture content calibrations for the Canadian Fine Fuel Moisture Code. *Int. J. Wildland Fire* **2007**, *16*, 463–472. [CrossRef]

36. Mao, W.X.; Tong, D.H.; Zhang, C.; Zhao, D.D.; Ding, Y.Y.; Jin, S. Spatial Heterogeneity of Moisture Content of Land Surface Dead Fuel in Larch Stand and Effects of Sampling Methods on Moisture Estimation. *J. Northeast For. Univ.* **2012**, *40*, 29–33. [CrossRef]

37. Bülent, S.; Bilgili, E.; KuUk, O.; Durmaz, B.D. Determination of surface fuels moisture contents based on weather conditions. *For. Ecol. Manag.* **2006**, *234*, S75–S81. [CrossRef]

38. Viegas, D.X.; Piñol, J.; Viegas, M.T.; Ogaya, R. Estimating live fine fuels moisture content using meteorologically-based indices. *Int. J. Wildland Fire* **2001**, *10*, 223–240. [CrossRef]

39. Van Wagner, C.E. Equilibrium Moisture Contents of some Fine Forest Fuels in Eastern Canada. *Inf. Rep. Petawawa For. Exp. Stn.* **1972**, *PS-X-36*.

40. Lawson, B.D.; Armitage, O.B.; Hoskins, W.D. Diurnal variation in the fine fuel moisture code: Tables and computer source code. *FRDA Rep.* **1996**, *245*. (Pacific Forestry Centre and Research Branch: Victoria, BC).

41. Sun, P.; Zhang, Y.L.; Sun, L.; Hu, H.H.; Guo, F.T.; Wang, G.Y.; Zhang, H. Influence of Fuel Moisture Content, Packing Ratio and Wind Velocity on the Ignition Probability of Fuel Beds Composed of Mongolian Oak Leaves via Cigarette Butts. *Forests* **2018**, *9*, 507. [CrossRef]

42. Byram, G.M.; Nelson, R.M. An Analysis of the drying Process in Forest Fuel Material. In *General Technical Report*; Southern Research Station, USDA Forest Service: Blacksburg, VA, USA, 1963; pp. 1–38.

43. Simard, A.J. The moisture content of forest fuels-1. In *A Review of the Basic Concepts*; Information Report FF-X-14; Canadian Department of Forest and Rural Development, Forest Fire Research Institute: Ottawa, ON, Canada, 1968; p. 47.

44. Liu, X.; Jin, S. Development of Dead Forest Fuel Moisture Prediction Based on Equilibrium Moisture Content. *Sci. Silvae Sin.* **2007**, *43*, 126–133. [CrossRef]

45. Viney, N.R. A Review of Fine Fuel Moisture Modelling. *Int. J. Wildland Fire* **1991**, *1*, 215–234. [CrossRef]

46. Luke, R.H.; Mcarthur, A.G.; Brown, A.G.; Mcarthur, A.G.; Hillis, W.E. *Bushfires in Australia*; Australian Government Publishing Service: Canberra, Australia, 1978.

47. Pixton, S.W.; Warburton, S. Moisture Content/Relative Humidity Equilibrium of Some Cereal Grains at Different Temperatures. *J. Stored Prod. Res.* **1971**, *6*, 283–293. [CrossRef]

48. Britton, C.M.; Countryman, C.M.; Wright, H.A.; Walvekar, A.G. The Effect of Humidity, Air Temperature, and Wind Speed on Fine Fuel Moisture Content. *Fire Technol.* **1973**, *9*, 46–55. [CrossRef]

49. Anderson Hal, E. Moisture Diffusivity and Response Time in Fine Forest Fuels. *Can. J. For. Res.* **1990**, *20*, 315–325. [CrossRef]

50. Zhang, H. Influencing Factors of Dead Surface Fuel Moisture Prediction in Great Xing'an Region. Ph.D. Thesis, Northeast Forestry University, Harbin, China, 2014.

51. Slijepcevic, A.; Anderson, W.R.; Matthews, S. Testing existing models for predicting hourly variation in fine fuel moisture in eucalypt forests. *For. Ecol. Manag.* **2013**, *306*, 202–215. [CrossRef]

52. Zhang, Y.L. Study on Influencing Factors and Prediction Model of Dynamic Change of Litter Moisture Content of Mongolian Oak and Pinus Koraiensis. Ph.D. Thesis, Northeast Forestry University, Harbin, China, 2019. [CrossRef]

53. Yu, H.Z.; Jin, S.; Di, X.Y. Prediction Models for Ground Surface Fuels Moisture Content of Larix gmelinii Stand in Daxing'anling of China Based on One-hour Time Step. *Chin. J. Appl. Ecol.* **2013**, *24*, 1565–1571.

54. Yu, H.Z.; Jin, S.; Di, X.Y. Models for Predicting the Hourly Fuel Moisture Content on the Forest Floor of Birch Stands in Tahe Forestry Bureau. *Sci. Silvae Sin.* **2013**, *49*, 108–113. [CrossRef]

55. Lu, X. Research on Dynamic and Prediction Model of Dead Fuel Moisture Content of Typical Stands in Great Xing'an Mountains. Ph.D. Thesis, Northeast Forestry University, Harbin, China, 2016.

56. Liu, X. Effects of Temperature and Humidity on Fuel Equilibrium Moisture Content. Master's Thesis, Northeast Forestry University, Harbin, China, 2007.

57. Carlson, J.D.; Bradshaw, L.S.; Nelson, R.M. Application of the Nelson Model to Four Timelag Fuel Classes Using Oklahoma Field Observations: Model Evaluation and Comparison with National Fire Danger Rating System Algorithms. *Int. J. Wildland Fire* **2007**, *16*, 204–216. [CrossRef]

 *forests*

Article

Modeling Drying of Degenerated *Calluna vulgaris* for Wildfire and Prescribed Burning Risk Assessment

Torgrim Log 

Fire Disasters Research Group, Department of Safety, Chemistry and Biomedical Laboratory Sciences, Western Norway University of Applied Sciences, 5528 Haugesund, Norway; torgrim.log@hvl.no; Tel.: +47-900-500-01

Received: 19 June 2020; Accepted: 10 July 2020; Published: 14 July 2020

Abstract: Research highlights: Moisture diffusion coefficients for stems and branches of degenerated *Calluna vulgaris* L. have been obtained and a mathematical model for the drying process has been developed and validated as an input to future fire danger modeling. Background and objectives: In Norway, several recent wildland–urban interface (WUI) fires have been attributed to climate changes and accumulation of elevated live and dead biomass in degenerated *Calluna* stands due to changes in agricultural activities, i.e., in particular abandonment of prescribed burning for sheep grazing. Prescribed burning is now being reintroduced in these currently fire prone landscapes. While available wildfire danger rating models fail to predict the rapidly changing fire hazard in such heathlands, there is an increasing need for an adapted fire danger model. The present study aims at determining water diffusion coefficients and develops a numerical model for the drying process, paving the road for future fire danger forecasts and prediction of safe and efficient conditions for prescribed burning. Materials and methods: Test specimens (3–6 mm diameter) of dead *Calluna* stems and branches were rain wetted 48 h and subsequently placed in a climate chamber at 20 °C and 50% relative humidity for mass loss recordings during natural convection drying. Based on the diameter and recorded mass versus time, diffusion coefficients were obtained. A numerical model was developed and verified against recoded mass loss. Results: Diffusion coefficients were obtained in the range 1.66–10.4 × 10^{-11} m^2/s. This is quite low and may be explained by the very hard *Calluna* "wood". The large span may be explained by different growth conditions, insect attacks and a varying number of years of exposure to the elements after dying. The mathematical model described the drying process well for the specimens with known diffusion coefficient. Conclusions: The established range of diffusion coefficients and the developed model may likely be extended for forecasting moisture content of degenerated *Calluna* as a proxy for fire danger and/or conditions for efficient and safe prescribed burning. This may help mitigate the emerging fire risk associated with degenerated *Calluna* stands in a changing climate.

Keywords: drying tests; humidity diffusion coefficients; wildfire; prescribed burning; modeling

---

1. Introduction

Wildfires represent an increasing threat to people and property in the wildland–urban interface (WUI) worldwide, in particular in the USA, Canada, Australia and the countries surrounding the Mediterranean Sea [1–3]. In Europe, the number of fires has in recent years decreased while the impact of the WUI fires has generally become more severe both with respect to the number of fatalities and the number of lost structures [4]. Recently, this has also been an issue in coastal Norway [5–7]. This area was part of the North-Western Europe cultural landscape that originated soon after the introduction of livestock husbandry, stretching from Portugal to the Arctic Circle. The heathlands of Western Norway was thus managed by anthropogenic fire regimes to increase pasture value and herbivore production [8], keeping the landscapes virtually free of severe fires. The combination of grazing and prescribed burning was vital for maintaining vegetation composition and successional dynamics in this semi-natural environment [9,10]. Fire removed old vegetation and prevented shrubs and trees from re-establishing in the habitat and was therefore of ecological importance [11]. Fire frequencies were set by the local landuse, and 10–20-year burning rotations were common, although regional and local variation occurred [12,13]. After the fire, a number of grasses and herbs established, including the *Calluna* regenerating from seeds and through resprouting from stems and subterranean organs [14]. Despite some years with difficult weather conditions, prescribed fire management generally worked well [15]. The *Calluna* stands in coastal Norway are adapted to these burning cycles to the extent that smoke-adapted germination is observed [16].

A lack of prescribed burning and fire suppression leaves the *Calluna* to grow larger. Today, this unique coastal landscape is therefore endangered through the lack of traditional management, resulting in older *Calluna* stands, much accumulated biomass, nature-type degeneration and a succession of bushes and shrubs [14]. This process also results in increasing amounts of dead material on the ground (as litter) as well as in the standing vegetation (dead plants). Due to these changes, the heathlands are now, according to the EC Habitat Directive 92/43/EEC, of international conservation importance [17–19]. Similar trends are also observed in Mediterranean *Calluna* stands, where scrub and woodland encroachment can be observed. There, combinations of prescribed fire, mowing and increased grazing/browsing will be necessary to achieve the long-term conservation of heathlands [20,21]. In old *Calluna* stands, dead branches constitute the lower canopy [22]. Spot and line fire ignited field burns revealed much more intense fires in old stands (more than 50 years since last burning), compared to young stands (approximately 10 years old) [23]. When unmanaged, Norwegian heathlands gradually develop vegetation compositions where species such as juniper, pine, spruce and birch enter the heathlands as part of a succession. This additionally contributes in the biomass build-up and in particular, junipers (*Juniperus communis* L.) with their highly flammable resinous foliage contribute considerably to rapid spread of fire [24].

In addition to the fire safety aspect, there are a number of motivations for the increased interest in resuming heathlands in coastal Norway. Among the most important motivations we find, e.g., biodiversity [17], cultural values [25], aesthetics [26], tourism [27] and concerns regarding the trend to increase local food production. Farmers may ask for some limited economic support for prescribed burning and the application success rate is currently high. Grazing is important to keep the heathland habitat. The Old Norse Sheep breed is particularly suited for being kept outside all year grazing the evergreen *Calluna*, i.e., a practice that stems back to the Viking Age. In the 1970s, the breed was reduced to about 1000 individuals at the islands of Austevoll community, Norway, but now counts 30,000+ individuals. The breed, which is locally branded as Norsk Villsau (Norwegian Wild Sheep), is even protected by a national regulation [28]. Many bird species depend on open landscapes, and the current local decline in, e.g., the majestic Eurasian eagle-owl (*Bubo bubo* L.) population is of national concern [17]. This comes along with early 19th century introduction of the black-listed Sitka spruce (*Picea sitchensis* L.) currently invading large areas of Norwegian heathlands [29,30]. Due to all these reasons, managing heathlands by fire to remove old *Calluna* plants and invasive species, and support *Calluna* regeneration, is increasingly popular. For resuming prescribed burning, the fire danger needs to be known.

Modern methods, such as remote sensing, may be used for assessing the current fire danger [31,32]. In contrast to remote sensing presenting the current fire danger, fire weather index systems have previously been developed to predict wildfire risk a few days into the future. Anderson and Anderson [33] refined the fine fuel moisture code (FFMC) of the Canadian Fire Weather Index System to predict the fine gorse (*Ulex europaeus* L.) shrub fuel moisture content, i.e., branches less than 5 mm diameter. The elevated dead fuel moisture content was poorly predicted by the FFMC. Their effort to improve the FFMC prediction accuracy through regression modeling was also unsuccessful. Another approach to predict fire behavior in such fuels is therefore required.

Fire behavior in heathlands is influenced by several factors, such as stand age influencing fuel load, structure and height [33], the fraction of dead fuel [34], fuel moisture content [35,36] and wind speed [37–39]. Recent results also indicate that critical differences in fire severity and fuel consumption, including possible destruction of soil seedbanks, can also be linked to the ground fuel layer flammability [30]. Davies et al. [22] revealed that for dry periods during wintertime, live *Calluna* stems showed no moisture gradient along the height. They also experienced rapid changes in live stems moisture content when the ground was frozen. During field studies in Scottish heathlands, Davies and Legg [40] for on-site spot and line fire ignition discovered a threshold value of lower canopy dead fine fuel moisture content for sustained burning. This shows that keeping track of the dead fuel moisture content is important for evaluating the fire danger of degenerated *Calluna* dominated heathlands. Being able to predict the fuel moisture content and the wind conditions, fire danger warnings may be issued, and when the fire danger is in a low range, conditions for safe prescribed burning may also possibly be predicted. This would require knowledge of the moisture diffusion coefficient in the critical fuel component, i.e., the elevated dead *Calluna* stems and branches acting as kindling fuel for engaging the whole *Calluna* plants in fire [40]. There is indeed a need for understanding and predicting how fast this biomass dries in order to issue valid fire danger ratings.

In Norway, there was no record of subzero temperatures wildfires until January 2014. However, due to decreasing snow cover during recent decades, in combination with accumulated live and dead biomass, the probability for such wildfires was on the rise, though unnoticed by the fire brigades. During January 2014, two surprising wildfires took place in the Atlantic heathlands at Flatanger and Frøya, Norway, 1.9° (210 km) south of the Arctic Circle. The first of these, the 15 km^2 Flatanger fire, destroyed more structures than any fire in Norway since 1923 [5]. The reasons for the destruction were many, such as a fire start in the darkness, storm strength wind, extremely rugged terrain, few hours of daylight, no access roads, frozen lakes, rivers and creeks limiting the access to fire water as well as the long distances covered by the fire [7]. On top of that, the climatic conditions had also resulted in the wooden structures, mainly homes, farm buildings, huts and boat sheds, being very susceptible to fire due to a period of dry air exposure [41]. These fires, as well as severe fires in South Western Norway heathlands in April 2019, one of these a result of lost control in prescribed burning, have demonstrated that there is a need for improved fire danger predictions.

In other areas, e.g., the Gulf of Mexico coastal region, prescribed fire is increasingly used as a management tool to restore declining native ecosystems. However, since treated sites are more susceptible to biological invasion of, e.g., Chinese tallow (*Triadica sebifera* L.) [42] this is a delicate balance. A review on the dynamics of prescribed burning, tree mortality and injury in managing oak natural communities to minimize economic losses in North America was done by Dey et al. [43]. This is indeed easier in the *Calluna* dominated heathlands, which represent an anthropogenic landscape that used to be managed by regular burning. However, given the current condition of the heathland, the fire risk is considerably higher than for managed heath and there is an urgent need for research to get a grip on the fire danger associated with degenerated heathlands in need of prescribed burning to (a) reduce the accumulated biomass representing the potential fire fuel and (b) return the landscapes to the previous farmland.

Sorption curves for most wildland fuel, *Calluna* stems and branches included, are not available, i.e., the sorption data for wood [44] currently represents the best available alternative. Diffusion coefficients are, however, established for some plant species. Pith diffusion coefficient for sunflower obtained by Sun et al. [45] was about 1–2.5 × 10^{-9} m^2/s. Diffusion coefficient for willow stems was reported by Gigler et al. [46] to be 3 × 10^{-10} m^2/s. The value reported for willow stems match with the typical diffusion coefficients for humidity transport in wood, i.e., 1–5 × 10^{-10} m^2/s as reported by Baronas et al. [47]. Studying dehydration kinetics of fermented cocoa beans, Adrover and Brasiello [48] found typical diffusion coefficients of 7.5 × 10^{-11} m^2/s at 25 °C. Domínguez-Pérez et al. [49] found for roasted cocoa beans moisture diffusion coefficients in the range 1.26–5.70 × 10^{-10} m^2/s. For lemongrass, Nguyen et al. [50] found effective diffusion coefficient in the range from 7.64 × 10^{-11} to 1.48 × 10^{-10} m^2/s. Schmalko and Alzamora [51,52] studied shrinking, apparent density, sorption curves and moisture diffusion coefficients for yerba mate. For the xylem, the values varied between 1.7 × 10^{-10} and 8.3 × 10^{-9} m^2/s. Faggion et al. [53] analyzed drying of yerba mate twigs ranging from 3.5 to 10 mm thickness and established that convection dominated the heat and mass transfer. Betie et al. [54] recorded and modeled the power and instrument transformers insulation paper drying process by thermogravimetric analyses to ensure safe transformer operations. Similar moisture diffusion coefficient studies have, however, not been found for *Calluna* stands.

The purpose of the present study is to analyze the drying processes of representative elevated dead biomass associated with degenerated (old) *Calluna* stands, i.e., similar to stands resulting in the January 2014 fires in Flatanger and Frøya [5–7], Trøndelag and the April 2019 fires in Tysvær and Sokndal, Rogaland. Based on drying tests, the moisture diffusion coefficient is obtained for selected test specimens at controlled ambient conditions. A numerical model for the involved mass and heat transfer is developed and the modeled results are compared to the results from the drying tests. The potential future outcome is that the model may later on be integrated in, e.g., the Canadian Fire Danger Rating Model for providing reasonable hour by hour fire danger modeling of the degenerated *Calluna* dominated heathlands currently representing an increasing fire threat in coastal Norway. It may also be used as a tool for assessing safe conditions for prescribed burning. The test specimens and climate chamber mass loss recordings are presented in Section 2. The results from drying tests and numerical modeling are presented and discussed in Section 3. Possible future impact and possible obstacles for future use is also presented in Section 3. Concluding remarks are presented in Section 4. The background theory and numerical model are outlined in Appendix A.

2. Materials and Methods

In heathland fires, the lower *Calluna* canopy fuel moisture content (FMC) is of high importance regarding sustained fire spread [40]. The drying experiments were therefore arranged to determine the drying rate of dead *Calluna* stems and branches. Test specimens were collected from *Calluna* stands in the degeneration phase, i.e., old *Calluna* stands, from an area south of Haugesund (H), N 59.362, E 5.325 and Ytstevika (Y), N 61.941, E 5.027. These *Calluna* stands, which are quite representative of the Norwegian coastal heath, had been left unmanaged and not been exposed to prescribed burning or natural fires over the last 50+ years. Some of the plants in these stands died through age while others died through extreme desiccation during the winter of 2014. Such die backs have also been observed in Scotland [55]. Since the *Calluna* plants may spread by sprouting, test specimens were collected with a minimum of 50 m separation to make sure that they were collected from different plants. The test specimens were stored dry (40–50% relative humidity (RH)) at room temperature prior to further treatment.

The test specimen diameter was recorded with a caliper at 10 different locations, and it was rotated to get a proper average reading. Test specimens of 3–7 mm diameter were cut to about 200 mm length, placed on a stainless steel grid and gently sprayed with deionized water for 48 h (rate 1 mm/h) prior to the drying tests. After wetting, each test specimen was then cut to 110–160 mm length. They were then equipped with short (3–4 mm) plastic lids to prevent axial drying, as seen in Figure 1.

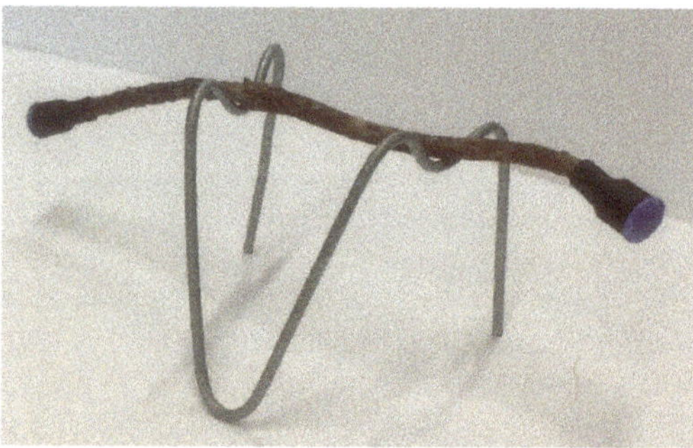

Figure 1. Sample holder and test specimen ready for climate chamber drying.

The drying tests were performed in a climate chamber (Binder KBWF 720) at 20 °C and 50% RH. To ensure vibrational noise dampening, the procedure as explained by Log et al. [7] was used. This included placing a 50 mm thick Styrofoam plate on the lowest grid shelf for noise dampening. A 5 mm thick glass plate rested on the Styrofoam plate and the balance (Sartorius CP324S, resolution 0.1 mg) was placed on top of the glass plate. To limit humidity accumulation in the vicinity of the drying test specimens, the balance side doors were kept open. The air drafts close to the test specimens were recorded to be in the order of 1 cm/s, i.e., practically stagnant air conditions. This set-up has been previously shown to give a noise ratio of as low as ± 0.3 mg root mean square and a ramping of less than 0.01 mg/h [7].

The test specimens were put on stainless steel tripod sample holders designed to minimize the contact area, and the mass loss was recorded every minute until equilibrium moisture content was achieved. This corresponded to a fractional mass loss of less than 10^{-7}/h (10^{-5}%/h). This could typically take a week of drying. After each drying test, the test specimen dry mass was recorded after oven drying for 48 h at 80 °C and then for 48 h at 105 °C. This allowed comparison with results obtained by Davies and Legg [40], who dried their *Calluna* test specimens at 80 °C to minimize the loss of volatile organic compounds, and for comparison with recommended drying at 105 °C [56].

Based on the recorded mass loss versus time it is possible to obtain the test specimen humidity diffusion coefficient. In the present study, this requires knowledge about the test specimen diameter and the time needed to dry half way ($t_{1/2}$) and three quarters of the way ($t_{3/4}$) towards the new equilibrium condition. The total number of test specimens was 12 and each drying experiment lasted for one week. All tests revealed the $t_{1/2}$ (s) and $t_{3/4}$ (s), i.e., respectively the time to reach half and three quarters of the final mass loss. The individual test specimen humidity diffusion coefficients were calculated using Equation (A14), as outlined in Appendix A.

A numerical model was developed for calculating the drying of similar rain wetted *Calluna* stems and branches. The theory behind the model and the numerical approach is outlined in Appendix A. The modeled mass loss during drying was compared to the results obtained in the climate chamber.

3. Results and Discussion

3.1. Drying Test Results

The mass loss versus time for a 6.01 mm (±0.11 mm SD) diameter and 129 mm long test specimen is first presented in detail. When put in the climate chamber, the mass was 2.8767 g. After 164 h (6.8 days) in the climate chamber the strict relative mass loss (based on the dry mass) requirement of 10^{-7}/h was met. The mass was then 2.0964 g, which corresponds to a water mass loss of 0.7803 g. The mass loss as a function of time for the first 96 h (4 days) is presented in Figure 2.

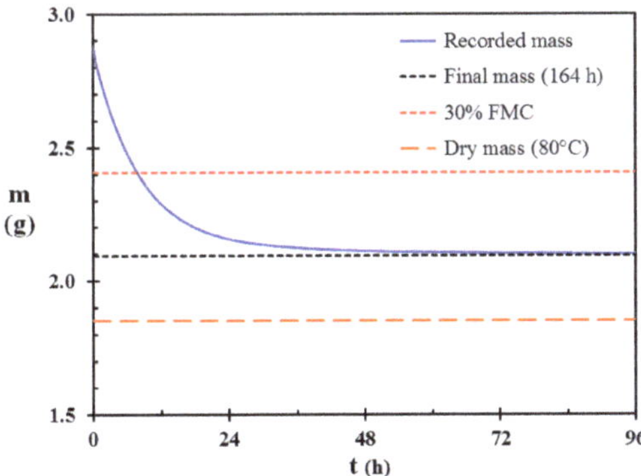

Figure 2. Recorded mass as a function of time for a 6.01 mm (±0.11 mm) diameter *Calluna* test specimen. The final mass after 164 h, the mass after oven drying at 80 °C and the mass at a 30% fuel moisture content (FMC) threshold value for sustainable fire in *Calluna* stands [34] are also marked on the figure.

Subsequent oven drying the test specimen at 80 °C for 48 h resulted in a dry mass of 1.8523 g. According to Davies et al. [34], 30% FMC for dead *Calluna* stems and branches represents the limit for sustained burning succeeding on-site spot and line fire ignition. For the presented test specimen this corresponds to a mass of 2.4080 g, which was reached at 27,970 s (7.77 h). This indicates that degenerated *Calluna* stands in 50% RH and 20 °C could change from rain wet to combustible within 8 h.

Drying in stagnant air, and starting from rain wet conditions, must be considered as very special conditions. In real conditions, the relative humidity may not be 100% during the nights or succeeding rainy conditions. Hence, the dead branches and stems may be partially dry when possibly exposed to sunlight, lower RH or windy conditions, or a combination thereof. Sunny weather usually generates thermal effects, which in elsewise stagnant conditions may initiate buoyantly driven wind. This would then significantly increase the associated drying rates. The fast drying under real conditions may explain the local prescribed burner groups experience that while not being able to get the *Calluna* dominated heather alight in the morning, it may burn at dangerous rates in the early afternoon.

The dimensionless mass of water being evaporated during drying is shown in Figure 3. The recorded values for $t_{1/2}$ and $t_{3/4}$, respectively 21,005 s (5.83 h) and 42,650 s (11.85 h), are also marked on the figure. Using the recorded values for $t_{1/2}$, $t_{3/4}$ and the test specimen radius, the moisture diffusion coefficient according to Appendix A, Equation (A14), was found to be 4.73×10^{-11} m^2/s. When the moisture diffusion coefficient is known, it may be used for the numerical model presented in Appendix A.4.

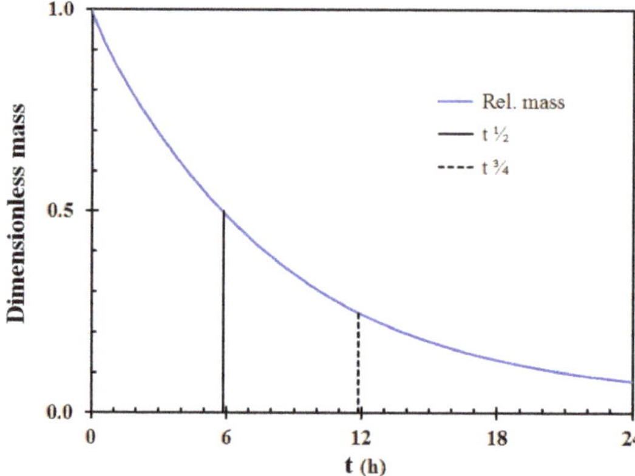

Figure 3. Dimensionless (normalized) mass as a function of time for a 6.01 mm (±0.11 mm) diameter *Calluna* twig. The time to reach $t_{1/2}$ and $t_{3/4}$, i.e., 5.83 h and 11.85 h, are also marked on the figure.

3.2. Variation in Recorded Diffusion Coefficients

A total of 12 degenerated *Calluna* branches were wetted and subsequently dried separately on the balance in the climate chamber at 50% RH as previously described. The test specimen data and the calculated moisture diffusion coefficient for these test specimens are given in Table 1.

It is quite clear from these drying experiments that the obtained moisture diffusion coefficients vary considerably, i.e., from 1.66×10^{-11} to 10.4×10^{-11} m^2/s. The average diffusion coefficient was 4.16×10^{-11} m^2/s, with a standard deviation as high as 2.69×10^{-11} m^2/s. Two of the test specimens displayed a moisture diffusion coefficient close to twice the value of the third highest one, respectively 8.96×10^{-11} m^2/s and 10.4×10^{-11} m^2/s. The drying time is according to Appendix A, Equation (A12), a function of radius squared. Normalizing the drying time by the radius squared may then be used to demonstrate the difference in drying time due to varying diffusion coefficients. This is shown in Figure 4 for selected drying tests involving low, medium and high diffusion coefficients.

Table 1. Growth place, diameter, length and resulting moisture diffusion coefficient of the degenerated *Calluna* test specimens.

Test Specimen	Location	D (mm)	L (mm)	$D_{w,s}$ (m^2/s)
1	Haugesund 1	6.01 ± 0.66	129	4.73×10^{-11}
2	Haugesund 1	5.47 ± 0.50	157	4.35×10^{-11}
3	Haugesund 1	6.90 ± 0.46	160	8.96×10^{-11}
4	Haugesund 2	5.00 ± 0.54	160	2.32×10^{-11}
5	Haugesund 2	5.32 ± 0.59	132	1.78×10^{-11}
6	Haugesund 2	6.07 ± 0.60	130	10.4×10^{-11}
7	Nerlandsøy 1	5.48 ± 0.35	150	2.33×10^{-11}
8	Nerlandsøy 1	5.21 ± 0.63	135	1.88×10^{-11}
9	Nerlandsøy 1	4.76 ± 0.24	130	1.66×10^{-11}
10	Nerlandsøy 2	6.45 ± 0.68	129	4.04×10^{-11}
11	Nerlandsøy 2	3.54 ± 0.66	130	3.56×10^{-11}
12	Nerlandsøy 2	6.43 ± 0.60	110	3.96×10^{-11}
Average		-	-	$4.16 \pm 2.69 \times 10^{-11}$

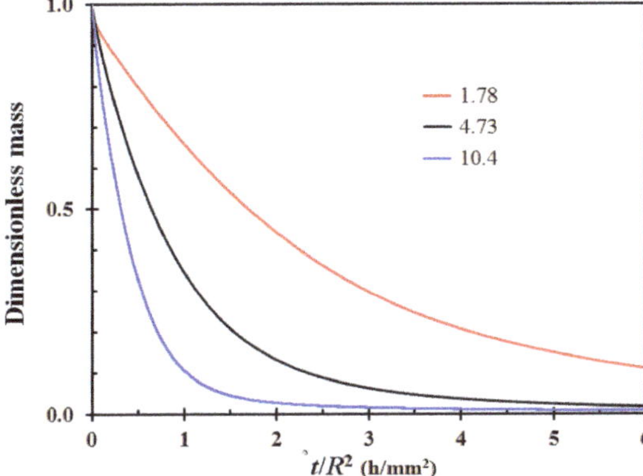

Figure 4. Dimensionless mass as a function of time normalized by the radius squared for selected low, medium and high diffusion coefficient test *Calluna* test specimens. The curves are labeled by their respective diffusion coefficients (in units 10^{-11} m^2/s).

The present study is, however, not the only study revealing large ranges in moisture diffusion coefficients. Baronas et al. [47] reports typical diffusion coefficients for moisture transport in wood in the range 1–5×10^{-10} m^2/s. For lemongrass, Nguyen et al. [50] found effective diffusion coefficient in the range from 7.64×10^{-11} to 1.48×10^{-10} m^2/s. Schmalko and Alzamora [51,52] recorded xylem moisture diffusion coefficients for yerba mate in the range from 1.7×10^{-10} to 8.3×10^{-9} m^2/s.

There may be several reasons for the large differences in observed moisture diffusion coefficients in the present study. To name a few: the *Calluna* plants may have grown in areas where there were differences in nutrients giving faster or slower growth rates and thereby softer or harder wood. The dead branches collected may have been dead for a varying number of years, and insect attacks may have made some of them more permeable than others. Cycles of freezing and thawing may in some locations have resulted in crack development and crack expansion thereby resulting in increased porosity for some of the test specimens. Additionally, not to forget, the diameter was varying along the quite tortuous branches, as seen in Figure 1, which in the present study were assumed to be perfect cylinders. It was, however, surprising for the author that the obtained diffusion coefficient varied nearly an order of magnitude from 1.66×10^{-11} to 10.4×10^{-11} m^2/s. Given the many factors making impact on the degenerated *Calluna* plants during years of exposure to the elements, as well as possible insect attacks, it may be assumed that collecting more samples could have revealed both higher and lower diffusion coefficients.

For future fire danger modeling, the observed variation makes it clear that one cannot assume one single value for the moisture diffusion coefficient. When modeling, the best approach may rather be to use a range of diffusion coefficients, ranging from 1×10^{-11} to 20×10^{-11} m^2/s, and probably putting more weight on the recorded average value, i.e., about 4×10^{-11} m^2/s.

3.3. Numerical Modeling Results

Recorded and modeled mass as a function of time for the 6.01 mm (±0.11 mm) diameter *Calluna* test specimen is shown in Figure 5. The modeled mass during drying generally follows the recorded values quite well. Some deviations between these two curves may, however, be observed.

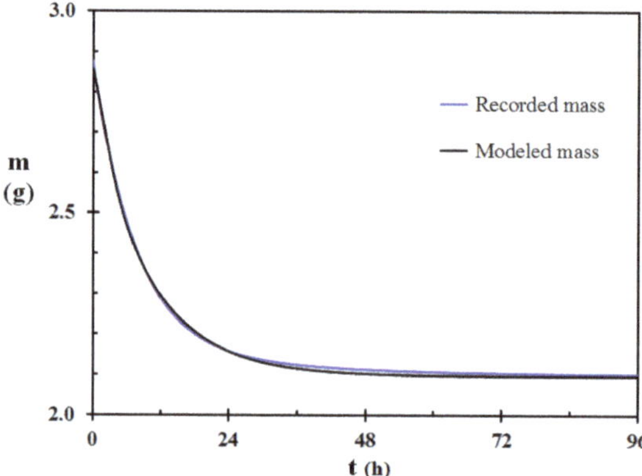

Figure 5. Recorded and modeled mass as a function of time for a 6.01 mm (±0.11 mm) diameter *Calluna* test specimen.

There may be several explanations for the deviation between the recorded and modeled mass loss as seen in Figure 5. It may partly be due to the moisture diffusion coefficient varying some with water concentration. The engineering expressions for the convective heat transfer coefficient may also be imprecise. The convective heat transfer to the test specimen may also be slightly increased due to minor air currents in the climate chamber, however, without being too severe. According to the modeling, the heat transfer to the test specimen was dominated by convection. This is in agreement with the findings by Faggion et al. [53]. The heat transfer by radiation is, however, not negligible. The assumption of constant emissivity of the *Calluna* twig surface may not be strictly correct since the emissivity may vary with the surface water content.

The modeling does, however, give a good indication about the mass loss during free convection drying, and the deviations shown in Figure 5 were quite representative also for the other test specimens. The main reason for the quite good fit is that the model uses a moisture diffusion coefficient established from the drying test, i.e., the modeling is therefore not strictly independent but validates that the modeling method is reliable. It may therefore be concluded that when the moisture diffusion coefficient is known, the modeling gives quite reliable results. If adjusted to the engineering equations describing forced convection drying, it is therefore likely that extending the model for forced convection may work well also for field conditions.

3.4. Future Possibilities

In the present study the drying tests were performed in a climate chamber with stagnant, or at least very close to stagnant ambient conditions, i.e., under free convection. With knowledge of the moisture diffusion coefficients for dead elevated parts of degenerated *Calluna* stands, modeling drying under forced convection should be possible. By collecting field data about the average branch thicknesses, modeling of drying of complete *Calluna* plant communities may be possible.

Anderson and Anderson [33] were not able to predict the elevated dead fuel moisture content of gorse (*Ulex europaeus* L.). By recording moisture diffusion coefficients of elevated dead branches and stems of gorse, it is quite likely that the approach presented in the present study can solve the difficulties encountered for gorse, as well as for similar potential fire fuels.

Based on hourly temperature, relative humidity, wind and insolation forecasts, the presented mathematical model may be expanded to predict the fire danger conditions of degenerated *Calluna* stands. Getting good estimates of the fire danger 48 h into the future would be very beneficial for the fire brigades. Being better prepared may assist in preventing losses as experienced in the Flatanger WUI fire [5–7]. Permanent manning of rural fire stations on high risk days, allowing for a significantly faster response in the case of a fire, could also be decided based on proper fire danger modeling. This is in line with fleet allocations, as suggested by Pérez et al. [57] according to the seasons, although for shorter periods, or even single days, when the fire risk is expected to be particularly high. The fire brigades could also ban prescribed burning on days of too high fire danger predictions. This would help reduce the fire disaster risk as suggested by Log et al. [58].

Modeling real conditions, e.g., 48 h ahead would also be of great value when planning to perform prescribed burning. Knowledge about days when prescribed burning can be done efficiently and knowledge about days when the risk of losing control is too high would be very valuable and may result in less contradiction regarding prescribed burning [59].

4. Conclusions

Drying of rain wet degenerated *Calluna* stems and branches under free convection condition at 50% relative humidity revealed moisture diffusion coefficients in the range from 1.66×10^{-11} to 10.4×10^{-11} mm^2/s. The mean value was $4.16 \pm 2.69 \times 10^{-11}$ mm^2/s. Numerical modeling of the natural convection drying process gave results close to the recorded mass loss. This is promising regarding future expansion of the numerical model to field conditions, i.e., forced convection, wind and insolation. Based on weather forecasts, this may pave the road for forecasting the dryness of the degenerated *Calluna* stands for safe prescribed burning as well as for alarming about high fire danger. The fire brigades may use such predictions to be better prepared if the conditions are likely to develop into high danger and may use this information to ban prescribed burning on high risk days.

Funding: The present work was supported by the Norwegian Research Council, grant no 298993 "Reducing fire disaster risk through dynamic risk assessment and management (DYNAMIC)" and the Fire Disaster Research Group, Western Norway University of Applied Sciences (HVL). The APC was funded by HVL.

Acknowledgments: The author appreciate the help from Liv Guri Velle for collecting *Calluna* branches at Nerlandsøy and the support from Gunnar Thuestad for wetting the test specimens and performing some of the drying tests while the author was on a mission to the Arctic areas of Norway. The improvements suggested by the two anonymous reviewers are highly appreciated.

Conflicts of Interest: The authors declare no conflict of interest.

Appendix A. (Theory and Model Description)

Appendix A.1. Water Vapor Concentration in Air

The exchange of humidity adsorbed, H_2O_{ad}, to water in the gas phase, $H_2O_{(g)}$, may be described by:

$$H_2O_{ad} = H_2O_{(g)} \tag{A1}$$

and the process may thermodynamically be described by:

$$\Delta G = \Delta H_{vap} - T \times \Delta S_{vap} \tag{A2}$$

where ΔG (J/mol), ΔH (J/mol) and ΔS (J/mol·K) are the Gibbs energy, enthalpy and entropy of the vaporization process described in Equation (A1). The drying process is entropy driven, and as the enthalpy of water evaporation is very high, the evaporation process results in temperature depletion during drying. The heat required for the drying process of a suspended stem or branch is supplied through convection and heat radiation.

The saturation vapor pressure of water is an almost exponential function of temperature, T_c (°C), and may be expressed by [60]:

$$P_{w,sat} = 610.78 \times e^{\left(\frac{17.2694 \cdot T_c}{T_c + 238.3}\right)} (\text{Pa}), \tag{A3}$$

The corresponding vapor concentration of water at a given temperature, T (K), may then be obtained by:

$$C_{w,sat} = \frac{P_{w,sat} \times M_w}{R \times T} \ (\text{kg/m}^3), \tag{A4}$$

where M_w (0.01802 kg/mol) is the molecular mass of water and R (8.314 J/K·mol) is the universal gas constant. The dryness of the air relative to the saturation conditions is usually expressed as the air relative humidity (RH), i.e., $RH = P_w/P_{w,sat} = C_w/C_{w,sat}$.

In Western Norway coastal areas, the relative humidity is generally quite high. In conditions of adiabatically heated air, e.g., due to high pressure subsidence or foehn wind, clear skies are often experienced. During the spring months of March and April, the sun's heat radiation adds significantly to the ambient air temperature increase. This further lowers the relative humidity of the ambient air. The sun also directly heats the wildland fuel, which thereby experience very low humidity air in the close vicinity of the fuel surface. This results in drying of the heather, and in particular, drying of the dead biomass fraction, which has no access to soil humidity. This drying renders the accumulated and elevated dead heathland biomass very prone to fires.

Appendix A.2. Wooden Fuel Equilibrium Moisture Content (EMC)

Dependent on the previous sorption history and the current conditions, dead cellulose based biomass, such as wood and dead *Calluna* stems and branches, adsorb humidity from or release humidity to, the surrounding air. Cellulosic materials consists of complicated molecular structures with free hydroxyl groups, which may result in hysteresis effects when exposed to cycles of dry and humid air [61,62]. However, given a very long time, i.e., $t \rightarrow \infty$, the corresponding equilibrium moisture content (EMC) for representative wooden materials may be calculated by [44]:

$$\text{EMC} = \frac{1800}{W} \left\{ \frac{K \cdot RH}{1 - K \cdot RH} + \frac{K_1 K \cdot RH + 2K_1 K_2 K^2 RH^2}{1 + K_1 K \cdot RH + K_1 K_2 K^2 RH^2} \right\} \tag{A5}$$

where

$$W = 349 + 1.29T + 0.0135T_c^2$$

$$K = 0.805 + 0.000736T_c - 0.00000273T_c^2$$

$$K_1 = 6.27 - 0.00938T_c - 0.000303T_c^2$$

$$K_2 = 1.91 + 0.0407T_c - 0.000293T_c^2$$

The wood EMC corresponding to Equation (A5) at 22 °C is presented in Figure A1 and the inverse curve is shown in Figure A2.

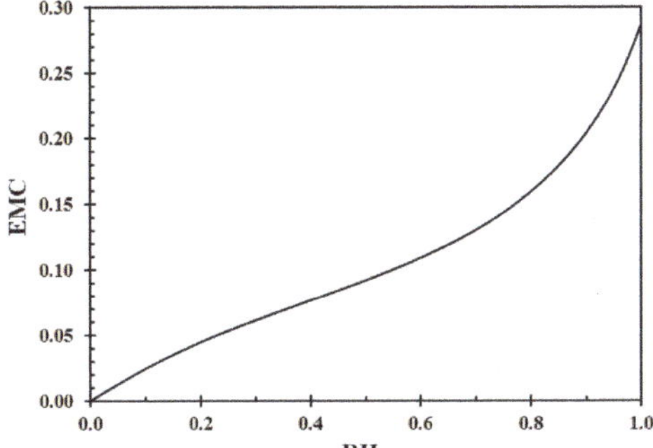

Figure A1. Equilibrium moisture content (EMC) at 22 °C as a function of relative humidity (RH). Data from Equation (A6) [44].

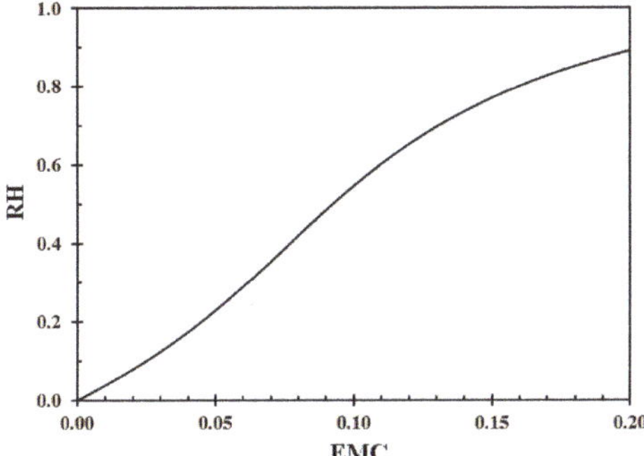

Figure A2. Relative humidity (RH) as a function of equilibrium moisture content (EMC) at 22 °C. Data from Equation (A6) [44].

For modeling purposes, the 4th order polynomials developed by Log [63] based on the data presented in Figures A1 and A2 is used in the present study, i.e.,

$$\mathrm{FMC} = 0.0017 + 0.2524 \times \mathrm{RH} - 0.1986 \times \mathrm{RH}^2 + 0.0279 \times \mathrm{RH}^3 + 0.167 \times \mathrm{RH}^4 \tag{A6}$$

$$\mathrm{RH} = 0.0698 + 1.258 \times \mathrm{FMC} - 125.35 \times \mathrm{FMC}^2 + 809.43 \times \mathrm{FMC}^3 + 1583.8 \times \mathrm{FMC}^4 \tag{A7}$$

The regression coefficients obtained for Equations (A6) and (A7) were very close to unity. For simplicity, assuming that the hysteresis does not dominate the sorption processes, Equations (A6) and (A7), can then be used to model transport of humidity between *Calluna* twig surfaces and the surrounding air. It should be noted that the "wood" in *Calluna* stems and branches do not necessarily follow exactly the same relationship as a representative wooden material. Since such data is missing for most wildland fuel, *Calluna* twigs and stems included, the data for wood [44] currently represents the best available alternative.

Appendix A.3. Transport of Humidity in Dead Calluna Stems and Branches

The transport of humidity in solids may be expressed by Fick's law of diffusion:

$$m'' = -D_{w,s} \times dC/dx \times \left(\mathrm{kg/m^2 \cdot s}\right) \tag{A8}$$

Twigs and branches of dead *Calluna* may mathematically be best described as cylindrical objects, where the internal humidity transport may be described by the associated heat equation:

$$\frac{\partial C}{\partial t} = \frac{1}{r}\frac{\partial}{\partial r}\left(rD_{w,s}\frac{\partial C}{\partial r}\right) \times \left(kg/m^3{\cdot}s\right) \tag{A9}$$

If the stem/branch has a uniform concentration at $t = 0$, for simplicity assumed to be unity, and for $t > 0$ the surface concentration is, due to the supply of dry air, kept at a new lower concentration, for simplicity zero, the dimensionless concentration within the cylinder is given by [64]:

$$C^* = -1 + 2\Sigma_{n=1}^{\infty}\exp\left\{-\beta_n^2 Fo\frac{J_0\left(\frac{r}{R}\beta_n\right)}{\beta_n J_1(\beta_n)}\right\} \tag{A10}$$

where $J_0(\beta)$ and $J_1(\beta)$ are Bessel functions of the first kind for integers 0 and 1, respectively, and where $\pm\beta_n$, $n = 1, 2, \ldots$ are the roots of $J_0(\beta) = 0$. The Fourier number, Fo, is in this case given by:

$$Fo = \frac{D_{w,s} \times t}{R^2} \tag{A11}$$

The dimensionless concentration given by Equation (A10) as a function of dimensionless radius is shown in Figure A3a for selected Fourier numbers. The average (integrated) dimensionless concentration as a function of Fourier number is shown in Figure A3b.

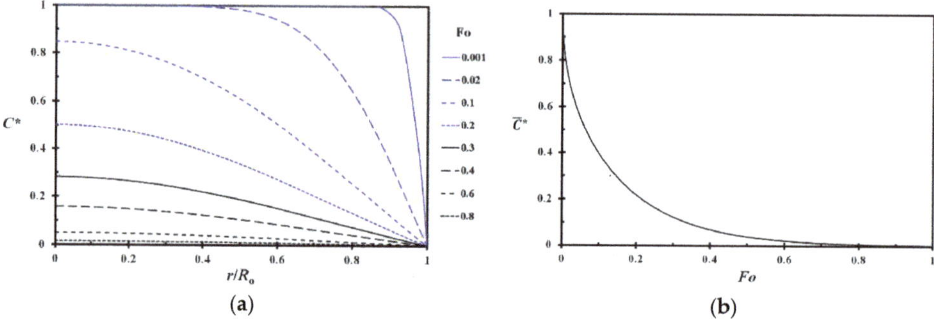

Figure A3. Dimensionless concentration (Equation (A10)) as a function of the relative radius and Fourier number (**a**) and dimensionless average concentration as a function of the Fourier number (**b**).

The time needed to achieve a certain level of average dryness, i.e., a relative mass loss as shown in Figure A3b, may then be calculated from Equation (A11) by:

$$t = \frac{R^2}{D_{w,s}}Fo \tag{A12}$$

As an example, at $Fo = 0.0630$, the average humidity, as shown in Figure A3b, has reached half the way towards the new EMC value. In a drying test, the heat required for the evaporation of water from the drying twig will, however, reduce the test specimen temperature. This results in a reduced water vapor concentration driving force to the bulk air in the vicinity of the drying stem or branch, i.e., a reduced drying rate. The temperature depletion is most conspicuous during the early phase of the drying process, i.e., when the drying rate is highest. However, during the last part of the drying phase, the test specimen temperature depletion is significantly reduced due to the low drying rate.

This part of an experimentally obtained drying curve may then be used for estimating the water diffusion coefficient, e.g., by:

$$D_{w,s} = R^2 \frac{Fo_2 - Fo_1}{t_2 - t_1} \tag{A13}$$

where the subscript refers to situations where a certain fraction of the mass loss towards the new equilibrium concentration is achieved. Taking the time and Fourier numbers from half the mass loss ($Fo_{1/2} = 0.0630$) to three quarters of the mass loss ($Fo_{3/4} = 0.1763$), the diffusion coefficient may simply be calculated by:

$$D_{w,s} = 0.1133 \frac{R^2}{t_{3/4} - t_{1/2}} \tag{A14}$$

where $t_{1/2}$ (s) and $t_{3/4}$ (s) are the time recorded to half and three quarters of the infinite time mass loss, respectively. As long as $t_{1/2}$ is on the order of hours, the temperature depletion below ambient conditions will not be significant from that time and onwards. This assumption may be verified numerically.

To make this technique work properly, the experimental boundary conditions need to be kept strictly constant. This is best achieved by performing drying tests in a carefully controlled atmosphere, e.g., a climate chamber, where both the temperature and relative humidity is kept constant through the complete drying period. Ideally, the test specimen should also be a perfect cylinder with no axial humidity transfer.

Appendix A.4. Numerical Model for the Drying Process

When a cylindrical test specimen is exposed to dry air, the heat consumption required for the water vaporization will decrease the test specimen temperature. Other objects in the vicinity will then radiate heat to the colder test specimen, and even in stagnant air, the temperature differences will trigger buoyant flow and thermal convection. The heat radiation may be expressed by:

$$\dot{Q}''_{rad} = \varepsilon_r \sigma \left(T_a^4 - T_s^4 \right) \tag{A15}$$

where ε_r is the resultant emissivity of the surface and the surroundings, T_a (K) is the ambient temperature and T_s (K) is the surface temperature. Sun radiation may also add to the received thermal radiation. In a dense *Calluna* stand, the objects surrounding a stem or a branch may to a large extent be other stems and branches at similar temperature as the one of interest. Unless exposed to sunlight, it may therefore be assumed that the heat radiation is of minor importance. In a climate chamber, it can, however, not be ignored since each drying test specimen is dried individually.

The convective heat transfer to the test specimen is given by:

$$\dot{Q}''_h = h(T_a - T_s) \tag{A16}$$

where the convective heat transfer coefficient h (W/m^2·K) may be estimated through engineering expressions for the Nusselt number:

$$Nu = \frac{h \times D}{k} \tag{A17}$$

where k (W/m·K) is the ambient air thermal conductivity and D (m) is the diameter. Considering free convection, the Nusselt number may be calculated by nondimensional expressions of the Grashof number and the Prandtl number. The Grashof number is an expression for the ratio between the buoyant force and the resisting viscous drag.

$$Gr = \frac{g\beta(T_a - T_s)D^3}{v^2} \tag{A18}$$

where g (m/s^2) is the acceleration of gravity, v (m^2/s) is the kinematic viscosity of air and $\beta = 1/T_a$ (K^{-1}). The Prandtl number describes the relationship between the between momentum diffusivity and thermal diffusivity a (m^2/s):

$$Pr = \frac{v}{a} \tag{A19}$$

The product of the Grashof number and the Prandtl number is often expressed as the Rayleigh number:

$$Ra = Gr \times Pr \tag{A20}$$

For $Ra < 10^{12}$, the following expression has been shown to be generally valid for average Nusselt numbers for cylinders [65]:

$$\overline{Nu} = \left(0.60 + \frac{0.387 Ra^{1/6}}{\left(1 + \left(\frac{0.559}{Pr} \right)^{9/16} \right)^{8/27}} \right)^2 \tag{A21}$$

In wood, the thermal diffusivity is in the range of 10^{-7} m^2/s while the water mass diffusion coefficient is 3–4 orders of magnitude lower. The transport properties involved are therefore not limited by thermal diffusion. Whether the heat flow to, and into, the test specimen is limited by convection or internal conduction may be evaluated by considering the Biot number:

$$Bi = \frac{h \times R}{k_w} \tag{A22}$$

where k_w (W/mK) is the thermal conductivity of the test specimen and R (m) is the test specimen radius. For dense wood, the thermal conductivity is typically in the range 0.14–0.17 W/mK. Given $h \approx 1$ W/m^{2}K (may be verified by Equation (A21) during the modeling), $k_w \approx 0.15$ W/m^{2}K and maximum radius of 0.0035 m gives $Bi \approx 0.02$. This clearly shows that modeling of internal heat transfer is not necessary, i.e., the heat transfer may be treated as a lumped thermal capacity analysis. This is also supported by the research done by Faggion et al. [53]. This is also in agreement with the results obtained by Mortensen [66] where the boundary layer only became a limiting parameter when studying humidity transfer from thin layers of paper to indoor air. It was also confirmed by testing for larger and smaller diffusion boundary layers within reasonable limits in the present study.

The mass transfer from the surface to the surroundings may then be expressed by:

$$\dot{m}_{\text{surface}} = 2\pi RL \frac{D_{w,a}}{\partial} \left(RH_{(t)} C_{sat(T_a)} - RH_{1(t)} C_{sat(T_1)} \right) \tag{A23}$$

where $RH_{(t)}$ is the ambient air relative humidity, $RH_{1(t)}$ is the relative humidity at the surface, $C_{sat(T_a)}$ is the ambient air water saturation concentration, $C_{sat(T_1)}$ is the water saturation concentration corresponding to the surface temperature of the cylinder, L (m) is the length of the cylinder, ΔC_a (kgm^{-3}) represents the difference in water concentration of the ambient air and the air in intimate contact with the solid surface and δ (m) is the boundary layer thickness.

The parameter $\frac{D_{w,a}}{\delta}$ (m/s) corresponds to the convective heat transfer coefficient, h, in heat transfer. In free convection, the hydrodynamic and thermal boundary layers are inseparable as the flow is created by buoyancy induced by the temperature boundary layer and the ambient air. Based on the thermal conductivity of air, i.e., 0.026 Wm^{-1}K^{-1}, the boundary layer thickness may be calculated.

In order to model the transport of moisture within the cylinder, the cylinder is sliced in N hollow cylinders of thickness Δr (m). For the surface layer, i.e., $n = 1$, Equation (A9) may be discretized as:

$$C_{1(t+\Delta t)} = C_{1(t)} + \frac{\Delta t \cdot \dot{m}_{\text{surface}}}{2\pi r_1 \Delta r} + \frac{D_{w,a} \cdot \Delta t}{2\pi r_1 \Delta r^2} \left(\frac{r_1}{2} + \frac{r_2}{2} \right) \left(C_{2(t)} - C_{1(t)} \right) \ (\text{kg/m}^3) \tag{A24}$$

For the internal layers, the heat equation may be discretized as:

$$C_{n(t+\Delta t)} = C_{n(t)} + \frac{D_{w,a}\cdot\Delta t}{\Delta r^2}\left(C_{n-1(t)} - 2C_{n(t)} + C_{n+1(t)}\right) + \frac{D_{w,a}\cdot\Delta t}{2\cdot\Delta r\cdot r_{(n)}}\left(C_{n-1(t)} - C_{n+1(t)}\right)(kg/m^3) \quad (A25)$$

For simplicity, a fictitious layer number $N + 1$ was introduced as a reflection plane, i.e., mirroring the concentration of layer N. This eliminated the need for a separate working equation for layer N.

Since the temperature would change as a result of the water evaporation process, and thereby the vapor equilibrium concentration at the cylinder surface, the heat required for water evaporation was taken into consideration by:

$$\dot{Q}_{vap} = \dot{m}_{surface}\Delta H_{vap,w} \quad (A26)$$

where $\Delta H_{vap,w}$ (2.444 MJ/kg) is the heat of vaporization of water.

During drying, the corresponding temperature depletions sets up a buoyant flow, whereas the corresponding Gr and Ra can be calculated for each time interval Δt, and the convective heat transfer coefficient determined through the calculated Nu. The heat flux to the cylinder surface can then be calculated by:

$$\dot{Q}_s = 2\pi RLh(T_a - T_1) + \varepsilon_r\sigma\left(T_a^4 - T_1^4\right) \quad (A27)$$

where σ (5.67 $\times$ 10^{-8} Wm^{-2}K^{-4}) is the Stefan–Boltzmann constant and ε_r is the resultant emissivity given by:

$$\varepsilon_r = \frac{1}{1/\varepsilon_a + 1/\varepsilon_s - 1} \quad (A28)$$

where ε_a is the emissivity of the surfaces surrounding the cylinder and ε_s is the emissivity of the cylinder surface.

For each time interval, Δt, the new test specimen temperature was calculated by:

$$T_{t+\Delta t} = T_t + \frac{\dot{Q}_s - \dot{Q}_{vap}}{\pi R^2 L\rho C_P}\Delta t \quad (A29)$$

It should be noted that the integration time interval, Δt, must comply with a Fourier number less than 0.5 to ensure numerical stability, where the Fourier number for the numerical calculations is given by:

$$Fo_{num} = \frac{D_{w,s} \times \Delta t}{\Delta r^2} \quad (A30)$$

References

1. Mowery, M.; Read, M.; Johnston, K.; Wafaie, T. *Planning the Wildland-Urban Interface*; Report 594, Planning Advisory Service; American Planning Association: Washington, DC, USA, 2019; ISBN 978-1-61190-202-0.
2. Bento-Gonçalves, A.; Vieira, A. Wildfires in the wildland-urban interface: Key concepts and evaluation methodologies. *Sci. Total Environ.* **2019**, *707*, 135529. [CrossRef]
3. Badia, A.; Pallares-Barbera, M.; Valldeperas, N.; Gisbert, M. Wildfires in the wildland-urban interface in Catalonia: Vulnerability analysis based on land use and land cover change. *Sci. Total Environ.* **2019**, *673*, 184–196. [CrossRef]
4. Rego, F.M.C.C.; Rodríguez, J.M.M.; Calzada, V.R.V.; Xanthopoulos, G. *Forest Fires—Sparking Firesmart Policies in the EU*; Publications Office of the European Union: Luxembourg, 2018; p. 53, ISBN 978-92-79-77493-5. [CrossRef]
5. DSB. *Brannene i Lærdal, Flatanger og på Frøya Vinteren 2014*; Norwegian Directorate for Civil Protection: Tønsberg, Norway, 2014; p. 55, ISBN 978-82-7768-342-3. (In Norwegian)
6. Steen-Hansen, A.; Bøe, G.A.; Hox, K.; Mikalsen, R.F.; Stensaas, J.P.; Storesund, K. Evaluation of Fire Spread in the Large Lærdal Fire, January 2014. In Proceedings of the 14th International Fire and Materials Conference and Exhibition, San Francisco, CA, USA, 2–4 February 2015; pp. 1014–1024.

7. Log, T.; Thuestad, G.; Velle, L.G.; Khattri, S.K.; Kleppe, G. Unmanaged heathland—A fire risk in subzero temperatures? *Fire Saf. J.* **2017**, *90*, 62–71. [CrossRef]

8. Webb, N.R. The traditional management of European heathlands. *J. Appl. Ecol.* **1998**, *35*, 987–990. [CrossRef]

9. Gimingham, C.H. *Ecology of Heathlands*; Chapman and Hall: London, UK, 1972; ISBN 10: 0412104601.

10. Gimingham, C.H. Biological flora of the British isles: Calluna vulgaris (L) hull. *J. Ecol.* **1960**, *48*, 455–483. [CrossRef]

11. Kaland, P.E. The origin and management of Norwegian coastal heaths as reflected by pollen analysis. In *Anthropogenic Indicators in Pollen Diagrams*; Behre, K.E., Ed.; Balkema: Rotterdam, The Netherlands, 1986; pp. 19–36. [CrossRef]

12. Velle, L.G.; Nilsen, L.S.; Vandvik, V. The age of *Calluna* stands moderates post-fire regeneration rate and trends in northern *Calluna* heathlands. *Appl. Veg. Sci.* **2012**, *15*, 119–128. [CrossRef]

13. Velle, L.G.; Vandvik, V. Succession after prescribed burning in coastal Calluna heathlands along a 340-km latitudinal gradient. *J. Veg. Sci.* **2014**, *25*, 546–558. [CrossRef]

14. Gimingham, C.H. *The Lowland Heathland Management Handbook*; English Nature: Peterborough, UK, 1992; ISBN 1-85716-077-0.

15. Allen, K.A.; Denelle, P.; Sánchez Ruiz, F.M.; Santana, V.M.; Marrs, R.H. Prescribed moorland burning meets good practice guidelines: A monitoring case study using aerial photography in the Peak District, UK. *Ecol. Indic.* **2016**, *62*, 76–85. [CrossRef]

16. Vandvik, V.; Töpper, J.P.; Cook, Z.; Daws, M.I.; Heegaard, E.; Måren, I.E.; Velle, L.G. Management-driven evolution in a domesticated ecosystem. *Biol. Lett.* **2014**, *10*, 20131082. [CrossRef]

17. Lindgaard, A.; Henriksen, S. *Norsk Rødliste for Naturtyper 2011*; Artsdatabanken: Trondheim, Norway, 2011; ISBN-13: 978-82-92838-29-7.

18. Prøsch-Danielsen, L.; Simonsen, A. Palaeoecological investigations towards the reconstruction of the history of forest clearances and coastal heathlands in southwestern Norway. *Veg. Hist. Archaeobotany* **2000**, *9*, 189–204. [CrossRef]

19. Thompson, D.B.A.; MacDonald, A.J.; Marsden, J.H.; Galbraith, C.A. Upland heather moorland in Great Britain: A review of international importance, vegetation change and some objectives for nature conservation. *Biol. Conserv.* **1995**, *71*, 163–178. [CrossRef]

20. Calvo, L.; Baeza, J.; Marcos, E.; Santana, V.; Papanastasis, V.P. Post-fire management of shrublands. In *Post-Fire Management and Restoration of Southern European Forests*; Moreira, F., Ed.; Springer: Dordrecht, The Netherlands, 2012; pp. 293–319. [CrossRef]

21. Borghesio, L. Can fire avoid massive and rapid habitat change in Italian heathlands? *J. Nat. Conserv.* **2014**, *22*, 68–74. [CrossRef]

22. Davies, G.M.; Legg, J.J.; O'hara, R.; MacDonals, A.J.; Smith, A.A. Winter desiccation and rapid changes in the live fuel moisture content of Calluna vulgaris. *Plant Ecol. Divers.* **2010**, *3*, 289–299. [CrossRef]

23. Nilsen, L.S.; Johansen, L.; Velle, L.G. Early stages of Calluna vulgaris regeneration after burning of coastal heath in central Norway. *Appl. Veg. Sci.* **2005**, *8*, 57–64. [CrossRef]

24. Diotte, M.; Bergeron, Y. Fire and the distribution of *Juniperus communis* L. in the Boreal Forest of Quebec, Canada. *J. Biogeogr.* **1989**, *16*, 91–96. [CrossRef]

25. Norwegian Bioeconomy Institute (NIBIO). Gammelnorsk Sau (Old Norwegian Sheep). Available online: www.nibio.no/tema/mat/husdyrgenetiske-ressurser/bevaringsverdige-husdyrraser/sau/gammelnorsk-sau (accessed on 13 June 2020).

26. Harstad, B.; Harestad, A.; Fludal, A. *Beitetrykk i Lynghei;—Produksjon av Kjøtt—Produksjon av Landskap (Grazing Pressure in Heathlands;—Production of Meat—Production of Landscape)*; Norsk landbruksrådgiving Rogaland: Haugesund, Norway, 2014; p. 35.

27. Værlandet. Meet the Old Norse Sheep. Hurtigruten, Inc.: Seattle, WA, USA. Available online: www.hurtigruten.co.uk/excursions/norway/varlandet-meet-the-old-norse-sheep/ (accessed on 23 March 2020).

28. Forskrift om Vern av Villsau frå Norskekysten/Villsau fra Norskekysten som Geografisk Nemning (Regulation on the Protection of Villsau from the Norwegian Coast/Villsau from the Norwegian Coast as a Geographical Naming); The Norwegian Ministry of Agriculture and Food: Oslo, Norway, 2010. Available online: https://lovdata.no/dokument/SF/forskrift/2010-11-04-1402 (accessed on 13 June 2020).

29. Hauglin, M.; Ørka, H.O. Discriminating between Native Norway Spruce and Invasive Sitka Spruce—A Comparison of Multitemporal Landsat 8 Imagery, Aerial Images and Airborne Laser Scanner Data. *Remote Sens.* **2016**, *8*, 363. [CrossRef]

30. Nygaard, P.H.; Øyen, B.-H. Spread of the Introduced Sitka Spruce (*Picea sitchensis*) in Coastal Norway. *Forests* **2017**, *8*, 24. [CrossRef]

31. Jurdao, S.; Chuvieco, E.; Arevalillo, J.M. Modelling Fire Ignition Probability from Satellite Estimates of Live Fuel Moisture Content. *Fire Ecol.* **2012**, *8*, 77–97. [CrossRef]

32. Yebra, M.; Dennison, P.E.; Chuvieco, E.; Riaño, D.; Zylstra, P.; Hunt, E.R.; Danson, F.M.; Qi, Y.; Jurdao, S. A global review of remote sensing of live fuel moisture content for fire danger assessment: Moving towards operational products. *Remote Sens. Environ.* **2013**, *136*, 455–468. [CrossRef]

33. Anderson, S.A.J.; Anderson, W.R. Predicting the elevated dead fine fuel moisture content in gorse (*Ulex europaeus* L.) shrub fuels. *Can. J. For. Res.* **2009**, *39*, 2355–2368. [CrossRef]

34. Davies, G.M.; Legg, C.J.; Smith, A.A.; MacDonald, A.J. Rate of spread of fires in *Calluna vulgaris*-dominated moorlands. *J. Appl. Ecol.* **2009**, *46*, 1054–1063. [CrossRef]

35. Baeza, M.J.; De Luís, M.; Raventós, J.; Escarré, A. Factors influencing fire behavior in shrublands of different stand ages and the implications for using prescribed burning to reduce wildfire risk. *J. Environ. Manag.* **2002**, *65*, 199–208. [CrossRef] [PubMed]

36. Sharma, S.; Ochsner, T.E.; Twidwell, D.; Carlson, J.D.; Krueger, E.S.; Engle, D.M.; Fuhlendorf, S.D. Nondestructive Estimation of Standing Crop and Fuel Moisture Content in Tallgrass Prairie. *Rangeland Ecol. Manag.* **2018**, *71*, 356–362. [CrossRef]

37. Santana, V.M.; Marrs, R.H. Flammability properties of British heathland and moorland vegetation: Models for predicting fire ignition. *J. Environ. Manag.* **2014**, *139*, 88–96. [CrossRef]

38. Morvan, D.; Dupuy, J.L. Modeling the propagation of a wildfire through a Mediterranean shrub using a multiphase formulation. *Combust. Flame* **2004**, *138*, 199–210. [CrossRef]

39. Davies, G.M.; Domènech, R.; Gray, A.; Johnson, P.C.D. Vegetation structure and fire weather influence variation in burn severity and fuel consumption during peatland wildfires. *Biogeosciences* **2016**, *13*, 389–398. [CrossRef]

40. Davies, G.M.; Legg, C.J. Fuel moisture thresholds in flammability of Calluna vulgaris. *Fire Technol.* **2011**, *47*, 421–436. [CrossRef]

41. Log, T. Cold Climate Fire Risk; A Case Study of the Lærdalsøyri Fire, January 2014. *Fire Technol.* **2016**, *52*, 1825–1843. [CrossRef]

42. Yang, S.; Fan, Z.; Liu, X.; Ezell, A.W.; Spetich, M.A.; Saucier, S.K.; Gray, S.; Hereford, S.G. Effects of Prescribed Fire, Site Factors, and Seed Sources on the Spread of Invasive Triadica sebifera in a Fire-Managed Coastal Landscape in Southeastern Mississippi, USA. *Forests* **2019**, *10*, 175. [CrossRef]

43. Dey, D.C.; Schweitzer, C.J. A Review on the Dynamics of Prescribed Fire, Tree Mortality, and Injury in Managing Oak Natural Communities to Minimize Economic Loss in North America. *Forests* **2018**, *9*, 461. [CrossRef]

44. Simpson, W.T. *Equilibrium Moisture Content of Wood in Outdoor Locations in the United States and Worldwide*; Research Note FPL-RN-0268; US Department of Agriculture: Washington, DC, USA, 1998; p. 14.

45. Sun, S.; Mathiasc, J.-D.; Toussainta, E.; Grédiaca, M. Hygromechanical characterization of sunflower stems. *Ind. Crops Prod.* **2013**, *46*, 50–59. [CrossRef]

46. Gigler, J.K.; Van Loon, W.K.P.; Meerdink, G.; Coumans, W.J. Drying Characteristics of Willow Chips and Stems. *J. Agric. Eng. Res.* **2000**, *77*, 391–400. [CrossRef]

47. Baronas, R.; Ivanauskasa, F.; Juodeikienéc, I.; Kajalavičiusc, A. Modelling of Moisture Movement in Wood during Outdoor Storage. *Nonlinear Anal. Model. Control* **2001**, *6*, 3–14. [CrossRef]

48. Adrover, A.; Brasiello, A. 3-D Modeling of Dehydration Kinetics and Shrinkage of Ellipsoidal Fermented Amazonian Cocoa Beans. *Processes* **2020**, *8*, 150. [CrossRef]

49. Domínguez-Pérez, L.A.; Concepción-Brindis, I.; Lagunes-Gálvez, L.M.; Barajas-Fernández, J.; Márquez-Rocha, F.J.; García-Alamilla, P. Kinetic Studies and Moisture Diffusivity During Cocoa Bean Roasting. *Processes* **2019**, *7*, 770. [CrossRef]

50. Nguyen, T.V.L.; Nguyen, M.D.; Nguyen, D.C.; Bach, L.G.; Lam, T.D. Model for Thin Layer Drying of Lemongrass (*Cymbopogon citratus*) by Hot Air. *Processes* **2019**, *7*, 21. [CrossRef]

51. Schmalko, M.E.; Alzamora, S.M. Modelling the drying of a twig of "yerba maté" considering as a composite material: Part I: Shrinkage, apparent density and equilibrium moisture content. *J. Food Eng.* **2005**, *66*, 455–461. [CrossRef]
52. Schmalko, M.E.; Alzamora, S.M. Modelling the drying of a twig of "yerba maté" considering as a composite material: Part II: Mathematical model. *J. Food Eng.* **2005**, *67*, 267–272. [CrossRef]
53. Faggion, H.; Tussolini, L.; Freire, F.B.; Freire, J.T.; Zanoelo, E.F. Mechanisms of heat and mass transfer during drying of mate (*Ilex paraguariensis*) twigs. *Dry. Technol.* **2016**, *34*, 474–482. [CrossRef]
54. Betie, A.; Meghnefi, F.; Fofana, I.; Yeo, Z. Modeling the Insulation Paper Drying Process from Thermogravimetric Analyses. *Energies* **2018**, *11*, 517. [CrossRef]
55. Hancock, M.H. An exceptional Calluna vulgaris winter die-back event, Abernethy Forest, Scottish Highlands. *Plant Ecol. Divers.* **2008**, *1*, 89–103. [CrossRef]
56. Matthews, S. Effect of drying temperature on fuel moisture content measurements. *Int. J. Wildland Fire* **2010**, *19*, 800–802. [CrossRef]
57. Pérez, J.; Maldonado, S.; López-Ospina, H. A fleet management model for the Santiago Fire Department. *Fire Saf. J.* **2016**, *82*, 1–11. [CrossRef]
58. Log, T.; Vandvik, V.; Velle, L.G.; Metallinou, M.-M. Reducing Wooden Structure and Wildland-Urban Interface Fire Disaster Risk through Dynamic Risk Assessment and Management. *Appl. Syst. Innov.* **2020**, *3*, 16. [CrossRef]
59. Farrar, A.; Kendal, D.; Williams, K.J.H.; Zeeman, B.J. Social and Ecological Dimensions of Urban Conservation Grasslands and Their Management through Prescribed Burning and Woody Vegetation Removal. *Sustainability* **2020**, *12*, 3461. [CrossRef]
60. Tetens, O. Uber einige meteorologische Begriffe. *Z. Geophys.* **1930**, *6*, 297.
61. Salin, J.G. Inclusion of the Sorption Hysteresis Phenomenon in Future Drying Models. Some Basic Considerations. *Maderas. Cienc. Tecnol.* **2011**, *13*, 173–182. [CrossRef]
62. Funk, M. Hysteretic moisture properties of porous materials: Part I: Thermodynamics. *J. Build. Phys.* **2014**, *38*, 6–49. [CrossRef]
63. Log, T. Modeling Indoor Relative Humidity and Wood Moisture Content as a Proxy for Wooden Home Fire Risk. *Sensors* **2019**, *19*, 5050. [CrossRef]
64. Carslaw, H.L.; Jaeger, J.C. *Conduction of Heat in Solids*, 2nd ed.; Oxford Science Publications: Oxford, UK, 1959; ISBN 0-19-853368-3.
65. Jaluria, Y. *Natural Convection. Heat and Mass Transfer*; Pergamon Press: Oxford, UK, 1980; p. 326, ISBN 9780080254326.
66. Mortensen, L.H. Hygrothermal Microclimate on Interior Surfaces of the Building Envelope. Ph.D. Thesis, Technical University of Denmark, Kongens Lyngby, Denmark, 2007; p. 166.

 *forests*

Article

Water-Soluble Inorganic Ions in Fine Particulate Emission During Forest Fires in Chinese Boreal and Subtropical Forests: An Indoor Experiment

Yuanfan Ma [1], Mulualem Tigabu [1,2], Xinbin Guo [1], Wenxia Zheng [1], Linfei Guo [1] and Futao Guo [1,*]

1   College of Forestry, Fujian Agriculture and Forestry University, Fuzhou 350002, China; 17746076057@163.com (Y.M.); mulualem.tigabu@slu.se (M.T.); guo18959151730@126.com (X.G.); ZWX15985703227@163.com (W.Z.); 18120894053@163.com (L.G.)
2   Southern Swedish Forest Research Centre, Swedish University of Agricultural Sciences, Box 49, SE-230 52 Alnarp, Sweden
*   Correspondence: guofutao@126.com; Tel.: +86-591-8378-0261

Received: 2 October 2019; Accepted: 4 November 2019; Published: 6 November 2019

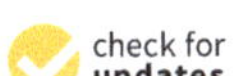

Abstract: Understanding of the characteristics of water-soluble inorganic ions (WSI) in fine particulate matter ($PM_{2.5}$) emitted during forest fires has paramount importance due to their potential effect on ecosystem acidification. Thus, we investigated the emission factors (EFs) of ten most common WSI from combustion of leaves and branches of ten dominant tree species in Chinese boreal and sub-tropical forests under smoldering and flaming combustion stages using a self-designed combustion unit. The results showed that EF of $PM_{2.5}$ was three times higher for the boreal (6.83 ± 0.67 g/kg) than the subtropical forest (1.97 ± 0.34 g/kg), and coniferous species emitted 1.5 times more $PM_{2.5}$ (5.35 ± 0.64 g/kg) than broadleaved species (3.45 ± 0.37 g/kg). EF of total WSI was 1.27 ± 0.08 g/kg for the boreal and 1.08 ± 0.07 g/kg for the subtropical forest and 1.28 ± 0.09 and 1.07 ± 0.06 g/kg for broadleaved and coniferous species, respectively. Individual ionic species also varied significantly between forest types and species within forest types, and K^+ and Cl^- were the dominant ionic species in $PM_{2.5}$, accounting for 25% and 30% for the boreal forest and 23% and 27% for the subtropical forest, respectively. Emissions of NO_2^- and SO_4^{2-} were the lowest, accounting for 3% and 5% for the boreal forest and 4% for each of the subtropical forests, respectively. Combustion of leaves emitted significantly more ionic species (1.29 ± 0.05 g/kg) than branches (1.05 ± 0.07 g/kg), and smoldering consistently emitted more ionic species (1.49 ± 0.09 g/kg) than flaming combustion (0.88 ± 0.03 g/kg). The cation to anion ratio was ≥ 1.0, suggesting that the particulate matter is neutral to alkalescent. As a whole, our findings demonstrate that forest fire in these regions may not contribute to ecosystem acidification despite the emission of a considerable amount of WSI during forest fires.

Keywords: acid rain; aerosol; biomass burning; forest fire; PM2.5

1. Introduction

Forest fire releases a large volume of smoke into the atmospheric environment, accounting for up to 42% of biomass-burning particulate emissions [1]. The smoke released from biomass-burning is mainly composed of particulate matter (PM) of different aerodynamic diameter, of which PM with a diameter less than 2.5 μm ($PM_{2.5}$) accounts for more than 90% of the total PM emitted from biomass combustion [2]. The PM derived from biomass-burning consists of condensed hydrocarbons, a mixture of elemental carbon and water soluble inorganic ionic species.

There is a growing concern about the emission of water-soluble inorganic ions, such as ammonium, nitrate, sulfate, and chloride, in PM emissions, as they modify the degree of acidity of the PM;

thereby resulting in acidification of ecosystems and contributing to the visibility reduction by light scattering [3,4]. Acid deposition and/or acid rain have a considerable impact on growth and productivity of forest and other ecosystems [5] by increasing foliar damage and soil acidity by acid water. The potential effects of acid rain on plants include foliar damage, disturbance to normal metabolic processes, inhibition of seed germination and seedling growth and predisposing the plants to other environmental stress factors [6,7]. Thus, characterizing water-soluble inorganic ions in atmospheric aerosol from mixed sources of emission has been a subject of many studies [3,4,8–10]. However, studies characterizing water-soluble inorganic ions in PM emitted from woody biomass burning are limited, particularly in China [11–13], where large forest fires occur annually. Given the predicted increase in the frequency and severity of forest fires in different parts of the globe as a consequence of climate change and vegetation encroachment, it is of paramount importance to understand the characteristics of water-soluble ions in PM emitted during forest fires.

Thus, the present study investigated the emissions of ten common water-soluble inorganic ions (Na^+, NH_4^+, K^+, Mg^{2+} and Ca^{2+} as well as F^-, Cl^-, NO_3^-, NO_2^- and SO_4^{2-}) in $PM_{2.5}$ from burning of leaves and branches of ten main tree species from boreal and subtropical forests of China under two combustion states, smoldering vis-à-vis flaming. The Chinese boreal and sub-tropical forest ecosystems are among China's four major forest management regions. The sub-tropical region, located in southern China, is an area that experiences high annual forest fire incidence, with nearly 15,000 forest fires occurring from 2000 to 2010 [14]. The Chinese boreal forest in northeastern China is prone to frequent wildfires [15,16] and has the largest average annual burned area in China, 1,300,000 ha between 1980 and 2005 [17].

The main objective of the study was to evaluate whether forest fire contributes to acidity of fine PM by characterizing the water-soluble composition of $PM_{2.5}$ emitted from forest fire according to combustion condition, fuel typologies, and forest type. The study specifically addressed the following questions: (1) Does emission of water-soluble ions vary among species and between forest types? (2) Does emission of water-soluble ions vary with fuel typologies, i.e., leaves versus branches, which are the common fuels during surface fire? (3) Does emission of water-soluble ions vary with combustion state, flaming versus smoldering? and (4) Is there a relationship between anions and cations in water-soluble ions? The emission of $PM_{2.5}$ has been shown to vary with tree species and combustion conditions, as more $PM_{2.5}$ is emitted during the smoldering stage, leaves release more $PM_{2.5}$ than branches, and conifer trees emit more $PM_{2.5}$ than broad-leaved trees [13]. Thus, it was hypothesized that the emission of water-soluble inorganic ions may vary among tree species, between forest type and combustion condition.

2. Materials and Methods

2.1. Samples and Collection of $PM_{2.5}$

A total of ten main tree species that are common in the boreal and subtropical forests of China, five from each forest type, were selected. The coniferous tree species included *Larix gmelinii* (Rupr.) Kuzen. and *Pinus sylvestris* var. mongolica Litv from the boreal forest region and *Pinus massoniana* L. and *Cunninghamia lanceolata* L. from the subtropical forest region. The broadleaved species from the boreal forest region included *Betula platyphylla* Suk., *Quercus mongolica* Fisch. ex Ledeb and *Populus davidiana* L. while those from the subtropical forest region were *Cinnamomum camphora* L., *Eucalyptus robusta* Smith, and *Phoebe bournei* (Hemsl.) Yang. Leaves and branches from these species were collected from the Nanying research station, Daxing'an mountains and the research forest of Fujian Agriculture and Forestry University in July and October, 2017, and the samples were evenly mixed and air-dried at a controlled relative humidity of 40% for three days until constant mass to avoid the confounding effect of seasonal variations in moisture and photosynthesis on the chemical composition of samples. All targeted trees were selected far away from the urban area to mitigate the influence of urban air pollutants. In addition, the branches and leaves of the same tree species were collected from different individual trees to ensure consistency of samples, and the samples were then mixed. The branch and leaf samples were then divided into three replicates (testing samples) using a 1/1000 electronic balance (50 g per testing sample).

The indoor combustion experiments were conducted by using a self-designed biomass combustion unit (Figure 1; [18]). The combustion device consists of a combustion chamber, temperature controller, flue gas analyzer, particle analyzer and sampler, and other additional components. The Flue Gas Analyzer (Testo350, Testo Instruments International Trading Co., Ltd., Lenzkirch, Germany) determines the pollutants in the flue gas based on infrared on-line monitoring and requires calibration with standard gas before each experiment. The calibrated instrument was used to identify the background concentrations of pollutants in the combustion chamber, which were recorded and then deducted from the final results. The instrument has a recording interval of 5 s, and the sensitivity of the instrument is 0.01% for CO_2 and 1 ppm for CO, CH and NO_x. The fine particulate matter ($PM_{2.5}$) was monitored using a particle analyzer (TSI8533, TSI Incorporated, Shoreview, MN, USA) throughout the combustion process, which works based on infrared on-line monitoring and requires calibration with standard gas. Background concentration identification is also required. The instrument has a recording interval of 5 s, and the background concentration of $PM_{2.5}$ for this experiment was 0.001 mg/m^3. The Deployable Particulate Sampler (DPS) System (SKC Incorporated, Eighty Four, PA, USA) is ideal for ambient and indoor air sampling of $PM_{2.5}$. The system includes a fully programmable Li-Ion-powered sample pump (SKC Ltd., Dorset, UK) for 24-hour sampling. The system pump provides a constant and accurate airflow with a flow rate of 10 L/min.

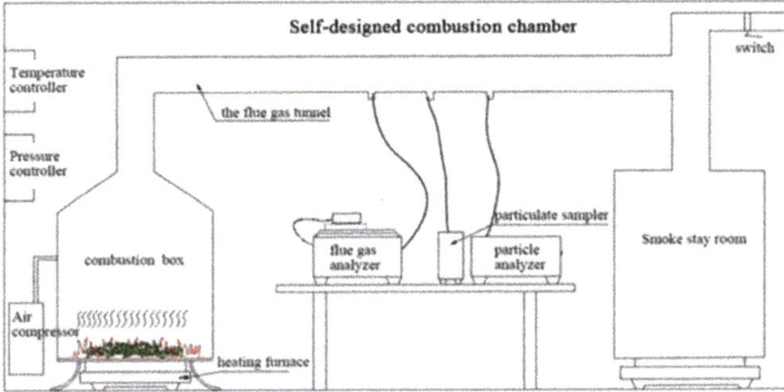

Figure 1. Schematic diagram of the self-designed biomass burning device.

Before starting the combustion test, the temperature controller was adjusted to control the temperature of the chamber to create either smoldering or flaming conditions. The combustion status was characterized by a Modified Combustion Efficiency (MCE), defined as the ratio of CO_2 to the change in CO and CO_2, calculated using the following formula:

$$MCE = \frac{\Delta CO_2}{\Delta CO + \Delta CO_2} \tag{1}$$

A combustion stage was considered as flaming when MCE reaches 0.99 and smoldering stage when MCE is between 0.65–0.85 [19,20]. After several preliminary experiments, the smoldering temperature was controlled at around 180 °C and the flaming temperature at around 270 °C. After adding samples into the combustion chamber, the flue gas analyzer and the particle analyzer were turned on for real-time monitoring, and the concentrations of various pollutants were measured and recorded within 5 s in order to calculate the emission factor of particulate matter and correct combustion efficiency. The particle sampler was then turned on when the combustion conditions remained stable and cooled down to room temperature. Teflon-polytetrafliuoroethylene (PTFE) membrane filters (SKC Ltd., Dorset, UK) were used to collect $PM_{2.5}$ during different combustion stages and were weighed before and after sample collection using a microbalance with 1 mg sensitivity. Control experiments were also performed using blank filters. Three replicates were conducted for each sample collection and analysis.

2.2. Extraction and Analysis of Water-Soluble Inorganic Ions

The concentrations of five anions (F^-, Cl^-, NO_3^-, NO_2^- and SO_4^{2-}) and five cations (Na^+, NH_4^+, K^+, Mg^{2+} and Ca^{2+}) were determined in aqueous extracts of the sample filters. To extract the water-soluble ions from the filters, the portions of the filters used for the gravimetric analysis were placed in separate 12 mL vials containing 10 mL of distilled–deionized water (18.2 MΩ resistivity). The vials were placed in an ultrasonic water bath and shaken with a mechanical shaker for 1 h to extract the ions. The extracts were filtered through 0.45 μm pore size microporous membranes, and the filtrates were stored at 4 °C in clean tubes before analysis.

A Dionex-1100 Ion Chromatograph (Dionex Inc., Sunnyvale, CA, USA) was used for determining both the cations and anions in the aqueous extracts of the air filters. For the cation analyses, the instrument was equipped with an IonPacCS12A column (20 mmol/L methanesulfonic acid as the eluent), while an ASRS-4num column (25 mmol/L KOH as the eluent) was used for anions. The measurements were taken under the following conditions: column temperature: 30 °C; flow rate: 1.0 mL/min; injection volume: 20 μL; flow precision $< \pm 0.1\%$; flow rate maximum error 0.1%. Detection limits were 4.5 mg L^{-1} for Na^+, 4.0 mg L^{-1} for NH_4^+, 10.0 mg L^{-1} for K^+, Mg^{2+} and Ca^{2+}, 0.5 mg L^{-1} for F^- and Cl^-, 15 mg L^{-1} for NO_2^- and NO_3^-, and 20 mg L^{-1} for SO_4^{2-}. Standard reference materials produced by the National Research Center for Certified Reference Materials (Beijing, China) were analyzed for quality control and assurance purposes. Data from blank samples were subtracted from the corresponding sample data after analysis [21].

2.3. Calculation of Emission Factors for Water-Soluble Ions in PM2.5

The emission factor (EF) of a given ion is defined as the amount of this ion emitted to the atmosphere per unit mass of fuel consumed by the fire. The emission factors of particulate matter and water-soluble ions were calculated in this study using the carbon mass balance method. The specific calculation procedure can be found in Zhang et al. [22]. The carbon mass balance method offers an advantage such that it is not necessary to collect all emitted pollutants, and the sampling position in the plume is adjustable [23].

2.4. Statistical Analysis

Two-way analysis of variance (ANOVA) was performed to determine the significant differences in EF of $PM_{2.5}$ and water-soluble inorganic ions (WSI) between forest types (boreal versus sub-tropical) and between species (conifer versus broadleaved), followed by least significant difference test for comparison of means. A t-test was performed to determine significant differences between species within each forest type and between combustion of leaves and branches as well as between smoldering and flaming stages of combustion. Correlation analysis between cations and anions was also conducted to examine the relationship between ionic species in $PM_{2.5}$.

3. Results

3.1. Emission of $PM_{2.5}$ and Water-Soluble Inorganic Ions

The EF of $PM_{2.5}$ varied significantly ($p < 0.05$) between forest types and species (Table 1). The EF of $PM_{2.5}$ was three times higher for the boreal (6.83 ± 0.67 g/kg) than the subtropical forest (1.97 ± 0.34 g/kg) when both coniferous and broadleaved species were burnt, and coniferous species emitted 1.5 times more $PM_{2.5}$ (5.35 ± 0.64 g/kg) than broadleaved species (3.45 ± 0.37 g/kg). However, combustion of coniferous species from the boreal forest emitted more $PM_{2.5}$ than the broadleaved species and species from sub-tropical forest. Similarly, the EF of WSI exhibited significant differences between forest types and species (Table 1). The EF total WSI was higher for the boreal forest (1.27 ± 0.08 g/kg) than for the subtropical forests (1.08 ± 0.07 g/kg), and combustion of broadleaved species emitted more WSI (1.28 ± 0.09 g/kg) than coniferous species (1.07 ± 0.06 g/kg). Combustion of broadleaved species from the boreal forest resulted in higher EF of WSI than the coniferous species as well as species from the subtropical forest. As a whole, the EF of WSI was higher for boreal than subtropical species. The heat of combustion of leaves and branches was 19,182 ± 294 kJ and 18,319 ± 264 kJ for coniferous and broadleaved species from the boreal region, respectively. For subtropical region, it was 19,115 ± 286 kJ and 18,054 ± 293 kJ for coniferous and broadleaved species, respectively (Table 1).

Table 1. Emission factors of $PM_{2.5}$ ($EF_{PM2.5}$) and water-soluble inorganic ions (EF_{WSI}) released during combustion of both leaves and branches of coniferous and broadleaved species from the boreal and sub-tropical forests together with heat of combustion of fuels.

Forest Type	Species	$EF_{PM2.5}$ (g·kg^{-1})	EF_{WSI} (g·kg^{-1})	Heat of Combustion (kJ)
	Conifer	9.03 ± 0.97a	1.03 ± 0.06a	19 182 ± 294a
Boreal	Broadleaved	4.62 ± 0.38b	1.51 ± 0.10b	18 319 ± 264b
	Mean	**6.83 ± 0.67A**	**1.27 ± 0.08A**	**18 751 ± 289A**
	Conifer	1.66 ± 0.31c	1.11 ± 0.06a	19 115 ± 286a
Subtropical	Broadleaved	2.28 ± 0.37c	1.05 ± 0.09a	18 054 ± 293b
	Mean	**1.97 ± 0.34B**	**1.08 ± 0.07B**	**18 584 ± 290B**

Means followed by different lower and upper case letter across the columns are significantly different between species and between the boreal and sub-tropical regions, respectively, at the 5% probability level.

With regard to the emission of individual ionic species, significant differences were observed between forest types and species for some ions (Table 2). The EF of Na^+, K^+, Mg^{2+}, Ca^{2+}, Cl^- and NO_3^- were higher for the boreal than subtropical forest while the EF of F^- was higher for the subtropical than boreal forest. Combustion of broadleaved species emitted more Na^+, Cl^-, SO_4^{2-} and NO_3^- than that of coniferous species. Comparison of the EF of WSI between species within each forest type (Table 3) revealed that broadleaved species from the boreal forest emitted significantly more ions, except K^+ and Ca^{2+}, than coniferous species. For the subtropical forest, combustion of coniferous species resulted in higher EF of NH_4^+, Mg^{2+} and F^- than broadleaved species. The dominant water-soluble ions in $PM_{2.5}$ released during burning of coniferous and broadleaved species were K^+ and Cl^-, accounting 25% and 30% for the boreal forest and 23% and 27% for the subtropical forest, respectively. Emissions of NO_2^- and SO_4^{2-} were the lowest, accounting 3% and 5% for the boreal forest and 4% each for the subtropical forests, respectively. The EF of individual ionic species was in the order of K^+ > Cl^- > Ca^{2+} > Na^+ > NH_4^+ > Mg^{2+} > NO_3^- > F^- > NO_2^- > SO_4^{2-} for coniferous species from the boreal region; Cl^- > K^+ > Ca^{2+} > Na^+ > Mg^{2+} > NH_4^+ > NO_3^- > F^- > SO_4^{2-} > NO_2^- for broadleaved species from the boreal region; Cl^- > K^+ > F^- > NH_4^+ > Ca^{2+} > Mg^{2+} > Na^+ > NO_2^- > SO_4^{2-} > NO_3^- for coniferous species from the subtropical region; and Cl^- > K^+ > Ca^{2+} > F^- > SO_4^{2-} > Na^+ > NH_4^+ > Mg^{2+}/NO_3^- > NO_2^- for broadleaved species from the subtropical region.

Table 2. Emission factor (g/kg) of water-soluble inorganic ions in $PM_{2.5}$ released during combustion of leaves and branches of coniferous and broadleaved species from the boreal and sub-tropical forests.

	Forest Type		Species	
Ions	Boreal	Subtropical	Conifer	Broadleaved
Na^+	0.101 ± 0.008a	0.061 ± 0.005b	0.067 ± 0.005A	0.091 ± 0.008B
NH_4^+	0.083 ± 0.009a	0.073 ± 0.006a	0.076 ± 0.006A	0.079 ± 0.008A
K^+	0.326 ± 0.012a	0.248 ± 0.015b	0.280 ± 0.015A	0.292 ± 0.013A
Mg^{2+}	0.085 ± 0.008a	0.052 ± 0.005b	0.061 ± 0.005A	0.073 ± 0.008A
Ca^{2+}	0.125 ± 0.008a	0.099 ± 0.009b	0.111 ± 0.008A	0.112 ± 0.008A
F^-	0.054 ± 0.004a	0.109 ± 0.008b	0.080 ± 0.009A	0.082 ± 0.006A
Cl^-	0.392 ± 0.027a	0.286 ± 0.010b	0.287 ± 0.011A	0.374 ± 0.023B
SO_4^-	0.043 ± 0.004a	0.062 ± 0.010a	0.033 ± 0.004A	0.066 ± 0.008B
NO_3^-	0.071 ± 0.009a	0.041 ± 0.004b	0.039 ± 0.003A	0.067 ± 0.008B
NO_2^-	0.041 ± 0.006a	0.046 ± 0.005a	0.038 ± 0.004A	0.047 ± 0.006A

Means followed by different lower and upper case letters across the rows are significantly different between the boreal and sub-tropical regions and between confer and broadleaved species, respectively, at the 5% probability level.

Table 3. Comparison of the emission factor (g/kg) of water-soluble inorganic ions in $PM_{2.5}$ released from combustion of conifer and broadleaved species within each region.

Ions	Within Boreal		Within Subtropical	
	Conifer	Broadleaved	Conifer	Broadleaved
Na^+	$0.078 \pm 0.006a$	$0.117 \pm 0.013b$	$0.055 \pm 0.008A$	$0.064 \pm 0.006A$
NH_4^+	$0.058 \pm 0.004a$	$0.099 \pm 0.014b$	$0.094 \pm 0.009A$	$0.058 \pm 0.007B$
K^+	$0.326 \pm 0.024a$	$0.326 \pm 0.012a$	$0.234 \pm 0.015A$	$0.257 \pm 0.022A$
Mg^{2+}	$0.057 \pm 0.004a$	$0.103 \pm 0.012b$	$0.065 \pm 0.009A$	$0.043 \pm 0.012B$
Ca^{2+}	$0.129 \pm 0.011a$	$0.122 \pm 0.011a$	$0.093 \pm 0.012A$	$0.103 \pm 0.011A$
F^-	$0.029 \pm 0.003a$	$0.070 \pm 0.006b$	$0.132 \pm 0.008A$	$0.093 \pm 0.011B$
Cl^-	$0.275 \pm 0.018a$	$0.470 \pm 0.037b$	$0.299 \pm 0.012A$	$0.278 \pm 0.016A$
SO_4^-	$0.017 \pm 0.001a$	$0.060 \pm 0.005b$	$0.048 \pm 0.006A$	$0.071 \pm 0.016A$
NO_3^-	$0.039 \pm 0.005a$	$0.092 \pm 0.013b$	$0.038 \pm 0.005A$	$0.043 \pm 0.006A$
NO_2^-	$0.022 \pm 0.003a$	$0.053 \pm 0.009b$	$0.053 \pm 0.006A$	$0.041 \pm 0.007A$

Means followed by different lower and upper case letter are significantly different between conifer and broadleaved species within the boreal and sub-tropical region, respectively, at the 5% probability level.

3.2. EF of Water-Soluble Ions in Relation to Combustion of Leaves and Branches

The EF of WSI varied significantly between combustion of leaves and branches of species group within each forest type (Figure 2). For coniferous species from the boreal region, combustion of leaves emitted more NH_4^+, K^+, Mg^{2+} and Ca^{2+} than combustion of branches, while the EF of the rest of the ions did not differ between leaves and branches. For broadleaved species from the boreal region, significantly higher NH_4^+, K^+, Ca^{2+} and Cl^- were emitted during combustion of leaves than branches. Combustion of leaves of coniferous species from the subtropical region emitted more Na^+, K^+, Mg^{2+} and Cl^- but less Ca^{2+} than combustion of branches. For broadleaved species from the subtropical region, combustion of leaves resulted in higher EF of all water-soluble ions than the combustion of branches.

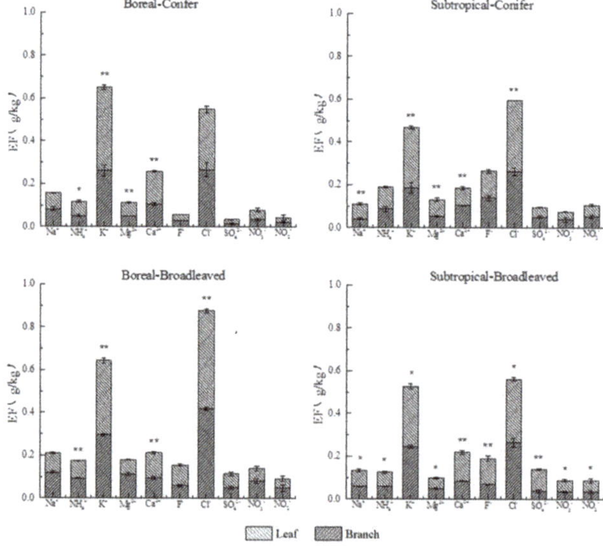

Figure 2. Comparison of the emission factor (g/kg) of water-soluble inorganic ions in $PM_{2.5}$ between combustion of leaves and branches of each group of species within each region. Bars with asterisks (** $p < 0.01$ and * $p < 0.05$) indicate significant differences between leaves and branches of conifer and broadleaved species within the boreal and sub-tropical region at the 5% probability level.

3.3. EF of Water-Soluble Ions in PM$_{2.5}$ Emitted During Smoldering and Flaming

For each species group within each forest type, significant differences were observed in the EF of WSI between smoldering and flaming stages of combustion ($p < 0.05$). Generally, smoldering released more WSI than flaming for most of the species (Figure 3). Smoldering resulted in significantly higher emissions of all ionic species, except K$^+$, than flaming for coniferous species from the boreal region while EFs for all WSI were higher during smoldering than flaming stages of combustion for broadleaved species from the boreal region. While smoldering resulted in higher EFs of all WSI, except Cl$^-$ and NO$_3{}^-$, it yielded higher EFs of all WSI for coniferous species from the subtropical region.

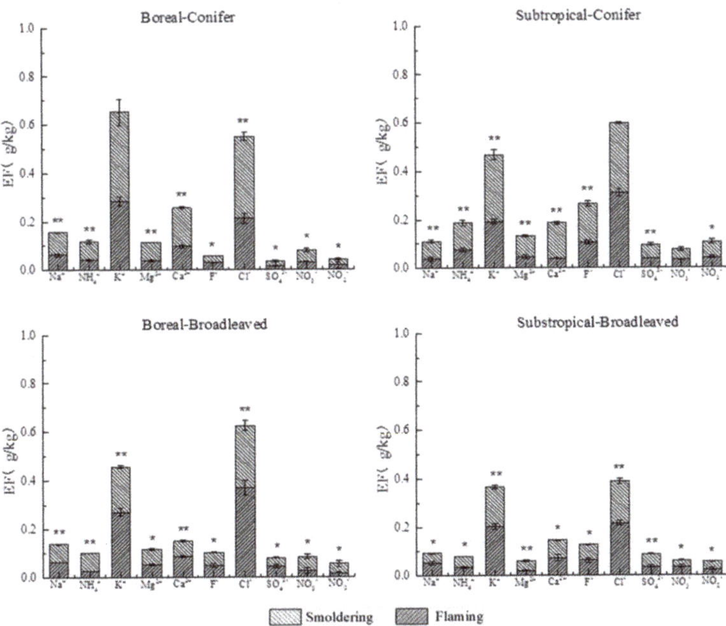

Figure 3. Comparison of emission factor (g/kg) of water-soluble inorganic ions in PM$_{2.5}$ between smoldering and flaming stages of combustion of leaves and branches of each group of species within each region. Bars with asterisk (** $p < 0.01$, * $p < 0.05$) indicate significant differences between leaves and branches of conifer and broadleaved species within the boreal and sub-tropical region at the 5% probability level.

3.4. Correlation Between Ionic Species in PM$_{2.5}$

Significant correlations were observed between ionic species in PM$_{2.5}$ (Table 4). For the boreal species, there were strong correlations (r > 0.75) between Na$^+$ and SO$_4{}^{2-}$ and NO$_2{}^-$ while the correlations between Na$^+$ and F$^-$ and NO$_3{}^-$ were moderate (r = 0.5–0.75). NH$_4{}^+$ and Mg^{2+} were strongly correlated with SO$_4{}^{2-}$, NO$_3{}^-$ and NO$_2{}^-$ but moderately correlated with F$^-$ and Cl$^-$. While K$^+$ had no correlation with any of the anions, Ca^{2+} had a moderate correlation with NO$_3{}^-$. For sub-tropical species, Na$^+$ was moderately correlated with NO$_3{}^-$ while NH$_4{}^+$ was strongly correlated with F$^-$, Cl$^-$ and NO$_2{}^-$. K$^+$ was moderately correlated with NO$_3{}^-$, Mg^{2+} was strongly correlated with F$^-$ but moderately correlated with Cl$^-$, and Ca^{2+} was strongly correlated with SO$_4{}^{2-}$ but moderately correlated with the rest of the anions.

Table 4. Correlations between cations and anions of water-soluble inorganic ions in $PM_{2.5}$ released during the burning of woody species.

	Boreal Species						**Sub-Tropical Species**				
	F^-	Cl^-	SO_4^{2-}	NO_3^-	NO_2^-		F^-	Cl^-	SO_4^{2-}	NO_3^-	NO_2^-
Na^+	0.58 *	0.18	0.82 **	0.53 *	0.81 **	Na^+	0.14	0.02	0.41	0.53 *	0.14
$NH4^+$	0.58 *	0.55 *	0.79 **	0.80 **	0.86 **	$NH4^+$	0.95 **	0.77 **	0.34	−0.10	0.77 **
K^+	−0.23	−0.06	0.18	0.19	0.07	K^+	0.21	0.21	0.41	0.71 **	0.23
Mg^{2+}	0.69 **	0.54 *	0.85 **	0.74 **	0.89 **	Mg^{2+}	0.78 **	0.57 *	0.46	−0.17	−0.61
Ca^{2+}	−0.28	0.36	−0.23	0.57 *	0.29	Ca^{2+}	0.55 *	0.69 **	0.91 **	0.57 *	0.70 **

* $p < 0.05$; ** $p < 0.01$.

4. Discussion

Our result showed regional differences in emission factors of $PM_{2.5}$ and total water-soluble ions; combustion of coniferous species from the boreal region emitted more $PM_{2.5}$ while combustion of broadleaved species emitted more WSI than the subtropical region (Table 1). This difference could be related to the flammability and surface area-to-volume ratio of the fuel and inter-species variation in chemical composition in the biomass. Coniferous fuels have high flammability and energy content and often result in high $EFPM_{2.5}$ [24,25]. Fuels with a large surface area-to-volume ratio require less heat for ignition, thereby favoring rapid and thorough flaming combustion. This is further evidenced in the present study where the heat of combustion was significantly higher for coniferous species (19,149 kJ) than the broadleaved species (18,187 kJ) (Table 1). Thus, rapid combustion of species from the boreal region emitted a larger proportion of water-soluble ions than the subtropical region. Furthermore, significant differences in the EFWSI in particulate matter could be related to the chemical composition of the species [26–29]. The chemical composition of the biomass is influenced by nutrient biogeochemistry, which in turn varies between sites [30]. Similar emission differences in ambient smoke plumes between vegetation types were observed during controlled burning of the Brazilian cerrado vegetation and tropical rainforest [31].

The dominant ionic species in $PM_{2.5}$ were K^+ and Cl^- irrespective of the forest type and species burnt (Tables 2 and 3), which is in line with previously reported studies, which sampled smoke particles in the field and indoors [13,31,32]. The presence of K^+ as the major cationic species in $PM_{2.5}$ is the result of primary production through volatilization of plant tissue material during the combustion process [13], and K^+ is often considered as a marker of biomass burning [11,12]. Emissions of all ionic species except F^- was higher for the boreal than subtropical regions and emissions of Na^+, Cl^-, SO_4^{2-} and NO_3^- were higher for broadleaved than coniferous species (Tables 2 and 3). This difference suggests that not only the species composition but also the geochemical differences between regions have an impact on the chemical composition of the fuel. Previous studies have shown that particulate matters released in smoke plumes during biomass burning were rich in K^+ and Cl^- [30,32] and the major water-soluble inorganic ions in $PM_{2.5}$ during forest fire were K^+, Na^+, NH_4^+, Ca^{2+}, Cl^-, NO_3^- and SO_4^{2-} [30,33].

Our result also showed significant differences in EFs of WSI between combustion of leaves and branches (Figure 2). Generally, combustion of leaves emitted more ionic species in $PM_{2.5}$ than branches. This could be related to high flammability of leaves [25] and structural differences between leaves and branches attributed to the denser structure and higher lignin content of branches than leaves [34]. The combustion stage had also a significant effect on emissions of ionic species in $PM_{2.5}$; smoldering generally emitted more WSI than flaming combustion (Figure 3). High concentrations of water-soluble ions associated with increased release of particulate matter in the smoke are emitted under limited oxidation during smoldering. Guo et al. [13] reported a similarly higher EF of ionic species in $PM_{2.5}$ during smoldering than flaming combustion while Liu et al. [12] demonstrated a higher content of total water-soluble ions in particulate matter released from the burning of dry branches during smoldering compared with flaming combustion.

There were strong correlations between Na^+ and SO_4^{2-} and NO_2^-, between NH_4^+ and SO_4^{2-}, NO_3^- and NO_2^- as well as between Mg^{2+} and SO_4^{2-}, NO_3^- and NO_2^- for species from the boreal region (Table 4). This suggests that the PM could be composed of sodium, ammonium and magnesium sulphate and nitrate during burning of boreal species, while ammonium fluoride, ammonium chloride, ammonium nitrite, potassium nitrate, magnesium fluoride and calcium sulphate were major constituents of PM during the burning of subtropical species. We also calculated the cation to anion ratio, as this is considered to be a good indicator of the acidity of the particulate matter [3]. The ratio was 1.2 for the boreal forest and 0.98 for the subtropical forest. Thus, the particulate matter released during forest fires is neutral to slightly alkaline rather than acidic. The slight increase in pH associated with wet and dry deposition of WSI on the soil surface could possibly improve availability of nutrients, particularly phosphorus which is limited in subtropical forest ecosystems due to the acidic nature of the soil and fixation by iron and aluminum [35]. This, in turn, may favor growth and productivity of forests. However, the relatively high emission of Cl^- compared to other anionic species calls for attention as wet deposition of chlorides may contribute to ecosystem acidification. As a whole, the study demonstrates that forest fire plays a minor role in the emission of acidic particulate matter.

5. Conclusions

Emission of $PM_{2.5}$ and its water-soluble ionic component was higher in the boreal forest region than in the subtropical forest region, and coniferous species emitted more than broadleaved species. EFs of WSI were generally higher during combustion of leaves than branches and consistently higher during smoldering than during flaming combustion. The water-soluble ionic component in $PM_{2.5}$ was dominated by K^+ and Cl^- irrespective of the forest ecosystem and species. As a whole, our findings demonstrate that forest fire could contribute a considerable amount of water-soluble ions to atmospheric emission depending on the forest type and species. However, the cation to anion ratio is higher than 1.0, particularly for the boreal forest, suggesting that the particulate matter is alkalescent. Thus, forest fire in these regions may not contribute to ecosystem acidification through emissions of water-soluble ions. Our data are crucial for understanding emissions of WSI during different phases of forest fires and how different typologies of biomass can affect the profile of speciation emissions.

Author Contributions: Conceptualization, F.G.; methodology, F.G., Y.M., X.G., W.Z. and L.G.; validation, M.T. and F.G.; formal analysis, M.T., Y.M., X.G., W.Z. and L.G.; investigation, Y.M., X.G., W.Z. and L.G.; data curation, F.G.; writing—original draft preparation, M.T.; writing—review and editing, M.T., F.G.; visualization, Y.M.; supervision, F.G., M.T.; project administration, F.G.; funding acquisition, F.G.

Funding: The study was financially supported by the National Natural Science Foundation of China (Grant No. 31770697), the Asia-Pacific Network for Sustainable Forest Management and Rehabilitation (APFNet Climate Research Project) Phase II, and the Fujian Agriculture and Forestry University Funds for Distinguished Young Scholar (xjq201613).

Acknowledgments: We thank colleagues at the forestry college, Fujian Agriculture and Forestry University for their support during sample collection and indoor experiment.

Conflicts of Interest: The authors declare no conflict of interest.

References

1. Ito, A.; Penner, J.E. Historical emissions of carbonaceous aerosols from biomass and fossil fuel burning for the period 1870–2000. *Glob. Biogeochem. Cycles* **2005**, *19*, 273–280. [CrossRef]
2. Deng, C.R. Identification of Biomass Burning Source in Aerosols and the Formation Mechanism of Haze. Ph.D. Thesis, Fudan University, Shanghai, China, April 2011.
3. Deng, X.-L.; Shi, C.; Wu, B.-W.; Yang, Y.-J.; Jin, Q.; Wang, H.-L.; Zhu, S.; Yu, C. Characteristics of the water soluble components of aerosol particles in Hefei, China. *J. Environ. Sci* **2016**, *42*, 32–40. [CrossRef] [PubMed]
4. Zhou, J.; Xing, Z.; Deng, J.; Du, K. Characterizing and sourcing ambient PM2.5 over key emission regions in China I: Water-soluble ions and carbonaceous fractions. *Atmos. Environ.* **2016**, *135*, 20–30. [CrossRef]

5. Elias, P.E.; Burger, J.A.; Adams, M.B. Acid deposition effects on forest composition and growth on the Monongahela National Forest, West Virginia. *Ecol. Manag.* **2009**, *258*, 2175–2182. [CrossRef]

6. Fana, H.B.; Wang, Y.H. Effects of simulated acid rain on germination, foliar damage, chlorophyll contents and seedling growth of five hardwood species growing in China. *Ecol. Manag.* **2000**, *126*, 321–329. [CrossRef]

7. Hu, Y.; Bellaloui, N.; Tigabu, M.; Wang, J.; Diao, J.; Wang, K.; Yang, R.; Sun, G. Gaseous NO2 effects on stomatal behavior, photosynthesis and respiration of hybrid poplar leaves. *Acta Physiol. Plant.* **2015**, *37*, 39. [CrossRef]

8. Wang, H.L.; Zhu, B.; Shen, L.J.; Kang, H.Q. Size distributions of aerosol and water-soluble ions in Nanjing during a crop residual burning event. *J. Environ. Sci.* **2012**, *24*, 1457–1465. [CrossRef]

9. Park, S.S.; Sim, S.Y.; Bae, M.S.; Schauer, J.J. Size distribution of water-soluble components in particulate matter emitted from biomass burning. *Atmos. Environ.* **2013**, *73*, 62–72. [CrossRef]

10. Contini, D.; Cesari, D.; Genga, A.; Siciliano, M.; Ielpo, P.; Guascito, M.R.; Conte, M. Source apportionment of size-segregated atmospheric particles based on the major water-soluble components in Lecce (Italy). *Sci. Total Environ.* **2014**, *472*, 248–261. [CrossRef]

11. Li, J.; Pósfai, M.; Hobbs, P.V.; Buseck, P.R. Individual aerosol particles from biomass burning in southern Africa: 2, Compositions and aging of inorganic particles. *J. Geophys. Res. Atmos.* **2003**, *108*, 347–362. [CrossRef]

12. Liu, G.; Huang, K.; Li, J.H.; Xu, H. Chemical composition of water-soluble ions in smoke emitted from tree branch combustion. *Environ. Sci. J. Integr. Environ. Res.* **2016**, *37*, 3737–3742, (In Chinese with English abstract).

13. Guo, F.; Ju, Y.; Wang, G.; Alvarado, E.C.; Yang, X.; Ma, Y.; Liu, A. Inorganic chemical composition of PM2.5 emissions from the combustion of six main tree species in subtropical China. *Atmos. Environ.* **2018**, *189*, 107–115. [CrossRef]

14. Zheng, H.Q.; Chen, J.P.; Zhang, X.; Zhang, C.A.; Zhang, C.G.; Chen, H. Study on the forecast system of forest fire weather ranks in Fujian. *Chin. J. Agrometeorol.* **2001**, *3*, 38–44, (In Chinese with English abstract).

15. Wu, Z.; He, H.; Yang, J.; Liu, Z.; Liang, Y. Relative effects of climatic and local factors on fire occurrence in boreal forest landscapes of northeastern China. *Sci. Total Environ.* **2014**, *493*, 472–480. [CrossRef] [PubMed]

16. Guo, F.; Su, Z.; Wang, G.; Sun, L.; Lin, F.; Liu, A.Q. Wildfire ignition in the forests of southeast China: Identifying drivers and spatial distribution to predict wildfire likelihood. *Appl. Geogr.* **2016**, *66*, 12–21. [CrossRef]

17. Guo, F.; Selvaraj, S.; Lin, F.; Wang, G.; Wang, W.; Su, Z.; Liu, A.Q. Geospatial information on geographical and human factors improved anthropogenic fire occurrence modeling in the Chinese boreal forest. *Can. J. For. Res.* **2016**, *46*, 582–594. [CrossRef]

18. Guo, F.T.; Jin, Q.F.; Yang, X.J.; Liu, A.Q. An Air Compression System that Simulates the Burning of Wild Biomass. China Patent 2016211196373, 2017.

19. McMeeking, G.R.; Kreidenweis, S.M.; Baker, S.; Carrico, C.M.; Chow, J.C.; Collett, J.L., Jr.; Hao, W.M.; Holden, A.S.; Kirchstetter, T.W.; Malm, W.C.; et al. Emissions of trace gases and aerosols during the open combustion of biomass in the laboratory. *J. Geophys. Res. Atmos.* **2009**, *114*, 1–20. [CrossRef]

20. Akagi, S.K.; Yokelson, R.J.; Wiedinmyer, C.; Alvarado, M.J.; Reid, J.S.; Karl, T.; Crounse, J.D.; Wennberg, P.O. Emission factors for open and domestic biomass burning for use in atmospheric models. *Atmos. Chem. Phys.* **2011**, *11*, 27523–27602. [CrossRef]

21. Zhang, T.; Cao, J.J.; Tie, X.X.; Shen, Z.X.; Liu, S.X.; Ding, H.; Han, Y.M.; Wang, G.H.; Ho, K.F.; Qiang, J.; et al. Water-soluble ions in atmospheric aerosols measured in Xi'an, China: Seasonal variations and sources. *Atmos. Res.* **2011**, *102*, 110–119. [CrossRef]

22. Zhang, J.; Smith, K.R.; Ma, Y.; Ye, S.; Jiang, F.; Qi, W.; Liu, P.; Khalil, M.A.K.; Rasmussen, R.A.; Thorneloe, S.A. Greenhouse gases and other airborne pollutants from household stoves in China: A database for emission factors. *Atmos. Environ.* **2000**, *34*, 4537–4549. [CrossRef]

23. Dhammapala, R.S. Evaluating Emission Factors of PM2.5, Selected PAHS and Phenols from Wheat and Kentucky Bluegrass Stubble Burning in Eastern Washington and Northern Idaho. Ph.D. Thesis, Washington State University, Pullman, WA, USA, August 2006.

24. Sandberg, D.V. Slash fire intensity and smoke emissions. In Proceedings of the Third National Conference on Fire and Forest Meteorology of the American Meteorological Society and the Society of American Foresters, Lake Tahoe, CA, USA, 2–4 April 1974.

25. Reid, A.M.; Robertson, K.M. Energy content of common fuels in upland pine savannas of the south-eastern US and their application to fire behavior modelling. *Int. J. Wildland Fire* **2012**, *21*, 591–595. [CrossRef]

26. Kauffman, J.B.; Cummings, D.L.; Ward, D.E. Relationships of fire, biomass and nutrient dynamics along a vegetation gradient in the Brazilian cerrado. *J. Ecol.* **1994**, *82*, 519–531. [CrossRef]

27. Alves, C.A.; Gonçalves, C.; Pio, C.A.; Mirante, F.; Caseiro, A.; Tarelho, L.; Freitas, M.C.; Viegas, D.X. Smoke emissions from biomass burning in a Mediterranean shrubland. *Atmos. Environ.* **2010**, *44*, 3024–3033. [CrossRef]

28. Hosseini, S.H.; Urbanski, S.P.; Dixit, P.; Qi, L.; Burling, I.R.; Yokelson, R.J.; Johnson, T.J.; Shrivastava, M.; Jung, H.S.; Weise, D.R.; et al. Laboratory characterization of PM emissions from combustion of wildland biomass fuels. *J. Geophys. Res. Atmos.* **2013**, *118*, 9914–9929. [CrossRef]

29. Urbanski, S.P.; Hao, W.M.; Baker, S. Chemical composition of wildland fire emissions. In *Developments in Environmental Science*; Bytnerowicz, A., Arbaugh, M., Riebau, A., Andersen, C., Eds.; Elsevier: Amsterdam, The Netherlands, 2009; Volume 8, pp. 79–107.

30. Allen, A.G.; Miguel, A.H. Biomass burning in the Amazon: Characterization of the ionic component of aerosols generated from flaming and smoldering rainforest and savannah. *Environ. Sci. Technol.* **1995**, *29*, 486–493. [CrossRef]

31. Yamasoe, M.A.; Artaxo, P.; Miguel, A.H.; Allen, A.G. Chemical composition of aerosol particles from direct emissions of vegetation fires in the Amazon Basin: Water-soluble species and trace elements. *Atmos. Environ.* **2000**, *34*, 1641–1653. [CrossRef]

32. Echalar, F.; Gaudichet, A.; Cachier, H.; Artaxo, P. Aerosol emissions by tropical forest and savanna biomass burning: Characteristic trace elements and fluxes. *Geophys. Res. Lett.* **1995**, *22*, 3039–3042. [CrossRef]

33. Vicente, A.; Alves, C.; Calvo, A.I.; Fernandes, A.P.; Nunes, T.; Monteiro, C.; Almeida, S.M.; Pio, C. Emission factors and detailed chemical composition of smoke particles from the 2010 wildfire season. *Atmos. Environ.* **2013**, *71*, 295–303. [CrossRef]

34. Schmidl, C.; Bauer, H.; Dattler, A.; Hitzenberger, R.; Weissenboeck, G.; Marr, L.L.; Puxbaum, H. Chemical characterisation of particle emissions from burning leaves. *Atmos. Environ.* **2008**, *42*, 9070–9079. [CrossRef]

35. Chen, H.J.; Li, Y.Q.; Chen, D.D.; Zhang, Y.; Wu, L.M.; Ji, J.S. Soil phosphorus fractions and their availability in Chinese fir plantations in south China. *For. Res.* **1996**, *9*, 121–126.

MDPI

St. Alban-Anlage 66

4052 Basel

Switzerland

Tel. +41 61 683 77 34

Fax +41 61 302 89 18

www.mdpi.com

Forests Editorial Office

E-mail: forests@mdpi.com

www.mdpi.com/journal/forests